Ecological Niches and Geographic Distributions

MONOGRAPHS IN POPULATION BIOLOGY

EDITED BY SIMON A. LEVIN AND HENRY S. HORN

A complete series list follows the bibliography

Ecological Niches
and Geographic Distributions

A. Townsend Peterson
Jorge Soberón
Richard G. Pearson
Robert P. Anderson
Enrique Martínez-Meyer
Miguel Nakamura
Miguel Bastos Araújo

PRINCETON UNIVERSITY PRESS
Princeton and Oxford

Copyright © 2011 by Princeton University Press

Published by Princeton University Press,
41 William Street, Princeton, New Jersey 08540
In the United Kingdom: Princeton University Press,
6 Oxford Street, Woodstock, Oxfordshire OX20 1TW
press.princeton.edu

Library of Congress Cataloging-in-Publication Data

Ecological niches and geographic distributions / A. Townsend Peterson . . . [et al.].
     p.   cm. — (Monographs in population biology ; no. 49)
    Includes bibliographical references.
    ISBN 978-0-691-13686-8 (hardcover : alk. paper) — ISBN 978-0-691-13688-2 (pbk. : alk.
paper) 1. Niche (Ecology)   2. Niche (Ecology)—Mathematical models.   3. Biogeography.
4. Biogeography—Mathematical models.   I. Peterson, A. Townsend (Andrew Townsend), 1964–
    QH546.3.E26 2011
    577.8'2—dc23   2011018009

British Library Cataloging-in-Publication Data is available

This book has been composed in Times Roman

Printed on acid-free paper. ∞

Printed in the United States of America

10 9 8 7 6 5 4 3 2 1

Table of Contents

Acknowledgments   ix

1. Introduction   1
    Practicalities   2
    This Volume   3

PART I
THEORY

2. Concepts of Niches   7
    Major Themes in Niche Concepts   9
    Grinnellian and Eltonian Niches   16
    Estimating Grinnellian Niches: Practicalities   19
    Summary   21

3. Niches and Geographic Distributions   23
    Relations between Environmental and Geographic Spaces   24
    The Ecological Equations   26
    The BAM Diagram: A Thinking Framework   29
    Ecological Niches and Geographic Distributions   31
    Estimating Geographic Areas and Ecological Niches   40
    Summary   46

PART II
PRACTICE

4. Niches and Distributions in Practice: Overview   51
    General Principles   52
    Steps to Building Niche Models   56

5. Species' Occurrence Data   62
    Types of Occurrence Data   62
    Occurrence Data Content and Availability   77
    Summary   81

6. Environmental Data   82
    Species-Environment Relationships   82
    Environmental Data for Ecological Niche Modeling   85
    Environmental Data in Practice   87
    Summary   95

7. Modeling Ecological Niches   97
    What Is Being Estimated?   98
    Modeling Algorithms   101
    Implementation   112
    Model Calibration   112
    Model Complexity and Overfitting   123
    Study Region Extent and Resolution Revisited   125
    Model Extrapolation and Transferability   126
    Differences among Methods and Selection of "Best" Models   128
    Characterizing Ecological Niches   131
    Summary   137

8. From Niches to Distributions   138
    Potential Distributional Areas   138
    Nonequilibrium Distributions   141
    Detecting and Processing Nonequilibrium Distributions   143
    Summary   149

9: Evaluating Model Performance and Significance   150
    Presences, Absences, and Errors   150
    Calibration and Evaluation Datasets   153
    Overfitting, Performance, Significance, and Evaluation Space   154
    Selection of Evaluation Data   156
    Evaluation of Performance   162
    Assessing Model Significance   167
    Future Directions   176
    Summary   180

PART III
APPLICATIONS

10. Introduction to Applications   185

11. Discovering Biodiversity   189
    Discovering Populations   190

Discovering Species Limits   191
Discovering Unknown Species   192
Connection to Theory   192
Practical Considerations   193
Review of Applications   195
Discussion   198

12. Conservation Planning and Climate Change Effects   200
Generalities   200
Connection to Theory   201
Practical Considerations   206
Review of Applications   208

13. Species' Invasions   215
Connection to Theory   216
Practical Considerations   216
Review of Applications   218
Caveats and Limitations   222
Future Directions and Challenges   224

14. The Geography of Disease Transmission   226
Connection to Theory   229
Practical Considerations   229
Review of Applications   230
Caveats and Limitations   235
Future Directions and Challenges   236

15. Linking Niches with Evolutionary Processes   238
Changes in the Available Environment   238
Niche Conservatism   240
Tests of Conservatism   243
Context   250
Learning More about Ecological Niche Evolution   250
Future Directions and Challenges   254

16. Conclusions   256

Appendices

Appendix A: Glossary of Symbols Used   261

Appendix B: Set Theory for G- and E-Space   266

Glossary   269

Bibliography   281

Acknowledgments

Acknowledging all of the people and institutions who have made the development of this book possible is not an easy task. That is, this book summarizes at least a decade of work by each of seven researchers, each of whom has seen significant support and encouragement from a number of sources. Certainly, then, we thank our home institutions—the University of Kansas; American Museum of Natural History; City University of New York; Instituto de Biología, Universidad Nacional Autónoma de México; Centro de Investigación en Matemáticas, A.C.; Museo Nacional de Ciencias Naturales; Consejo Superior de Investigaciones Científicas of Spain; and Comisión Nacional para el Uso y Conocimiento de la Biodiversidad. Without this firm institutional basis, this book surely would not have come to exist.

On a more proximate basis, we thank the following organizations for their support in terms of funding: Microsoft Research (grant to Peterson and Soberón), Fundação de Amparo à Pesquisa do Estado de São Paulo (grant to support research visit by Peterson to Brazil), the "Rui Nabeiro" Chair of Biodiversity at the University of Évora (Araújo), the Spanish Ministry of Science and Innovation (grant CGL2008-01198-E to Araújo), Proyectos de Investigación e Innovación Tecnológica UNAM (grant IN221208 to Martínez-Meyer), the U.S. National Science Foundation (grant DEB 0717357 to Anderson; DEB 0641023 and IPY 0732948 to the American Museum of Natural History), U.S. National Aeronautics and Space Administration (NNX09AK19G and NNX09AK20G to the American Museum of Natural History), and U.S. National Atmospheric and Oceanic Administration (NA04OAR4700191, NA05SEC4691002 to the American Museum of Natural History). We thank the Global Biodiversity Information Facility (GBIF) for supporting the series of global training courses (2004–2007) that assembled most of the authors of this book and initiated our discussions. We thank the Centro de Referência em Informação Ambiental (Brazil) and Landcare Research (New Zealand) for hosting Peterson on research visits during which significant portions of the book were prepared.

More personally, we thank numerous colleagues for intriguing discussions and debates that helped to clarify and solidify these ideas. In particular, in this vein, we thank Narayani Barve, Bastian Bentlage, Catherine Graham, Robert Guralnick, Alberto Jiménez-Valverde, Daniel Lew, Jorge Lobo, Sean Maher,

Andrés Lira-Noriega, Yoshinori Nakazawa, Adolfo G. Navarro-Sigüenza, Monica Papeş, Steven Phillips, Christopher Raxworthy, Víctor Sánchez-Cordero, Jeet Sukumaran, Volker Bahn, and Dan Warren. Thanks to Robert Holt for detailed reviews of the manuscript, to Terry Dickert for careful and detailed editing of the literature cited, and to Alison Kalett, Stefani Wexler, Dale Cotton, Lauren Lepow, Jason Alejandro, and Jennifer Harris of Princeton University Press for expert assistance with final preparation of the manuscript for publication. Finally, we thank our spouses and families for their kind and gentle patience during the preparation of this book.

Ecological Niches and Geographic Distributions

CHAPTER ONE

Introduction

The fields of historical biogeography and ecological biogeography have long been paradoxically disparate and distant from one another, with different terminologies, different concepts, and almost nonoverlapping sets of researchers. Ecological biogeography focuses on spatial pattern in the composition and functioning of ecological communities, while historical biogeography attempts to reconstruct the history of areas and their biotas. Although some recent steps have narrowed gaps a bit, the two fields have long been quite distinct and disconnected, and spatial understanding of biodiversity has suffered as a consequence.

Differences between the two biogeographies are manifold: certainly, spatial scale is an important one, with most of ecological biogeography focusing at regional scales and most of historical biogeography at continental or even global scales. Another important difference is in treatment of temporal dimensions, with ecological biogeography focused chiefly over time spans that are geologically instantaneous (i.e., in the present), but historical biogeography looking from the present back over evolutionary time, sometimes many millions of years. Although not without significant exceptions (e.g., MacArthur 1972), these homonymous fields both have important insights to offer regarding the geography of biodiversity, yet have developed in large part independently until quite recently.

In recent years, however, an emerging body of work has begun to bridge between the two, building toward a more synthetic biogeography. Ecologists looking over broader spatial extents and into history, and systematists thinking about environmental dimensions and interactions among species, have come to understand that species' distributions are a function of phenomena from both realms. Detailed thinking regarding areas of distribution has also provided a fascinating reawakening of interest in another classic concept, that of the ecological niche. In effect, understanding areas of distribution of species in terms of their ecological requirements across multiple scales of space and time has provided an arena for a meeting of these two disparate disciplines. The impressive impact on both fields of just the first few years of this interaction suggests that their integration will have a bright future.

PRACTICALITIES

The past few years have witnessed substantial increases in availability of species' occurrence data, also termed "primary biodiversity data" or "presence data" or "occurrence data" (Soberón et al. 1996, Graham et al. 2004a, Soberón and Peterson 2004). This trend results from large-scale efforts to digitize and reference to geographic coordinates ("georeference") the estimated 1–3 billion specimens held in world museums and herbaria (Chalmers 1996, Krishtalka and Humphrey 2000), as well as efforts to improve access to large observational data stores, at least for certain taxonomic groups (chapter 5). Presence data, as we will see in coming chapters, form the basis for most efforts to estimate ecological niches. Publicly accessible Internet portals now allow access to on the order of 300 million primary biodiversity data records (Edwards 2004).

Information regarding environmental variables is now similarly abundant. Petabytes (i.e., millions of gigabytes) of environmental information about climate, topography, soils, oceanographic variables, vegetation indices, land-surface reflectance, and so on, are available across almost the entire planet, and at increasingly finer resolutions (chapter 6). These datasets are being generated by agencies such as the European and U.S. space agencies, by the United Nations, by university researchers (e.g., Hijmans et al. 2005), and by many national institutions (e.g., CONABIO 2009, INPE 2009, NRSC 2009).

Finally, powerful software allowing estimation of both areas of distribution and theoretical objects related to niches has been implemented. The work of pioneers like Grinnell, Hutchinson, and Austin suddenly became thriving research areas linking ecological and historical dimensions of biogeography. In particular, these tools enable what has been termed "species distribution modeling" (SDM; Guisan and Zimmermann 2000, Hirzel et al. 2002, Guisan and Thuiller 2005, Araújo and Guisan 2006), as well as the related (but by no means equivalent) endeavor called "ecological niche modeling" (ENM; Peterson et al. 2002d, Soberón and Peterson 2005, Soberón 2007). These fields—the subject of this book—center on application of niche theory to questions about real and possible spatial distributions of species in the past, present, and future. In a very real sense, the availability of large quantities of data, technological developments like geographic information systems (GIS), and several computational tools are enabling a multitude of applications that are not only of biological importance, but that also can often be of extreme practical utility.

Nevertheless, many carefully pondered decisions are necessary before it is possible to turn these data and tools into interesting analyses and useful knowledge with full scientific rigor. Many crucial methodological issues remain to be explored and resolved, like the types of variables to be included, whether to use them in raw or transformed forms, whether to reduce dimensionality prior to analysis, what spatial and temporal resolution to use, and how best to assess model performance; all of these points are questions affecting niche modeling exercises. The answers to these practical questions depend on a rigorous conceptual framework. After a period of development in which conceptual and methodological rigor took a back seat to rapid development of software and data resources, the time has come to take stock of the advances and propose a conceptual reorganization. All of this thinking is the subject of this book.

THIS VOLUME

We offer this volume as a first synthesis of concepts in this emerging field. In spite of hundreds of research contributions and increasing numbers of reviews and commentaries, no rigorous and quantitative conceptual framework and synthesis has been presented. This lack of synthesis is nowhere more notable than in the debate between groups of researchers using the same tools to address the same questions, yet employing—whether knowingly or not—very distinct conceptual frameworks in the development of their analyses and conclusions. Such misunderstandings can be avoided, if a common language and thinking framework are available. This book represents the crystallization of years of thinking and work by a diverse suite of coauthors, all interested in the ecological, geographic, and evolutionary dimensions of geographic distributions of species.

We do not intend this book to serve as an exhaustive review of the burgeoning literature on ecological niches and geographic distributions. Quite simply, just in the time in which we have been preparing this manuscript, hundreds of new papers have been published, making the idea of an exhaustive review a moving target that is probably impossible to achieve. Moreover, recent publication of a book by J. Franklin (2010), *Mapping Species Distributions*, does a commendable job of reviewing and synthesizing the vast literature on this topic. Our approach is more conceptual: we aim to offer a body of terminology and schemes by which to understand and discuss phenomena of distributional ecology; a common language is badly needed in a field so rife with ill-defined jargon and loosely defined terms.

This book focuses on the complex relationships between ecological niches and geographic distributions of species, both across space and (perhaps to a lesser degree) through time. We provide a conceptual overview, which we hope will be of broad interest to researchers interested in diverse aspects of ecology, biogeography, and other related fields. However, we focus much of this book on how that conceptual framework links to the emerging fields of ecological niche modeling and species distribution modeling, both correlative approaches to understanding ecological niches and geographic distributions.

We do not make any broad attempt to provide a similar overview or deep understanding of process-based, physiological approaches to estimating ecological niches that other researchers are exploring—rather, we see the two approaches as complementary. The process-based approaches show considerable promise, and in some situations offer the only view possible into the fundamental ecological niches of species. However, such approaches remain in early stages of development and exploration. This book focuses on the correlative approaches, which we see as most broadly applicable to diverse questions regarding the ecology and geography of biodiversity phenomena.

The reader should not imagine that this process of finding a common language has been easy. Indeed, even among the seven authors of this book, who have worked together for years, some strong differences of opinion still exist regarding terminologies and concepts. We have, however, achieved a high degree of concordance, and have been willing to look past our colleagues' different views in striving for synthesis. Many of the insights that emerged from these debates concerned the relevance of ideas central to each of our respective backgrounds and skill sets, including integration of ideas from field biology, morphology, systematics, genetics, theoretical ecology, evolutionary biology, statistics, climatology, and geospatial science.

The result is this volume. We begin with a conceptual framework for thinking about and discussing the distributional ecology of species, which has involved considerable exploration of the field of population ecology, and has required revisiting several "sacred" texts, such as the fundamental early works of Hutchinson and Grinnell. A second section addresses the data and tools that have been marshalled in the early development of this field. We avoid carefully the temptation to review and assess specific software tools, as we consider these to be transitory and less important than the base concepts. It is much more important that the field have a consistent terminology and thinking framework than to "know" that such and such program is the "best." (Besides, as the reader will see in the chapters that follow, what appears to be the "best" frequently is not what it appears, and is certainly context- and scale-dependent.) In the final section, we provide a relatively brief overview of real-world situations to which these tools have been applied, to illustrate the promise of this new field. Our hope is to move the discourse in this field to a new level, once a common platform of ideas is established.

Part I

THEORY

CHAPTER TWO

Concepts of Niches

It has often been pointed out that the term "niche" disguises several concepts under a single label (Whittaker et al. 1973, Colwell 1992, Leibold 1995, Chase and Leibold 2003, Odling-Smee et al. 2003). Some authors, perhaps overwhelmed by the broad variety and subtle shades of meaning assigned to the word, have advised that "niche is perhaps a term best left undefined" (Bell 1982). We disagree: in science, arguments benefit from precise and consistent usage of key concepts; otherwise, clear thinking is hindered. Using the same word to refer to different ideas leads to confusion, as the picturesque history of the word "niche" amply demonstrates. Besides, different senses of niche are appropriate to deal with different biological problems. As a consequence, several niche concepts exist, and our first task is thus to clarify and specify which concept we will be using and why. We will choose terminology and ideas best suited to the problem in which we are interested—namely, that of estimating and understanding areas of distribution of species.

One reason why "niche" has acquired a veritable bush of meanings is that, since the first time it was used, ecologists have applied the term to analyze a very complex question: what combinations of environmental factors allow a species to exist in a given geographic region or in a given biotic community, and what effects does the species have on those environmental factors? Not only does the preceding statement refer to an intrinsically complex set of problems, but several of its terms can be interpreted and measured in a variety of ways. Moreover, the concept has been used at both geographic and local scales, most often assuming that the ensuing complications and differences should be obvious. For example, Grinnell (1917), studying the niche of the California Thrasher (*Toxostoma redivivum*) in relation to its area of distribution (this idea will be discussed later as the "existence" of the species), meant by the term niche the thrasher's climatic and habitat requirements (the environmental factors) expressed geographically (figure 2.1). In a contrasting interpretation, Elton (1927) viewed the niche as the functional role of an animal on a community (its local effects): the *existence* was taken for granted, but the emphasis was placed on the "impacts." These two early views of niche illustrate one of the main

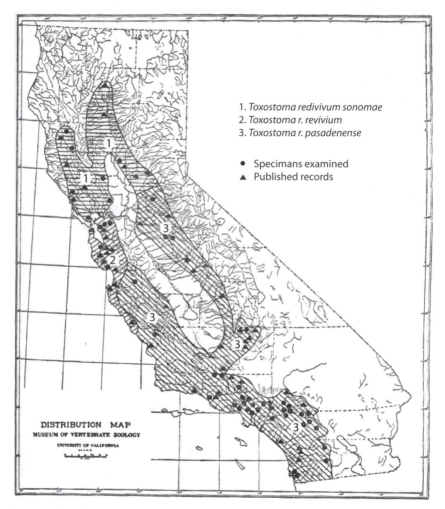

FIGURE 2.1. Distribution of the California Thrasher (*Toxostoma redivivum*) in California, from Grinnell (1917). Approximate distributional limits of the three subspecies are shown with different shadings, and occurrences are shown as dots (for specimens) or triangles (for published records).

causes of ambiguity of concept: stress on requirements at geographic scales versus stress on impacts at local scales (Chase and Leibold 2003).

After reviewing the history of niche concepts, Chase and Leibold (2003) concluded that much of the confusion surrounding the term results because "previous authors have not consistently distinguished between the responses of organisms to their environment and the effects of organisms on their environment." Indeed, this difference has both deep implications about the actual mathematical form of a multivariate niche definition and serious operational consequences, since certain variables related to requirements can be measured easily, whereas most variables related to impact require ad hoc experimental efforts.

Although attempting an exhaustive classification of niche concepts based on the preceding ideas would be interesting, that task is not our purpose in this chapter. Rather, we intend to propose a formal and operational definition of a particular niche concept (which is naturally related to the problem of estimating areas of distribution), offer approaches to characterize and measure it, and use it as a conceptual and terminological basis for describing and understanding much of the related practices of ecological niche modeling and species distribution modeling (Peterson 2006c). In this chapter, we explain the reasons for our choice of emphases, leading to a particular meaning and usage of niche. To accomplish this goal, we review briefly the themes most important in understanding niche concepts, highlighting the meanings most appropriate to the purpose of this book.

MAJOR THEMES IN NICHE CONCEPTS

Recall Hutchinson's (1957) definition of the fundamental niche of a species: a hypervolume of environmental variables, "every point of which corresponds to a state of the environment which would permit the species to exist indefinitely." Most differences in niche concepts depend on the formulation and relative importance given to three interrelated points, considered in turn later: (1) the meaning of "exist indefinitely," (2) what kinds of variables constitute the hypervolume, and (3) the nature of feedback loops between a species and the variables composing the hypervolume. Our definitions of niche are based on operational specifications of the preceding three points.

The "Existence" of a Species

As highlighted by Holt (2009), several ambiguities complicate the term "existence" of the species, as used in niche definitions. From a population ecology point of view, a species may exist for a period of time $t_0 < t < t_1$ in a place g of the world, if its total instantaneous growth rate $dx_g(t)/dt$ is on average nonnegative during $t_0 < t < t_1$ (Vandermeer 1972, Maguire 1973, Hutchinson 1978). However, and disregarding stochastic factors and evolutionary change, the instantaneous growth rate is composed (see next chapter) of (1) an intrinsic growth rate r_g that is by definition density-independent, (2) density-dependent factors and the results of interactions with other species, and (3) the population structure that determines arrival of dispersing individuals to the locality g and that may allow existence of the species in places where $r_g < 0$ (sink populations; Pulliam 1988 and 2000).

It is clear that quite-different combinations of the preceding factors may lead to a species "existing" [i.e., $dx_g(t)/dt > 0$] or not in a given region over a given period of time. Holt (2009) pointed out that some of the major issues requiring clarification are (1) the existence of Allee effects (reduced per-capita growth rate at small population densities) that will make some regions where $r_g > 0$ impossible to invade because the flow of migrants is below the Allee threshold; (2) the possibility that, in regions of high instantaneous growth rate, population growth will lead to irreversible alterations of the environment because the population impacts its niche (Chase and Leibold 2003); and (3) situations may exist in which the instantaneous growth rate fluctuates owing to stochastic variations, leading to a long-run growth rate (Lande et al. 2003) $\bar{r}$ that differs from the physiologically defined r.

In view of the preceding, a unequivocal niche definition requires a great degree of detail about how the environment in g affects different components of the growth rates. One can define a variety of niches (Hutchinson 1957, Pulliam 2000, Soberón 2007, Holt 2009) by focusing on the effects that the environment has on different components of the growth rate. In the next chapter, we will define niche concepts unequivocally based on how environmental variables affect some components of the instantaneous growth rate in the cells of a grid. However, first, we need to discuss what kinds of variables are used to build the environmental niche space.

TYPES OF VARIABLES IN NICHE SPACE

Since Hutchinson's seminal work, researchers have referred routinely to multivariate niche spaces. However, diverse types of niche variables are almost invariably mixed, often ignoring deep differences in their properties. For example, Hutchinson (1957) defined the niche as "the set, in a multidimensional space, of environmental states within which a species is able to survive." Here, "environmental states" are loosely represented as values along the axes of the n-dimensional niche hypervolume. However, the "environment" of a species consists of radically different factors, some of which cannot be represented simply as static axes. In particular, it is critical to distinguish environmental factors that are linked dynamically to the population of the focal species from those that are not; in other words, variables describing environmental aspects that are impacted by the species (by consumption or other modifications) must be distinguished from those that may affect the fitness of the species without being consumed or changed by it. The reason for this distinction is simple: the position of a point describing an environmental state in a noninteractive (non-linked) space of variables remains fixed regardless of the changes in numbers of a population of a species. When dynamic (linked) interactions exist, however, an initial point in a space of linked environmental variables (like consumed resources) actually traces a trajectory that depends on the changing numbers in the population of the species. This difference requires distinct types of mathematical objects to describe niche space.

The first author to notice the need to distinguish between dynamically linked variables and unlinked "conditions" was Hutchinson (1978), in a seldom-quoted chapter. He drew attention to deep differences between *scenopoetic* (from the Greek roots for "setting the stage") variables, as he called them, which are not consumed and for which no competition occurs, and others that can be dynamically consumed and may be the object of competition, which he termed *bionomic* variables (an unfortunate choice of term; see the following). Explicitly or implicitly, other authors have made this or parallel distinctions between linked and nonlinked variables (Austin 1980, James et al. 1984, Austin et al. 1990, Jackson and Overpeck 2000, Begon et al. 2006), although they may differ in emphasis. For example, Austin (1980) and Austin and Smith (1989) referred to "direct gradients" as those variables having direct physiological impacts on a population but that are not consumed, and offered as an example pH (although pH can be modified directly by the presence of a population without being consumed; Hinsinger et al. 2003). Likewise, Begon et al. (2006) defined "conditions" as abiotic environmental factors that influence the functioning of living organisms but are not "consumed," in contrast to resources; they also used pH as an example.

To define multidimensional environmental spaces in this book, we will make the crucial distinction between variables that are dynamically modified (linked) by the presence of the species versus those that are not. In other words, we follow what we consider to be the spirit of Hutchinson's (1978) chapter, which is an important point that merits reiteration. The distinction that we use is not about biotic versus abiotic variables, or consumed resources versus not consumed conditions. Instead, we construct the multivariate environmental spaces for our definitions based on variables that are not dynamically affected by the species, like climate, topography, and perhaps some habitat features, in contrast to variables that are dynamically modified (linked), such as consumed resources (Harper 1977, Austin 1980, Austin and Smith 1989, Begon et al. 2006) or those that are subject to modification by niche construction (see Odling-Smee et al. 2003, for a review) or niche destruction (Holt 2009). We use the term "dynamically linked" in the sense of terms that appear as parameters in population equations versus appearing as dynamic state variables (Meszéna et al. 2006), as we will see in the next chapter. Following the spirit of Hutchinson's (1978) chapter, and considering more recent usage (Jackson and Overpeck 2000), to avoid confusion, we call nonlinked variables "scenopoetic."

Several good reasons exist for choosing this criterion for distinguishing variables: first, scenopoetic, unlinked variables can be used to construct multivariate environmental spaces in which different niches are simple subsets. As we will see later, niches defined using nonscenopoetic, linked variables require much more elaborate mathematical definitions. Second, at coarse spatiotemporal resolutions (e.g., defined in years, and on the order of 1 km^2 or coarser), huge databases are available that summarize important scenopoetic variables for the entire planet; nonscenopoetic variables, on the other hand, will generally have to be generated for each particular situation.

Scenopoetic variables are typically measured at broad spatial extents (i.e., broader than ~10^4 km^2) and low spatial resolutions (e.g., 10^0 to 10^3 km^2; chapter 6). Obviously, the same factor (light availability for plants, for example), may be regarded as scenopoetic at a geographic scale (e.g., solar radiation), but dynamically linked at another scale (e.g., light competition among plants in light gaps). Still, we explore the feasibility and utility of this simple distinction between variables, to the extent that it is useful.

In terrestrial environments, the most obvious examples of scenopoetic variables are climatic and geomorphological variables, which are abiotic and coarsegrained. Austin's (1980) distinction between direct and indirect gradients (see chapter 6) remains valid: elevation, for example, affects species indirectly through atmospheric pressure, temperature, UV radiation, and other effects (Körner 2007), but whether the effect is direct or indirect, scenopoetic variables are not dynamically linked to changes in the population of a species. Clearly, these variables change over time, but following dynamics not related directly to the numbers of one or even a few species. Hence, on ecological timescales relevant to individuals and populations, they may be regarded as more or less constant, and therefore can be represented meaningfully by static sets of numbers.

James et al. (1984) used vegetation structure variables in the same sense: these variables may be biotic in nature, but they are neither consumed nor modified dynamically (linked) by populations of Wood Thrushes (*Hylocichla mustelina*) in the eastern United States. The structure of a forest is probably unaffected or affected only very slowly, by numbers of most forest-dwelling species. The ever-existing exceptions can be cited—for instance, moth outbreaks may defoliate a forest completely (Elkinton and Liebhold 1990).

The term "bionomic" that Hutchinson (1978) used to refer to nonscenopo-etic variables is not precise enough in our context, because we do not focus on whether the variable is biotic. Rather, we propose to focus on whether variables are linked dynamically to the population levels of the species in question. Some abiotic environmental factors can be consumed as resources (nutrients for ex-ample) or modified dynamically by the activities of individuals in a population (Harper 1977). Having said that, however, most examples of scenopoetic vari-ables we will be using are indeed abiotic, and most nonscenopoetic examples are related to biotic interactions. It may be possible to regard some very high-resolution variables as scenopoetic, in the sense of their not being impacted by the species in question (e.g., vegetation structure), but these exceptions are probably rare.

The multidimensional spaces of scenopoetic variables that exist in a given region, at a given time, we henceforth call "E-spaces"; in most cases, E-spaces are composed of coarse-grained variables, if only owing to data availability (see G-space discussions later in this chapter). An example appears in figure 2.2, and in chapter 6 we will discuss in detail what variables form E-spaces and several problems of constructing them. In conclusion, and anticipating more detailed discussions later, our selection of environmental variables to define niche spaces will be based on *scenopoetic variables that influence, at a geo-graphic scale, the intrinsic and instantaneous growth rates of a species.*

Niche as Requirements versus Niche as Impacts

The previous section developed one aspect of the idea that a major source of confusion about niche concepts is the disregard of the "impacts" part of the interactions between a species on its environment (Chase and Leibold 2003). By choosing to build environmental spaces using noninteractive variables that affect the density-independent component of population growth rates, we place ourselves firmly in the "niches as requirements" school for the purposes of this book, but this choice does not imply that a more comprehensive definition of niche including impacts is not needed.

Inclusion of consideration of impacts is fundamental (Leibold 1995), since at local scales species consume resources and interact with one another. At geographic scales, separating the requirements and the impacts components of niche may be a valid simplification (the Eltonian Noise Hypothesis, see discus-sion in chapter 3; see Odling-Smee et al. 2003, for what may be a dissenting view). At a more local scale, however, requirements and impacts cannot be separated easily, and even at geographic scales, local effects may be felt. For example, Leathwick and Austin (2001) presented evidence that *Nothofagus*

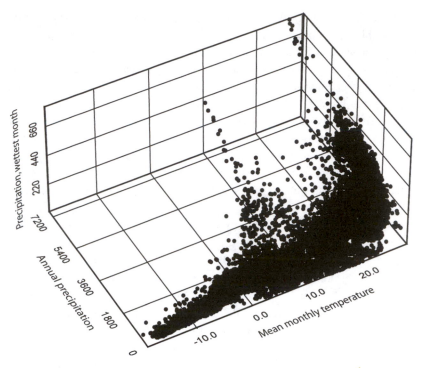

FIGURE 2.2. Projection of North and South America in an environmental space composed of three bioclimatic dimensions (annual precipitation in mm, mean monthly temperature in °C, precipitation of wettest month in mm). Each point corresponds to the environmental combination represented in one of 68,000 grid cells at 0.17° spatial resolution (climate data from Hijmans et al. 2005).

beeches have competitive effects on other species. Such effects, although local in mechanism, may at times have geographically discernible results.

Recently, a discussion of niche in the context of dynamically linked factors has been offered by Chase and Leibold (2003). They envisioned environmental spaces with resources and predator numbers as axes (sometimes also including static "stressors," related to scenopoetic variables) and defined subsets of these spaces as different classes of niches. However, since within a community, resources and consumers often show dynamic relationships, niches cannot be defined without mention of the dynamic effects that consumers have on the abundance and potentially on the spatial distribution of the resource and vice versa. To represent the environmental requirements of a population and the effects of that population on the environment simultaneously, a mechanistic model of the interaction is required (Leibold 1995). Using mechanistic consumer-resource models (MacArthur 1972, Tilman 1982), Chase and Leibold (2003) formalized a definition of niche in terms of both the species' environmental requirements and the species' impacts on the environment. Given the analytical complexities of the population dynamics of multiple species and multiple resources for each, they used a graphical device based on plotting simplified zero-net-growth isoclines for consumers in a two-resource space, together with resource-supply points and impact vectors.

In other words, even for the simplest cases, the niches defined by Chase and Leibold (2003) require knowledge of the full instantaneous growth rate (as opposed to just the intrinsic growth rate) in a space that is being modified continuously by the species of interest. This approach is theoretically very fruitful, but obtaining data about the dynamic (linked) parts of these niche spaces is seriously complicated, since knowledge of population dynamics is required. For all the preceding arguments, therefore, we see a fundamental advantage in defining an environmental space in terms only of noninteractive requirement variables. We suggest that this approach is valid at the very least at coarse spatial resolutions.

SCALING OF NICHES

For practical reasons, as discussed earlier and in chapter 6, most scenopoetic variables are measured at coarse spatial resolutions. As such, they are naturally suited to addressing macroecological and biogeographic problems. Whittaker et al. (1973), discussing the "place" and "role" meanings of niche, proposed a multidimensional environmental space composed of both "niche" (meaning role) and "habitat" (meaning place) variables that correspond roughly to the bionomic and scenopoetic variables of Hutchinson (1978). Whittaker et al. (1973) called this space the "ecotope." Leaving aside the problem mentioned earlier of dynamic (linked) variables requiring more complicated representations of niche space than unlinked (scenopoetic) variables, a problem emerges regarding the spatial scales at which different variables are measured. As mentioned by Araújo and Guisan (2006), the problem of obtaining detailed data about species' interactions over geographic scales is daunting owing to the very high spatial turnover in the details of the interactions, which may even change sign or direction over short distances. For instance, the interaction between the moth *Greya politella* and the wildflower *Lithophragma parviflora* varies from mutualistic to antagonistic over the span of their distribution (Thompson 2005). Therefore, we suggest that a natural scaling exists for spatial resolutions, in which many important scenopoetic variables are manifested at coarse spatial resolutions, while details of resource consumption and species' interactions are manifested at local scales commensurate with the activities of individuals.

In this vein, Pickett and Bazzaz (1978), Silvertown (2004), and Silvertown et al. (2006) developed the idea of α, β, and γ niches. These authors propose a hierarchy of niche definitions based on the scales at which different ecological processes operate: the α niche is the region of a species' realized niche corresponding to the local scale where interactions among species occur. The β niche is the region of a species' niche corresponding to the habitat where it is found (Silvertown et al. 2006). Finally, the γ niche is the geographic range of a species. These ideas are closely related to those of Whittaker et al. (1973), in that they acknowledge a hierarchy of processes related to community-level interactions, and habitat- and geographic-level determinants, and also in that they address niche definitions by defining a multidimensional space composed of mixtures of types of variables. Although we think that these ideas are headed in a fruitful direction, for the reasons already discussed, we find theoretical and practical problems with including both scenopoetic and dynamically linked variables in the same niche hypervolume, especially since they are now confounded additionally by the geographic variables of the γ niche (e.g., the spatial configuration of geographic barriers to dispersal).

Rather, it is probably more sensible to define two environmental hypervolumes. One is composed of scenopoetic variables, at the scales and resolutions at which this definition seems more reasonable (i.e., the γ niche, and perhaps the β niche, if scenopoetic variables can define habitats). The other, complementary, hypervolume comprises bionomic variables *sensu* Hutchinson (1978), at the local scales at which these dimensions are more meaningful and that correspond to the α niche. This approach will be our framework in this book. Therefore, in the terminology of Silvertown et al. (2006), this book is about γ and β niches.

GRINNELLIAN AND ELTONIAN NICHES

The preceding discussion aimed to clarify the meaning of "niche" underpinning this book. The main meaning is explicitly geographic in nature, and is based on E-spaces composed of scenopoetic variables taken as conditions or requirements. These niches have been called "Grinnellian" (James et al. 1984) or "environmental" (Austin 1980, Austin and Smith 1989, Jackson and Overpeck 2000). We retain the term "Grinnellian niche" in view of the emphasis that Grinnell placed on niche as a tool in understanding geographic distributions of species. We are aware that Grinnell mixed to some extent features of the "role" and the "place" interpretations of the niche (Whittaker et al. 1973) and scenopoetic and bionomic variables, and that in its "role" part, he emphasized the "requirements" view (Leibold 1995). However, since "Grinnellian niche" has already been used in applications to geographically oriented situations with nonlinked environmental variables, we retain it throughout this book.

Of course, other extremes of niche meaning are possible and important; in particular, as we saw, niche concepts exist that are oriented toward community-ecology questions, defined at local scales, and including models of resource consumption and impacts. We will refer to this scale and meaning as "Eltonian niches." One of the best and most synthetic expositions of modern Eltonian niche theory is that by Chase and Leibold (2003). Ideally, to understand fully the geographic distributions of species, both Grinnellian and Eltonian niche elements are needed.

Do Empty Niches Exist?

The preceding discussion about niches as properties of the environment versus niches as properties of the species has led in the past to the question of whether such a thing as "empty niches" can exist (Colwell 1992). Unless rigorous definitions of terms are provided, of course, this question remains merely semantic. Grinnellian niches, defined as subsets of scenopoetic environmental spaces, are entirely different entities in this sense than Eltonian niches, defined in terms of zero-growth isoclines, impact vectors, and supply points. As such, the answer to the question depends, at the very least, on which of these two meanings is involved.

It is clear that—strictly speaking—neither Grinnellian nor Eltonian niches can be operationalized without reference to a particular species. Indeed, in the Grinnellian case, niches represent subsets of elements of the E-space defined in terms of the effects of certain variables on population growth rates of the species in question. In the Eltonian case, on the other hand, it is also only in terms of a given species in a given community context that resources can be specified, a consumption model proposed, isoclines obtained, impact vectors estimated, and the niche thus identified. In this sense, no "empty" Grinnellian or Eltonian niches can exist.

However, it is easy to produce examples of realistic and biologically meaningful E-spaces using variables known to be important to distributions of large classes of species, without specific mention of any species in particular. Figure 2.2 is an example in which all combinations of three important bioclimatic variables across the entire Western Hemisphere are presented. The Grinnellian niche (defined in terms of these three variables) of any species living in the Americas will be a subset of that E-space. Hence, in a very real sense, figure 2.2 displays the domain available for Grinnellian niches in the Americas for the variables considered; some sectors of this space (even if suitable) may indeed be unoccupied by a species in the region for a variety of reasons (see the following and chapter 8).

In contrast, it is very hard to illustrate a realistic zero-growth isocline Eltonian niche in resource space without reference to a particular species. (Presenting a hypothetical Eltonian niche, however, based only on verbal definitions of niches as roles or functions in a community, is not difficult.) This difference is a consequence of the fact that Grinnellian niches may also be regarded as an attribute (measured through scenopoetic variables, and certainly with heritable components related to the species' physiology) of the area occupied by the species across its geographic range, whereas Eltonian niches are attributes of *interactions* between a species and its local resources and other species. Hence, while it is intuitive and reasonable to refer to empty Grinnellian niches, it is probably meaningless to discuss empty Eltonian niches (Colwell 1992).

HUTCHINSONIAN IDEAS APPLIED TO GRINNELLIAN AND ELTONIAN NICHES
Hutchinson (1957) defined two subtypes of his multidimensional niches—the fundamental and the realized. For Hutchinson, the "fundamental niche" was the set of "all the states of the environment which would permit the species S to exist" (Hutchinson 1957). The "realized niche," on the other hand, is the subset of the fundamental niche corresponding to environmental conditions under which species S is a superior competitor and can persist (Hutchinson 1957) in an interacting environment. Chase and Leibold (2003) provided rigorous definitions of Eltonian requirement-impact niches, and found that fundamental and realized niches can be defined in their terms. It is also possible and useful to analyze this dichotomy in relation to Grinnellian niches, as we will see in chapter 3.

An important side note, however, is that other sorts of niches not considered by Hutchinson (1957) can be conceptualized for Grinnellian niches defined in terms of E-spaces and considering geographic extents. Hutchinson originally discussed niches in abstract and nonspatial ways, without providing actual examples and mostly ignoring the importance of scale and geography. Later, in his "What Is a Niche?" chapter, Hutchinson (1978) provided real-life examples of niches, but still refrained from outlining how consideration of broad spatial and temporal scales might introduce new possibilities.

Jackson and Overpeck (2000), working with scenopoetic environmental spaces and following the main ideas of Hutchinson, defined the fundamental niche as the "subset of the *environmental space* defined by the n dimensions, consisting of the suite of combinations of variables that permit survival and reproduction of individuals." This definition is perfectly compatible with Hutchinson's ideas; however, it begs the question of how can one estimate "the suite of

combinations" of favorable environments that composes the fundamental Grinnellian niche. This fundamental niche can, after solving several technical difficulties, be estimated via physiological experiments or biophysical first principles (Sutherst 2003, Kearney and Porter 2004, Crozier and Dwyer 2006,
Buckley 2008). The fundamental niche therefore is an expression of the physiology and behavior of an organism, and its definition can be obtained independently of the localities where a species is observed.

An important contribution to Grinnellian niche theory is the realization by
Jackson and Overpeck (2000) that the fundamental niche may include combinations of environmental variables currently missing in the existing E-space,
which is constrained by geography, and changes continuously across evolutionary time periods (Manning et al. 2009). As a consequence, E-space may
include large regions of biologically sensible, but currently lacking, combinations of variables. For example, in figure 2.2, we can observe a large gap corresponding to high precipitation and intermediate temperatures—areas holding
these conditions simply do not presently occur in the Western Hemisphere.
However, nothing is absurd in considering that a species may have physiological tolerances that permit maintenance of populations under some of these nonexistent environmental combinations. Consider, for example, species distributed
in archipelagos of small islands with limited sets of environments represented,
even though the species' distributional possibilities are much broader. Therefore, it makes sense to define a "potential niche" (Jackson and Overpeck 2000)
that is quite simply the intersection of the existing E-space (or the "existing
environmental space") with the fundamental niche—in other words, the portion
of the fundamental niche that actually exists somewhere in the study region at
the time of analysis. As we will discuss later, the term "potential" is somewhat
unfortunate in this context, but the concept is very useful, so in chapter 3 we
rechristen this concept as the scenopoetic existing fundamental niche. Also, as
discussed in chapter 3, other perfectly sensible Grinnellian niches not considered by Hutchinson can be defined by considering dispersal and movement in
spatially explicit analyses.

ESTIMATING GRINNELLIAN NICHES: PRACTICALITIES

Key to our exposition of many ideas in this book is a relationship between environmental and geographic spaces. By defining a geographic space (G-space),
at a given resolution, and at a particular point in time, and by intersecting that
area with digital data layers of environmental data, it is possible to extract subsets of the existing E-space that correspond to different regions of G-space.

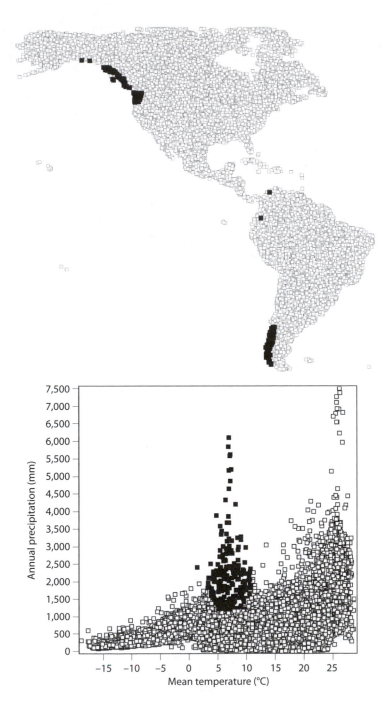

Conversely, GIS operations can be used to map in G-space all the regions that correspond to a given subset of E-space. Note that, for every point in G-space, one and only one point exists in E-space, whereas for each point in E-space more than one point may exist in G-space. This dual-space correspondence is illustrated in figure 2.3. E-space is changing all the time (Jackson and Overpeck 2000) according to the dynamics of global geology, climate, and physical environment in general.

The dramatic recent improvements in availability of data have enabled the work of pioneers like Grinnell, Hutchinson, and Austin to develop into a thriving area of macroecology and biogeography. In particular, as mentioned earlier this field includes what has been termed "species distribution modeling" (SDM; Guisan and Zimmermann 2000, Hirzel et al. 2002, Araújo and Guisan 2006), as well as the related (but by no means equivalent) endeavor named "ecological niche modeling" (ENM; Soberón and Peterson 2005, Soberon 2007). These fields—the subject of this book—consist of application of niche theory to questions about the observed and potential spatial distributions of species in the past, present, and future. In a very real sense, the availability of large quantities of data, technological developments like GIS, and several computational tools are enabling a multitude of applications that are not only of biological importance, but also can often be of extreme practical utility.

SUMMARY

The general idea of "niche" refers to the ecological conditions that a species requires to maintain populations in a given region, together with the impacts that the species has on its resources, other interacting species, habitat, and environment. The different emphases that various schools of thought have placed on different components of this central idea have led to a variety of particular niche concepts. For understanding of geographic distributions of species, two interpretations of the idea are important. The first, with roots in the work of Joseph Grinnell, emphasizes noninteractive unlinked variables measured mostly

---

FIGURE 2.3. Correspondence of geographic and environmental spaces, showing variation in mean monthly temperature and annual precipitation across the Americas in geographic and environmental dimensions. Cell resolution is 0.25°, or roughly 27 km at the Equator. The climatic data are drawn from the WorldClim dataset (Hijmans et al. 2005). An example of environments showing moderate temperature and high precipitation is shown in black in both spaces.

at coarse spatial resolutions. These dimensions are the scenopoetic variables of G. Evelyn Hutchinson and are very important in determining the broad aspects of species' distributions. Here, niches and distributions are estimated and visualized in associated geographic (G) and environmental (E) spaces. These niche dimensions (the Grinnellian niche, in effect) can be measured using data available in large quantities and estimated operationally. Another important perspective on the niche concept emphasizes impacts of a species on its environment, via linked variables and phenomena that are generally measured at finer spatial resolutions, which can be termed the "Eltonian niche." In this book, we focus on Grinnellian niches, as our interests lie primarily at geographic extents and resolutions.

CHAPTER THREE

Niches and Geographic Distributions

In chapter 2, we began developing and exploring a concept of niche that emphasizes multidimensional spaces of scenopoetic variables, typically measured at coarse spatial resolutions and over broad geographic extents. Such a niche concept not only has had a long and fruitful tradition in ecology, but also provides a natural connection to the study of geographic distributions of species and the broader field of biogeography. In this chapter, we develop this idea in greater detail.

First, we must consider the concept of the geographic distribution, or range, of a species, and the approaches available by which to measure it. Ranges are usually represented as maps, but maps of what? In a thought experiment, we could make all individuals of a species fluoresce and observe them from space (Brown 1995). We could then define a grid on the surface of the planet, and agree that the set of all cells within which fluorescent dots can be detected during a certain time interval constitutes the distribution of the species. Even such a detailed, pointillist map, however, would be a static simplification of the complex spatial and temporal footprint within which individuals of a species are distributed on Earth (Brown 1995).

It is not possible to follow each individual at the spatial resolution of its movements, except, perhaps, for a very few large-bodied and high-profile species (Maurer and Taper 2002). As a consequence, retreat to coarser resolutions becomes necessary, as well as adoption of conventions about which individuals and populations will be regarded as comprising the species' distribution—e.g., breeding versus migratory populations, or source versus sink populations (Udvardy 1969). The concept of the distribution is then related intrinsically to the resolution (grain) at which the grid is defined (Erickson 1945, Mackey and Lindenmayer 2001, Maurer and Taper 2002, Gaston 2003), although no single grid resolution can be taken as the "right" one (Gaston 2003): quite simply, distributions are scale-dependent. Therefore, we can define the distribution of a species as the set of all grid elements in which, within a given sampling time period, the probability of recording an individual of that species exceeds some given threshold. In some situations, to be more specific, it may be more important to emphasize "reproductive populations," rather than mere presence of individuals.

In sum, the concept of the distribution of a species includes the ideas of presence of individuals, presence of reproductive populations, and probability of detection, all considered at spatial and temporal resolutions and extents that are normally relatively coarse and broad, respectively. In what follows, then, the operational concept of distributional area of a species will be *subsets of geographic space in which the presence of individuals or populations of a species can be detected.* Some other areas, lacking observable populations or individuals but otherwise suitable, can also be defined. Relationships between areas of distribution and niches therefore depend on how ecological properties of species delineate subregions of the world for each species.

RELATIONS BETWEEN ENVIRONMENTAL AND GEOGRAPHIC SPACES

Let us go back to the two spaces introduced in chapter 2. One is a "geographic space," denoted by $\mathbf{G}$, which is composed of cells (= grid cells or pixels) covering a particular region. This space is usually two-dimensional (and possibly three-dimensional in some future applications), and the grid is characterized by a particular extent and a particular resolution (= grain). In most applications discussed later in this book, the grain is typically 1 km^2 per cell or more. The corresponding "environmental space" of environmental variables, at a given time, is denoted by $\mathbf{E}$. This space is defined by a suite of environmental attributes, such as climate, solar radiation, topography, and so on, all of which are generally characterized in relatively coarse environmental variables (see chapter 6). These dimensions are the scenopoetic variables of Hutchinson (1978). Hutchinson (1957) called $\mathbf{G}$ the "biotope." The linkage between environmental niches and the corresponding biotopes has been termed "Hutchinson's Duality" (Colwell and Rangel 2009). Now, we will explore this set of concepts in more formal terms.

Since one can take measures of v scenopoetic variables in each cell g in $\mathbf{G}$ (or in symbols, $g \in \mathbf{G}$), we can define vectors denoted by $\vec{e}_g = (e_1, e_2, \ldots e_v)_g$, one for each cell in $\mathbf{G}$, which describe the environmental characteristics of cell g. The space of all existing values of $\vec{e}_g$ comprises the environmental space denoted as $\mathbf{E}$. Jackson and Overpeck (2000) called $\mathbf{E}$ the "realized environmental space," noting that in geological time the set of environmental combinations existing is dynamic, and can change significantly through time. We refer to it simply as $\mathbf{E}$.

Particular environments may be repeated in different geographic regions (Aspinall and Lees 1994), so the number of elements (i.e., the "cardinality," denoted by vertical bars) of $\mathbf{E}$ and $\mathbf{G}$ may not be the same. Whether environments are repeated or not depends on the degree of heterogeneity of the environmental gradients, the resolution of the grid, the number of variables involved, and the precision with which they are measured. When many variables measured with high precision at relatively high resolutions are used to characterize the grid, in general, $|\mathbf{E}| = |\mathbf{G}|$.

Varying grid resolution creates an instance of the Modifiable Areal Unit Problem (MAUP; Fotheringham et al. 2000), a difficult conceptual problem in geography. Indeed, by changing grid resolution, different values of spatial statistics may be obtained (Openshaw and Taylor 1981). In the context of this book, this issue means that changing the grid may lead to distinct estimates of the niche of a species. The problem may not be very serious (Guisan et al. 2007), but neither is it likely to be trivial. In any case, the resolution and position of the grid, the precision at which variables are measured, and the methods for any changes in resolution should be stated explicitly. Otherwise, comparisons among results lack essential information.

Figure 2.3 presents an example of these two linked spaces. In that figure, a geographic space $\mathbf{G}$ is depicted with an extent covering the Western Hemisphere and resolution of cells of $0.25°$ (roughly 27 km on a side at the Equator). The corresponding space $\mathbf{E}$ is represented here by two climatic variables (mean annual temperature and annual precipitation) drawn from the WorldClim dataset (see chapter 6). In this case, because only two variables are used, and with only three significant digits, repeated environmental combinations are present, and $|\mathbf{E}| < |\mathbf{G}|$. In this example, the number of cells, $h = |\mathbf{G}| = 138,223$, which exceeds the number of distinct environmental combinations $|\mathbf{E}| = 98,432$. If three or more WorldClim environmental variables are used to characterize $\mathbf{E}$, however, no repeated environments are present, and $|\mathbf{G}| = |\mathbf{E}| = 138,223$.

Note that, with suitable tools such as GIS, for every spatial subset $\mathbf{G}' \subset \mathbf{G}$, it is simple to find the environmental subset of $\mathbf{E}$ associated with $\mathbf{G}'$. In other words, a function exists $\eta: \mathbf{G} \to \mathbf{E}$ that maps geography onto environment, or, in Hutchinson's (1957) terminology, biotopes onto environments. Implicitly, one assumes that it is the biotope of a particular species that is mapped by η; in this case, the environment would be a Grinnellian niche. The environment at a specific location g can be denoted $\eta(g) = \vec{e}_g$. We can denote the environments of sets of cells in symbols as $\eta(\mathbf{G}') = \{\eta(g) | g \in \mathbf{G}'\}$, which is termed in mathematics the "direct image" of $\mathbf{G}'$. The inverse operation maps sets of environments onto sets of geographic cells, and is defined as $\eta^{-1}(\mathbf{E}') = \{g \in \mathbf{G} | \vec{e}_g \in \mathbf{E}'\}$. In other words, $\eta^{-1}(\mathbf{E}')$ maps particular environments onto biotopes, or environmental space onto geography. Both functions, once again, can be implemented conveniently via GIS software, as was done to produce figure 2.3. However, several subtleties related to mapping continuous variables in discretized space need to be considered, such as mapping many-dimensional spaces onto fewer-dimensional subspaces, and issues related to the extremely different topologies (i.e., configurations) of **E** and **G** (Aspinall and Lees 1994, Stockwell 2007). Some of these problems will be discussed in detail in chapter 8.

THE ECOLOGICAL EQUATIONS

We have already discussed how distributional areas of species relate to the way in which individuals of the species are themselves distributed. We can begin exploring how spatially explicit population growth patterns can be related to variation in the ecological characteristics of species. Several approaches have been explored to achieve this relationship. For the sake of clarity, we will approach the problem first using a classic Lotka-Volterra phenomenological formulation. However, as will be shown later, for a fuller characterization, the Lotka-Volterra formulation should be replaced by a mechanistic consumer-resource model (MacArthur 1972, Tilman 1982, Chase and Leibold 2003, Meszéna et al. 2006).

THE PROBLEM OF MULTIPLE RESOLUTIONS

We consider the situation of $i = 1, 2, \ldots, s$ interacting species that inhabit **G**, which in turn is partitioned into a grid of h cells, $g = 1, 2, \ldots, h$, of a certain resolution. In practice, as we will see in chapter 6 and in many examples explored in this book, the resolution of the grid is relatively coarse. Indeed, given limitations of data availability and storage, as well as of computation, we will seldom find cells smaller than 1 km^2 in continental applications; indeed, in many cases, the resolution in such studies is on the order of 10–100 km^2 or coarser.

From a theoretical standpoint, if models are to be fully relevant, it is important to acknowledge that the resolution of the grid over which models are fitted should ideally be commensurate with the scale of the resource-consumer interactions characteristic of the Eltonian niche, as outlined in chapter 2. Essentially, this assertion means that at least two resolutions are relevant to understanding the relationship between niches and distributional areas. The first is the spatial scale best suited to studying resource consumption and the results of biotic interactions. Typically, this resolution is that at which Eltonian niches are expressed, which for many terrestrial vertebrates and plants is around 10^{-6}

to 10^{-2} km^2. However, such fine resolution is impractical when attempting to express areas of distribution, so one must aggregate thousands of the cells at the Eltonian resolution to reach grain sizes that are meaningful for describing geographic ranges. This rescaling of resolutions is one of the main theoretical problems challenging efforts to link Grinnellian and Eltonian niches. Of course, we are highlighting the extremes of a continuum, and the details will depend on the species and its autecology; still, the distinction is instructive, and so we explore it.

PHENOMENOLOGICAL EQUATIONS RELATING AREAS AND NICHES

Let us then explore factors affecting population growth rates of species in a spatially explicit grid of Eltonian resolution, using ideas first presented by Vandermeer (1972), and then developed for simplified environmental gradients by Holt et al. (2005 and 2009), and Pulliam (2000). The idea is to use spatially explicit population equations to represent the distribution of a species in space and then calculate the niches on the basis of the environments occurring in the region that the species is capable of occupying, actually or potentially.

Let $(1/x_{i,g})(dx_{i,g}/dt)$ be the per-capita growth rate of species i in cell g, where $x_{i,g}$ is the density of species i in cell g. Ignoring movements of individuals of the species among cells, the growth rate is the difference between an intrinsic growth rate (r) and a regulatory term φ that depends on the densities of all other species:

$$\frac{1}{x_{i,g}} \frac{dx_{i,g}}{dt} = r_{i,g}(\vec{e}_g) - \varphi_{i,g}(\vec{e}_g, \vec{R}_{i,g}; \vec{x}_g). \qquad (3.1)$$

The vector $\vec{e}_g$ represents the values of the ν scenopoetic variables that affect the growth rates of all species under consideration. The vector $\vec{R}_{i,g}$ represents equation parameters related to interactions with other species (including resources), which we will call, for lack of a better term, "biotic parameters." Later, we present a more explicit interpretation in a mechanistic resource-consumer model. The function $\varphi_{i,g}(\vec{e}_g, \vec{R}_{i,g}; \vec{x}_g)$ represents the regulation term (described later) of equation 3.1 (Meszéna et al. 2006). This overall function is directly related to the Eltonian niche. We now proceed to explore $\vec{x}_g$.

In the classic niche literature, the only population interactions considered are competitive and predator-prey interactions. The inclusion of positive (mutualistic) interactors, however, represents an important gap in niche theory (Colwell and Fuentes 1975, Araújo and Guisan 2006). Although we acknowledge this gap, we restrict our discussion to negative interactions, since available theory regarding Eltonian niches has disregarded mutualism almost entirely (Chase and Leibold 2003; but see chapter 8). Including such classes of interactions will in all likelihood alter many concepts in niche theory. In general terms, then, $\varphi_{i,g}$ is a function of the vector $\vec{x}_g$, which represents the population densities of all species occurring at a given time in cell g.

The still-not-spatial equation 3.1 expresses the Grinnellian and the Eltonian niche effects in contrasting ways: the intrinsic growth rate $r_{i,g}(\vec{e}_g)$ depends only on scenopoetic parameter values, and is directly related to Grinnellian niches (although, as we will see later, it should include effects of resources that are manifested at resolutions much finer than those of the climatic data usually considered in niche modeling applications). On the other hand, Eltonian niches, which depend on interactions with other species, require full consideration of the regulation term $\varphi_{i,g}(\vec{e}_g, \vec{R}_{i,g}; \vec{x}_g)$, which determines equilibrium values (or limit cycles, or even strange attractors) and also the effects of resources and dynamically linked species. Eltonian niches may also require consideration of initial conditions, when stable equilibria have nonglobal basins of attraction, or for unstable equilibria, as may happen in the case of two-species Lotka-Volterra competition (Begon et al. 2006).

Now, consider the situation when species can move around within **G**. This possibility introduces a third term ψ into equation 3.1:

$$\frac{1}{x_{i,g}} \frac{dx_{i,g}}{dt} = r_{i,g}(\vec{e}_g) - \varphi_{i,g}(\vec{e}_g, \vec{R}_{i,g}; \vec{x}_g) + \psi(\mathbf{T}_i; \vec{x}_i). \tag{3.2}$$

$\mathbf{T}_i$ denotes a transition matrix (Vandermeer 1972) expressing the instantaneous probabilities of all intercell movements for species i. The vector $\vec{x}_i$ represents the entire metapopulation of species i, or the population density of species i in every cell in **G** at a given time. If $\mathbf{T}_i$ is not irreducible (Bailey 1964), then it may include submatrices that represent mutually inaccessible regions. Any large geographic extent is likely to have inaccessible subsets, given the dispersal potential of a particular species, which might correspond to isolated areas such as islands, or areas on opposite sides of barriers like mountain ranges or large rivers. Although, strictly speaking, the movement term $\psi(\mathbf{T}_i; \vec{x}_i)$ should include also the corresponding matrices and metapopulation structure of the competitors and predators of i, this addition only complicates an already seriously complex model and does not add any substantial insights to the discussion that follows, so we will dispense with such terms.

As has been discussed already by many authors (Grinnell 1917, Good 1931, Hutchinson 1957, MacArthur 1972, Holt and Keitt 2000, Pulliam 2000, Pearson and Dawson 2003, Araújo and Guisan 2006), interactions between the three sets of factors (density-independent growth rate, biotic interactions, and dispersal capacity and movements) determine the area of distribution of a species (Soberón and Peterson 2005). In fact, we are also ignoring evolutionary factors and perturbations, which would complicate the problem even more (Kirkpatrick and Barton 1997). Equation 3.2 therefore represents an attempt to reduce the problem of defining distributional areas to the fundamentals of population dynamics. As discussed later, considerations of intrinsic and instantaneous growth rates, and of regions in which population growth rates are positive, allow rigorous definition of subsets of **G**, which represent areas of particular interest. Conversely, the environments present in these geographic subspaces allow us to define already-described niches, as well as new types of niches, in a rigorous way.

THE BAM DIAGRAM: A THINKING FRAMEWORK

The generally defined model in equation 3.2 allows not only for local equilibria of species, but also for limit cycles and strange attractors; however, for the purposes of this book, we do not need a full solution, and it is possible to use a static representation of equation 3.2 to discuss extreme situations and define niches. Initial discussions and verbal models were provided by Pulliam (2000) and Soberón and Peterson (2005), who published Venn diagrams describing the simultaneous influence of environmental conditions, biotic interactions, and dispersal in shaping species' geographic distributions. The spatial structure also allows for traveling waves and other complex forms of behavior (Solé and Bascompte 2006). This wealth of spatiotemporal behavior is very interesting from the point of view of population dynamics, but is not necessary to define Grinnellian niches.

We use a simple diagram (figure 3.1 and table 3.1) to display the joint fulfillment of the three sets of conditions that appear in dynamic form in equation 3.2. Set **A** represents regions in geographic space where (mainly abiotic) scenopoetic conditions (and existing resources) allow intrinsic growth rates to be positive, i.e., $r_{i,g} > 0$. Another set **B** represents the geographic regions where the interacting factors (mainly biotic interactions with other species) are favorable for the presence of the species. Finally, a third set, **M** (relating to movements of individuals of the species) corresponds to the geographic regions that have been accessible to the species within a given time span [e.g., dispersal over recent generations or since the Last Glacial Maximum (LGM)]. $\mathbf{G}_O = \mathbf{A} \cap \mathbf{B} \cap \mathbf{M}$ is then defined as the "occupied distributional area." It is the subset of the accessible region in which both scenopoetic and biotic conditions permit the species to maintain populations, and is synonymous with the "realized range" of Gaston (2003).

$\mathbf{G}_I = \mathbf{A} \cap \mathbf{B} \cap \mathbf{M}^C$, on the other hand, is the "invadable distributional area" that the species could occupy if present distributional constraints were to be

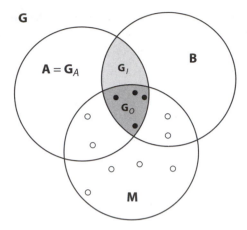

FIGURE 3.1. A simplified heuristic device termed the "BAM diagram," which depicts the interaction between biotic (**B**), abiotic (**A**), and movement (**M**) factors. Four areas are depicted: **G** the geographic space within which analyses are developed, $\mathbf{G}_A$ = the abiotically suitable area, $\mathbf{G}_O$ = the occupied distributional area, and $\mathbf{G}_I$ = the invadable distributional area. Circles indicate occurrence data: solid circles indicate presences, and open circles indicate absences. Note that the only data relevant in calibrating niche models are those within **M** and **B** (see Barve et al. 2011); for this reason, no absences are depicted outside of **M**, although the species is indeed absent here.

overcome. ($\mathbf{M}^C$ is the complement of **M**, or all of the areas which to the species is currently incapable of sending migrants.) The union of occupied and invadable areas $\mathbf{G}_P = \mathbf{G}_O \cup \mathbf{G}_I$ can be defined as the "potential distributional area" (Gaston 2003) of the species.

The types of populations occurring across space can also be displayed using the BAM diagram framework. Source populations, by definition, can occur only

TABLE 3.1. Summary of distributional areas (in geographic space, **G**) and corresponding Grinnellian niches (in environmental space, **E**), as defined in this book using scenopoetic (noninteractive) variables.

Distributional areas		Grinnellian niches	
Symbol	Name	Symbol	Name
$\mathbf{G}_A$	Abiotically suitable area	$\mathbf{E}_A$	Existing fundamental niche
$\mathbf{G}_P$	Potential distributional area	$\mathbf{E}_P$	Biotically reduced niche
$\mathbf{G}_I$	Invadable distributional area	$\mathbf{E}_I$	Invadable niche
$\mathbf{G}_O$	Occupied distributional area	$\mathbf{E}_O$	Occupied niche

See figure 3.1 for a visual representation of relationships among these distributional areas. Note that $\mathbf{E}_A$, the existing fundamental niche present in a given **G**, will likely represent only a subset of the full scenopoetic Grinnellian fundamental niche, $\mathbf{N}_F$.

within $\mathbf{G}_O$, whereas sink populations may occur anywhere outside $\mathbf{G}_O$ but within $\mathbf{M}$. Observations of "presence" of a species can originate anywhere within $\mathbf{M}$, while "absences" can come from anywhere within $\mathbf{G}$, either because the species is genuinely absent, or resulting from one of many factors that could cause nondetection (see chapter 5).

More complicated situations can also create difficult-to-interpret "presences." For instance, Holt (2009) described a case in which scenopoetic conditions are physiologically favorable, but Allee effects create a threshold of invasibility, leading to "establishment" and "persistence" niches: a population may be found within an area with favorable scenopoetic conditions (i.e., inside the establishment niche), but at a density below its Allee threshold (i.e., outside its persistence niche); this effect is mostly Eltonian in nature. In chapters 5 and 7, we will discuss effects of different types of presences and absences, and their effects on the process of modeling ecological niches.

ECOLOGICAL NICHES AND GEOGRAPHIC DISTRIBUTIONS

MacArthur (1972) and Tilman (1982), assuming a simple substitutable resources model, have both shown that, in a linear, resource-consumer model, intrinsic growth rates are simple functions of an environmentally determined death rate d_i and the availability of resources at their equilibrium level (in absence of consumers), which we denote by R_l^*, as follows:

$$r_{i,g} = -d_i(\vec{e}_g) + \sum_{l=1}^{n} w_{i,l} R_l^* a_{i,l}.$$ (3.3)

The term $a_{i,l}$ represents the per-unit time probability of finding the resource l, assuming that it exists at its equilibrium level R_l^*, and therefore that $a_{i,l} R_l^*$ represents the amount of resource encountered per unit time; the $w_{i,l}$ factors are conversion parameters to transform resource-encounter rates to units of population growth rate. Equation 3.3, therefore, can be used to define region $\mathbf{A}$ precisely, as

$$\mathbf{G}_A = \{g | -d_i(\vec{e}_g) + \sum_{l=1}^{n} w_{i,l} R_l^* a_{i,l} > 0\}.$$ (3.4)

In words, $\mathbf{G}_A$ (or $\mathbf{A}$ for short) is the region (equivalent to Hutchinson's biotope) where, in the absence of competitors and other negatively interacting species, and Allee effects (and implicitly given unlimited dispersal abilities), species i will be able to establish populations (Holt et al. 1997, Pulliam 2000). $\mathbf{G}_A$ represents the set of all cells in which the scenopoetic environment is favorable for species i and resources are available to sustain at least a small population, and we term it the "abiotically suitable area." The environment associated with $\mathbf{G}_A$

is $\eta(\mathbf{G}_A) = \mathbf{E}_A$, obtainable by standard GIS operations. We suggest that $\mathbf{E}_A$ corresponds very closely to the "potential niche" of Jackson and Overpeck (2000). The term "potential niche" may be somewhat unfortunate, however, since it represents the currently existing manifestation of the fundamental niche (see chapter 2) that is in reality available at the moment, rather than the species' potential. Moreover, other subregions of E-space may much more appropriately be termed "potential," as we will see later. For this reason, we decided to rebaptize Jackson and Overpeck's (2000) "potential niche" as the "existing fundamental niche," to stress the fact that it represents the existing subset of $\mathbf{E}$ that is within the tolerance limits of a species (its fundamental niche). This niche is the intersection of the fundamental niche, which, as we said before, may be obtainable only by mechanistic methods (Kearney and Porter 2004), with the available set of environmental conditions. In symbols, if $\mathbf{N}_F$ represents a fundamental niche estimated independently, $\mathbf{E}_A = \mathbf{E} \cap \mathbf{N}_F$ (Soberón and Nakamura 2009). Conceptually, the fundamental niche $\mathbf{N}_F$ is a physiological characteristic of a species, defined independently of the existing environment and the existing fundamental niche $\mathbf{E}_A$ may be estimated using nonphysiological techniques, and perhaps even correlative modeling (see the following).

The concept of fundamental niche of Hutchinson (1957), defined in terms of favorable conditions and lack of competition, ignored complications owing to nonexistence of conditions favorable for the species across the specific landscape of interest (Jackson and Overpeck 2000). Moreover, recall that in his famous *Concluding Remarks*, Hutchinson (1957) did not distinguish between bionomic and scenopoetic variables. This distinction is critical in our definition of Grinnellian scenopoetic niches. Recent authors have published concepts akin to Hutchinson's fundamental niche that are based on scenopoetic variables, such as the fundamental niche of Jackson and Overpeck (2000), the Grinnellian niche of Pulliam (2000) and James et al. (1984), and the environmental niche of Austin et al. (1990). Since in theory it is feasible to define Eltonian fundamental niches (Chase and Leibold 2003), to avoid confusion, we henceforth refer to the Grinnellian fundamental niche $\mathbf{N}_F$ as the "scenopoetic fundamental niche."

SPATIAL RESOLUTION OF SCENOPOETIC VARIABLES

In the literature of niche and distribution modeling, scenopoetic environmental variables have been used at coarse resolutions, mostly for reasons of data availability. It is possible, however, to model using some higher-resolution variables. In a recent example, Heikkinen et al. (2007) modeled distributions of several species of owls as a function not only of the customary habitat and climatic variables, but also of the presence of an important biotic component:

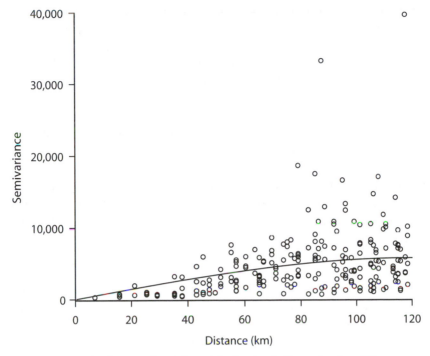

FIGURE 3.2. Semivariogram of annual precipitation (mm) climate data from Hijmans et al. (2005), based on 300 random points distributed across the central
Great Plains of the United States. The range (i.e., the distance over which points
are not likely to be independent owing to spatial autocorrelation) calculated from
this analysis was 118.5 km.

woodpeckers that excavate tree holes that the owls use as nesting sites. Both of
the latter suites of variables may be regarded as scenopoetic, in the sense that
the owl populations do not affect either climate or numbers of woodpecker
holes, but their resolutions are different. The finer-resolution variable represented by availability of nesting sites is unaffected by owl populations (assuming that the two do not compete for holes and that owls do not eat woodpeckers
commonly), and in this sense constitutes a fine-resolution scenopoetic variable. Most of the examples we will present in this book, however, focus on
relatively coarse-resolution scenopoetic variables—that is to say, their semi-
variograms or autocorrelograms show broad spatial lags, in typical situations
with vertebrates on the order of 10^2 to 10^3 km (figure 3.2). This resolution is a
natural one at which to represent the coarse-grained features of the distributional area of a species, but it also means that a single coarse-grained cell in the

grid may contain thousands of the Eltonian cells used in equation 3.4. Implications of this assertion are discussed later.

ESTIMATION OF THE FUNDAMENTAL AND EXISTING FUNDAMENTAL NICHES
How can one estimate scenopoetic niches? Different avenues are available to estimate different niches (Holt 2009). In the first place, N_F can be studied mechanistically by means of process-based physiological models (Sykes et al. 1996, Porter et al. 2002, Sutherst 2003, Kearney and Porter 2004, Buckley 2008). This approach requires resorting to first principles of energy transfer and function to describe the ecological physiology of a species, and has been applied to a variety of species of animals. For example, Kearney and Porter (2004) studied *Heteronotia binoei*, an Australian gecko, and produced a model of its thermal limits, water tolerance, and time needed for egg development. These values reveal, at least partially (Godsoe 2010), the species' scenopoetic fundamental niche, N_F. The results were integrated with a microclimatic model that delineated above-ground temperature profiles and relative humidity across Australia. The interaction of these two models leads to predictions about the spatial extent of the existing fundamental niche E_A, or $G_A = \eta^{-1}(E \cap N_F)$, which the authors expressed in terms of the probability of observing the species. The maps generated had a spatial resolution of 0.05°, or about 5 km on a side.

A related technique involves laboratory experiments or greenhouse studies to determine the ranges of conditions under which a species can live. It has been used, for example, to estimate two variables (temperature and moisture) in the fundamental niche of flour weevils (*Calandra* spp.; Birch 1953) and pH and Ca^+ concentration constraints for the daphnia *Daphnia magna* (Hooper et al. 2008). These techniques are potentially quite powerful, as they allow estimation of N_F directly, at least for some variables (Godsoe 2010). It also has the major theoretical advantage of being based on direct information about the biology of a species, which can be obtained independently of the correlative methods that are the main focus of this book.

The main drawback of these mechanistic approaches to measuring N_F is that they require laborious experiments, measurements, and/or calculations, and therefore are not readily applicable to large numbers of species. A related problem is that most climatic data are not in the form of variables related directly to limiting physiological mechanisms. For example, derived climatic information such as degree days needed for development in ectotherms, or growing days to budburst important to certain trees, must be obtained from available climate data (Sykes et al. 1996), which are normally available only at low spatial resolutions. Variables related to the physiology of organisms require some sort of modeling, as exemplified by Kearney and Porter (2004). However, in general at least, as the example cited and others show, direct, mechanistic estimation of N_F is possible, thus providing a potentially powerful and independent means of estimating the **A** region in the BAM diagram, since by hypothesis $G_A = \eta^{-1}(E \cap N_F) = A$.

A second method would estimate E_A via G_A. In a thought experiment, imagine that individuals of a species could be introduced at low to moderate densities in all cells in **G**, in experimental settings suitably replicated, and protected from competitors, predators, and diseases. After some appropriate length of time, the set of cells where the species shows a positive intrinsic growth rate $\{g|r_{i,g} > 0\}$ is an estimate of G_A; then, standard GIS operations can be used to perform the operation $E_A = \eta(G_A)$. To our knowledge, experimental estimation of G_A using methods of this sort has never been attempted, even though estimating G_A empirically is what farmers have been doing since the Neolithic, and transplants of individuals to check their population responses are indeed feasible (Angert and Schemske 2005).

The third method relies on correlative methods, using the approaches here referred to as "ecological niche modeling," to estimate E_A from observations of the presence of a species in relation to environmental variation. The feasibility of estimating the existing fundamental niche using distributional data has been contentious, since it is likely that the occupied distribution of a species already includes reductions owing to biotic interactions and dispersal limitations. As we will see later, under particular configurations of the BAM diagram, such correlative estimates of environments associated with the occupied area of a species may coincide with E_A; however, without assumptions and hypotheses about the configuration of the BAM diagram, outputs of ecological niche modeling exercises cannot be equated simply and directly to E_A.

THE BIOTICALLY REDUCED NICHE

In the preceding thought experiment, G_A was estimated by introducing propagules (protected from competitors, predators, and other negative interactors) over all of **G**, and observing which introductions were able to establish initial populations. Now, consider a second thought experiment: Imagine that competitors and other negative interactors are allowed contact with the populations of the focal species that were introduced across **G** in the first experiment. After some steady state has been reached, we remove every cell in which the negative interactors were capable of excluding the populations of the species of interest. This new area, by hypothesis, will be smaller than or equal to G_A (but see the following; such is the case because positive interactions are neglected in these treatments)—we can refer to this area as $\mathbf{G}_P$, or the "potential distributional area." Why potential? Because $\mathbf{G}_P$ is the area that the species could occupy if it were able to disperse there.

This thought experiment has given us information about a potential distributional area $\mathbf{G}_P$, which is composed of two regions, $\mathbf{G}_O$ and $\mathbf{G}_I$, so $\mathbf{G}_P = \mathbf{G}_A \cap \mathbf{B}$ $= \mathbf{G}_O \cup \mathbf{G}_I$ (figure 3.1). We called $\mathbf{G}_O = \mathbf{G}_A \cap \mathbf{B} \cap \mathbf{M}$ the "occupied distributional area" or occupied range: the area that is suitable from both abiotic and biotic perspectives *and* that has already been reached by individuals of the species capable of founding populations. Only $\mathbf{G}_O$ is actually amenable to being studied using observational data. Most maps of species' distributions are geographically realistic attempts to depict $\mathbf{G}_O$ (as opposed to abstractions, like the BAM diagram).

The "invadable distributional area" $\mathbf{G}_I$ is suitable to a species from the scenopoetic and biotic perspectives, but has not been accessible to individuals of the species in question. The identification of such invadable areas has seen considerable study in the world of invasive species biology (NAS 2002), although not in a spatially explicit manner until the advent of ecological niche modeling (Peterson 2003a). $\mathbf{G}_I$ can be estimated by transferring (i.e., applying across space) models of the environments in $\mathbf{G}_O$ and assuming that effects of biotic interactions on $\mathbf{G}_A$ are parallel in the two regions. More direct approaches would require something along the lines of the second thought experiment (wherein competitors and other negative interactors are allowed to interact with the populations of the focal species), which is almost impossible in practice. Chapter 13 treats the special challenges and problems involved in estimating $\mathbf{G}_I$.

Kearney (2006) suggested that $\mathbf{G}_I$ could be estimated by means of competition experiments conducted in tandem with physiological measurements. This approach may prove feasible if the interactor species are few and have all been identified—in this case, experiments can be performed, if biotic interactions do not induce complicated dynamics (see the following), and if they are sufficiently known. However, much of the theory on competitive interactions stresses that spatial heterogeneity plays a strong determinant role in the outcome of interactions (Tilman 1982, Chesson 2000, Amarasekare 2003, Hirzel and Le Lay 2008). At the large extents characterizing most biogeographic work, performing experiments that would cover the entire set of conditions exhaustively will often prove unfeasible.

What are the subsets of E-space that correspond to $\mathbf{G}_O$ and $\mathbf{G}_I$? We call the subset of environmental space defined by $\mathbf{E}_P = \eta(\mathbf{G}_P)$ the "biotically reduced niche." Conceptually, it is closely related to the realized niche of Hutchinson (1957), since it represents the part of the existing fundamental niche $\mathbf{E}_A$ that remains habitable after reductions by competitors and other negative interactors. Still, since this niche is defined using only scenopoetic variables, and because the distinction between occupied and potential distributional areas (i.e., the effects of $\mathbf{M}$) remains unclear, we will avoid using the term "realized" niche to refer to $\mathbf{E}_P$. Similarly, the sets in environmental space corresponding to $\mathbf{G}_O$ and $\mathbf{G}_I$ are termed the "occupied niche" and the "invadable niche," respectively, and are denoted as $\mathbf{E}_O$ and $\mathbf{E}_I$. Since $\mathbf{G}_P = \mathbf{G}_O \cup \mathbf{G}_I$, it follows from the niche-space operations in appendix B that $\eta(\mathbf{G}_P) = \eta(\mathbf{G}_O) \cup \eta(\mathbf{G}_I)$, or $\mathbf{E}_P = \mathbf{E}_O \cup \mathbf{E}_I$.

An important question is whether $\mathbf{E}_O \approx \mathbf{E}_I$, which is often assumed to be the case (Guisan and Thuiller 2005), and may frequently be *mostly* true. The success of numerous examples of predictions of the geographic potential of species' invasions strongly suggests that the assumption is correct (Peterson 2003a), but this question is explored further in chapter 13. Appendix A summarizes the notational and terminological orgy presented in this chapter.

Caveats about Reduction of Grinnellian Niches and the Eltonian Noise Hypothesis

Although the preceding arguments appear to be straightforward, the biotic reduction of $\mathbf{E}_A$ to $\mathbf{E}_P$ is a complicated issue, as we now discuss. In figure 3.1, the region $\mathbf{B}$ represents the spatial subregion of $\mathbf{G}$ in which the aggregated biological milieu (McGill et al. 2006) permits the species in question to maintain populations. It is the intersection of $\mathbf{B}$ and $\mathbf{A}$ that Hutchinson (1957), at his community-level scale, regarded as yielding the reduced "biotope" that corresponds to the realized niche. Although this situation appears simple, two issues require clarification (Soberón 2007), one related to the predictability of the outcome of interactions, and the other related to the scaling of observations.

Outcomes of interactions. Consider the simplest model, with no considerations of limited migration and with only competitors and a "mean field" (Rescigno and Richardson 1973, Molofsky et al. 1999) biotic parameter $c_{i,g}$. (A "mean field" competition model assumes that the competitive effects can be summarized as an average of competitive effects by all species pooled together, which is denoted by $c_{i,g}$.) In this extreme case, only one species can survive per cell, as in the competitive exclusion principle attributed to Gause (1936), and $\mathbf{B}$ can be defined in terms of parameters of equation 3.2: $\mathbf{B} = \{g | r_{i,g}(\vec{e}_g)/c_{i,g}$ is a maximum}. Although information about the spatial distribution of $c_{i,g}$ is needed (which would be a daunting task), finding $\mathbf{B}$ under such a hypothetical situation is not, in principle at least, impossible.

However, under a more comprehensive, resource-consumer model (Tilman 1982, Chase and Leibold 2003), defining $\mathbf{B}$ in terms of parameters requires knowledge of the precise resource-exploitation model, the initial conditions, resource supply points, and their location in relation to the impact vectors. Therefore, for a more realistic resource-consumer model, it usually would be impractical to measure parameters and define **B** operationally, in terms of sets of cells that fulfill the conditions. Even worse, in certain realistic scenarios, results of complex competitive interactions cannot be predicted *even with* full knowledge of the parameters (Huisman and Weissing 2001), so definition of **B** in terms of the parameters of the equations may be fundamentally impossible. It is likely, then, that direct estimation of **B** from experimental data may be possible only in extreme situations in which competitive interactions are simple and apparent, or phenomenologically, a posteriori from observations (e.g., Anderson et al. 2002b). In any case, the problem of obtaining the experimental data regarding the presence of negative interactors across broad spatial extents remains a formidable practical problem (Araújo and Guisan 2006).

Scaling and biotic reduction of niches.   The second problem with defining **B** in terms of Eltonian processes is one of spatial scale, in particular of spatial resolution. As we have seen, **B** often has a fine-grained structure, since it is defined mainly by the outcomes of the dynamics of equation 3.2, and so requires knowledge of resource consumption and biotic interactions. We illustrate this point in figure 3.3, which is a BAM diagram modified to show explicitly that **A** and **B** are manifested at different spatial resolutions (Pearson and Dawson 2003).

Although competition for resources and indirect competition via shared predators are manifested at the scale of single grid cells, **A** is defined at the scale of clusters of cells sharing similar values of the (usually coarser-grained) scenopoetic variables. A wealth of ecological theory and experience suggests that, at broad spatial extents (i.e., large enough to include spatial heterogeneity), competitors can coexist for long periods of time (Tilman 1982, Chesson 2000, Amarasekare 2003, Hirzel and Le Lay 2008). In this framework, by aggregating many local cells to compose the coarser-grained cells that define species' geographic distributions, different outcomes of competitive interactions are averaged, or at least muted, and the coarse-grained manifestations of fine-grained processes may not even be noticeable (Soberón 2010). For example, assuming the most extreme scenario, in which many local populations are extirpated by competitors, the coarse-scale pattern is not affected unless the extirpation takes place in all of the local cells in **B** that together compose a coarse-grained cell in **A** (figure 3.3, right side).

Even if competitors *are* capable of excluding populations of a given species in the entirety of a coarse-resolution grid cell, the set $\mathbf{E}_A$ would not be reduced

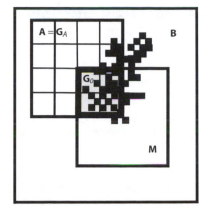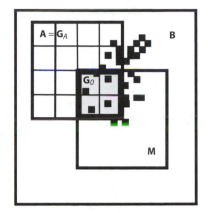

FIGURE 3.3. A BAM diagram redrawn to stress differences in spatial resolution of **B** versus **A** and **M**. The large box in each panel indicates **G**, the geographic space under consideration. **A** and **M** are shown as boxes, with coarse-grained cells in **A** to indicate the broad autocorrelation structure of most scenopoetic variables. **B** is shown as finer-resolution cells (small dark squares indicate biotically suitable sites) to emphasize the more restricted autocorrelation structure of bionomic variables. The right-hand panel shows a reduced **B**, perhaps owing to more intense competition, that nonetheless does not affect the coarse-resolution distribution of the species in terms of cells of **A**.

unless the cell in question presents a unique combination of niche conditions (Pearson and Dawson 2003). The possibility of competitive exclusion taking place (in local cells), but not reducing E_A (as a consequence of the different scales at which the two niches are defined, and possibly owing to geographic heterogeneity in the distributions of the competitors) is in stark contrast with the case of the Eltonian niche and Hutchinson's (1957) analysis. In the case of the latter niche, competitive exclusion at local levels always implies both alteration of the fundamental niche and reduction of occupied local-level cells (Chase and Leibold 2003).

Examples of local competition leading to spatially broad extirpation of populations have been documented, meaning that local effects can indeed affect geographic distributions (Bullock et al. 2000, Leathwick and Austin 2001). Such can be the case when broad-scale overlap between competing species is prevented by competitive interactions at the edges of species' ranges. Hence, no need exists for extirpation to occur in all of the local cells: it simply needs to occur in the areas of intersection between the competing species' ranges. Examples of such situations include the creation of "suture zones" in areas where populations of closely related and competing species come into contact, e.g., after dispersing out of glacial refugia after cold periods (Hewitt 2000). Other examples of co-occurrence and co-exclusion patterns being driven by biotic interactions include mutualistic interactions at local scales that affect patterns at macroscales (e.g., Heikkinen et al. 2007, Hampe 2004).

However, given the preceding arguments, the details of Eltonian processes probably are not frequently dominant in determining the broad brush-stroke characteristics of species' geographic distributions (Pearson and Dawson 2003, but see Anderson et al. 2002b). Another point of view is that, for many species, $\mathbf{A}$ and $\mathbf{B}$ may coincide broadly—that is, interactive effects may not influence the scenopoetic fundamental ecological niche at geographic extents and resolutions. Of course, the idea that $\mathbf{A} \approx \mathbf{B}$ for a particular species is a matter for empirical verification; the issue also will depend on the spatial extent of $\mathbf{G}$ and the spatial resolution of the grid that partitions it.

The fact that models based only on scenopoetic variables at relatively coarse resolutions are frequently capable of powerful predictions, and particularly in cases in which testing is transferred among regions or time periods (e.g., Iguchi et al. 2004, Araújo et al. 2005a), suggests that in many cases $\mathbf{A} \approx \mathbf{B}$. We call this idea—which once again is an issue for empirical testing—the "Eltonian Noise Hypothesis," because, if $\mathbf{A} \approx \mathbf{B}$, Eltonian processes would act principally to generate noise in correlations between scenopoetic variables and patterns of occurrence of species. We return to this hypothesis at numerous points in the remainder of this book.

ESTIMATING GEOGRAPHIC AREAS AND ECOLOGICAL NICHES

In the final topic of this chapter, we introduce conceptual aspects of the main theme of this book: how to estimate areas of distribution and their associated environmental subspaces (i.e., ecological niches). This discussion is an elaboration of Hutchinson's duality of niche and biotope (Hutchinson 1957, Colwell and Rangel 2009). Chronologically speaking, the first problem that researchers attempted to address using niche modeling (ENM and SDM) was estimation of $\mathbf{G}_O$, the current area of distribution of a species (Busby 1991, Guisan and Zimmerman 2000, Peterson 2001). This initial challenge can be separated into two sorts of problems: "spatial interpolation" and "spatial transferability" (see chapter 10). As used here, spatial interpolation refers to extension of known distributions into areas that are not statistically independent of the areas used for model calibration (i.e., spatially autocorrelated), whereas spatial transferablity refers to inferences regarding areas that are statistically independent of those used for model calibration (i.e., areas not spatially autocorrelated, usually different landscapes or different time periods with different climates).

Note that having obtained (by whatever means) an estimate of the occupied distributional area $\hat{\mathbf{G}}_O$ (the ^ indicates an estimate), the simple GIS-based operation $\eta(\hat{\mathbf{G}}_O)$ produces an estimate of the occupied niche $\hat{\mathbf{E}}_O$. This idea is illustrated in figure 3.4, which depicts the relationship between a species' distribution in geographic and environmental spaces. Here, the crosses indicate known occurrences, and the area of interpolation is represented by the gray region surrounding the crosses. Spatial interpolations can be performed based on nothing more than the spatial arrangement of observed presences, because of the strongly autocorrelated structure of most spatial data (Bahn and McGill 2007). However, often, we are interested not only in areas around known occurrences, but also in predicting *beyond* known regions of occurrence, via "spatial transferability," represented by areas A and B in figure 3.4. Such problems call for algorithms capable of generalizing to regions beyond that used for calibrating the models, and require more than spatial interpolation because the only common link between the different areas is the environmental conditions that they manifest (see chapter 10).

When data documenting "true absences" (absences due to the species not being present, rather than to insufficient exploration, see chapter 5) are available, estimation of $\mathbf{G}_O$ is best achieved by means of conventional statistical techniques that contrast explicitly populations of presence and absence points. Approaches such as generalized linear modeling (GLM), generalized additive modeling (GAM), regression trees, and others are well suited to this task (e.g., Guisan et al. 2002). Leaving aside the statistical methods, another distinction lies with the integration of true absences in the calibration process, which introduces effects other than those of the environmental variables in $\mathbf{E}$, such as geographic barriers to dispersal. The suite of applications that use true absence data in addition to presence records is what should be termed species distribution modeling (SDM) *sensu stricto*, and will generally fall in the realm of SDM applications, which are fairly well explored and documented in the literature (Ferrier et al. 2002, Thuiller et al. 2004a, Guisan et al. 2006).

Most biodiversity datasets, however, lack true absences (see chapter 5 for detailed discussion of some of the terms that follow). Lacking true absence data, presence-only data (as well as presence/pseudoabsence data and presence/background data; see chapter 5) alone are not enough to estimate probabilities of presence of a species (and therefore $\mathbf{G}_O$), unless certain configurations of the BAM diagram are known or assumed (Phillips and Dudík 2008, Phillips et al. 2008, Anderson and Raza 2010). Rather, models (binary or continuous) indicating relative suitability can be generated. In fact, as explained in chapter

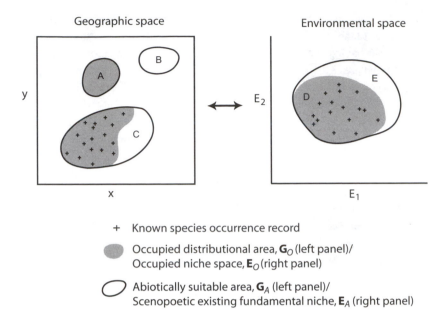

FIGURE 3.4. Geographic and environmental spaces for a hypothetical species. Observed presences $\mathbf{G}_+$ are shown as +'s; the occupied distributional area $\mathbf{G}_O$ and the occupied niche space $\mathbf{E}_O$ are shown with gray shading; and the abiotically suitable area $\mathbf{G}_A$ and scenopoetic existing fundamental niche $\mathbf{E}_A$ are shown as open outlines. Notice that some parts of $\mathbf{G}_O$ may be unknown (e.g., area A is occupied, but the species has not been detected there) and, similarly, that observed presences may not identify the full extent of $\mathbf{E}_O$ (e.g., the shaded area immediately around label D does not include any known occurrence localities). Also, notice that some regions of $\mathbf{G}_A$ may not be inhabited by the species: for example, area B may be beyond the dispersal range of the species, while the nonshaded area around label C may be uninhabited due to competition with another species. Hence, some parts of both $\mathbf{G}_A$ and $\mathbf{E}_A$ are unoccupied, yielding the invadable distributional area $\mathbf{G}_I$ (e.g., area B) and invadable niche $\mathbf{E}_I$ (the nonshaded area around label E, drawn to represent environments found in area B). Redrawn from Pearson (2007).

7, niche modeling, which aims to estimate $\mathbf{E}_A$ or perhaps $\mathbf{E}_P$, may be best performed without true absence data, which essentially introduce in the modeling process the effects of interactive species, Allee effects, dispersal limitations, and other factors not directly related to Grinnellian niches. Since we concentrate primarily on presence-only situations (although we review presence/absence methods as well for completeness), the obvious question is what quantity is being estimated by ENM algorithms.

It is probably correct to state that niche modeling estimates niche-related objects along a continuum between the existing fundamental niche $\mathbf{E}_A$ and the

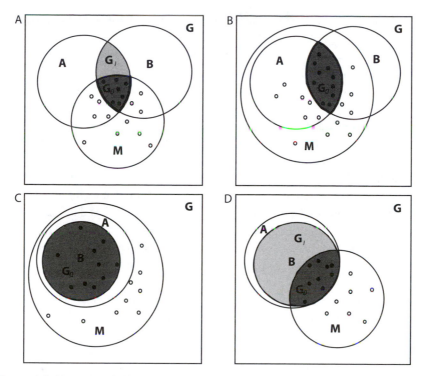

FIGURE 3.5. Examples of different configurations of the BAM diagram. Panel A shows an intuitive hypothetical configuration, as in figure 3.1. Panel B shows a situation in which all of the abiotically suitable area G_A is accessible, so the invadable distributional area G_I is null and all parts of the potential distributional area G_P are inhabited. Panel C shows a situation in which A and B are almost coincident, and the entire area is accessible to the species, so neither biotic nor movement considerations reduce the distributional potential of the species substantially. Finally, panel D depicts a situation similar to C, except that substantial restriction of dispersal exists, such that not all suitable potential distributional areas are inhabited. In all panels, open circles denote absences of the species, solid circles denote presences of the species, light shading indicates G_P, and darker shading indicates G_O.

occupied niche E_O (Jiménez-Valverde et al. 2008), and that the position along this spectrum depends on the particular configuration of the BAM diagram (figure 3.5) in the universe chosen as the study region, the availability (and use or not) of true absence information, and the particular niche-modeling method used (Anderson and Raza 2010). The essence of the problem lies in the fact that, although estimations of G_O and G_P present very different problems (Soberón and Peterson 2005, Peterson 2006c), as can be seen from figure 3.1, their associated environments may be quite similar. Indeed, a fundamental (but often only implicit) assumption in many SDM applications is that $\mathbf{E}_P = \mathbf{E}_O$. Confusing estimation of $\mathbf{G}_P$ with that of $\mathbf{G}_O$ is easy, and has led to some model evaluation and comparison exercises that have not distinguished between the two challenges in the best manner possible (e.g., Elith et al. 2006). The bottom line is that, without information about true absences, the predictions that any ENM produces represent a hypothesis about environments similar to those where the species has been observed, and these environments are likely to be found between the extremes of $\mathbf{E}_A$ and $\mathbf{E}_O$. Similarly, the associated biotopes or geographic ranges can be hypothesized to lie between $\mathbf{G}_A$ and $\mathbf{G}_O$ (Jiménez-Valverde et al. 2008).

In figure 3.5, we illustrate how different configurations of the BAM diagram correspond to different estimation problems. First, we denote the set of observations (presences and, if existing, true absences) as $\mathbf{G}_{\text{data}}$, and the set of presence-only data as $\mathbf{G}_+$. The operation of applying any algorithm to $\mathbf{G}_{\text{data}}$, given an environmental space $\mathbf{E}$, to estimate subsets of $\mathbf{G}$ like $\mathbf{G}_O$, $\mathbf{G}_P$, or $\mathbf{G}_P$, is symbolized with $\mu(\mathbf{G}_{\text{data}}, \mathbf{E})$. Note that we are not including information about the biotic environment $\mathbf{B}$; we should not confuse this modeling step with η, which simply maps $\mathbf{G}$ space onto $\mathbf{E}$ space. This operation represents the set of steps required to obtain a prediction of an area from occurrence data and environmental variables. Many methods produce an estimate of some subset of $\mathbf{E}$ that is then expressed geographically. Therefore, the operation $\mu(\mathbf{G}_{\text{data}}, \mathbf{E})$ identifies subsets of $\mathbf{G}$, although internally the first step is generally to identify some subset of $\mathbf{E}$. Many ways of implementing $\mu(\mathbf{G}_{\text{data}}, \mathbf{E})$ are available, at times based on completely different mechanisms, as we will discuss in chapter 7. In this section, we discuss not the details of methods, but rather the general framework and interpretation.

The researcher always should ask first what region in the BAM framework is being modeled, given occurrence data sampled from what area of the BAM diagram, and with what set of biological assumptions. Then, at least, we expect $\mathbf{G}_+ \subseteq \mu(\mathbf{G}_{\text{data}}, \mathbf{E}) \subseteq \mathbf{G}$; a point in $\mathbf{G}_+$ that lies outside of the model prediction $\mu(\mathbf{G}_{\text{data}}, \mathbf{E})$ represents an error of omission, whereas areas predicted by the model that are outside of the true $\mathbf{G}_P$ represent errors of commission (see chapter 9 for a much more detailed treatment of these ideas). An algorithm that predicts only the known occurrence points $\mathbf{G}_+$ as suitable has no commission error (meaning that no known absence points are incorrectly predicted; see chapters 5 and 9), but has the maximum possible omission error when confronted with an independent evaluation dataset. On the other hand, an algorithm that predicts all of $\mathbf{G}$ as suitable incurs all possible commission error, but no omissions. Obviously, both extremes are useless. We desire an algorithm that improves predictions at least to $\mathbf{G}_O \subseteq \mu(\mathbf{G}_{\text{data}}, \mathbf{E}) \subseteq \mathbf{A}$.

Niche-modeling algorithms, albeit using diverse methods, seek sets of cells that are environmentally similar to $\mathbf{G}_+$, trying to balance omissions of the available data (usually only presences) and measures of commission, often based on different types of pseudoabsences or samples from the full background (Hirzel and Le Lay 2008, Phillips et al. 2008; see chapter 7). The question is how to achieve the correct balance between errors of commission and errors of omission (Anderson et al. 2003). The fundamental problem of most niche modeling efforts is the fact that one is modeling without any reliable information about absences. As a consequence, estimating commission error is difficult (Anderson et al. 2003), as will be seen later, specifically in chapter 9.

To clarify the question of what area of the BAM diagram is being modeled, we examine some extreme cases (see figure 3.5). In the simplest case, assume that the species in question is an excellent disperser, and has been capable of colonizing all suitable regions of $\mathbf{G}$; as a consequence, in such a case, $\mathbf{M}$ includes the entirety of the intersection of $\mathbf{A}$ and $\mathbf{B}$ (see panels B and C of figure 3.5). This broad dispersal potential implies that occupied and potential areas coincide, or $\mathbf{G}_P = \mathbf{G}_O$, and similarly for the associated environments $\mathbf{E}_P = \mathbf{E}_O$. Assume further that the Eltonian Noise Hypothesis is true, in which case $\mathbf{A} \approx \mathbf{B}$, as is illustrated in panel C. In this situation, if the presence data have been drawn uniformly from across environmental space, one would expect a reasonable algorithm to give an estimate of sets that coincide: $\mu(\mathbf{G}_+, \mathbf{E}) = \mathbf{G}_A \approx \mathbf{G}_O$. In this situation, the environments $\mathbf{E}_A$ and $\mathbf{E}_P$ (corresponding to the existing fundamental niche and the biotically reduced niche, respectively) coincide, and many algorithms will estimate them approximately. The two types of problems defined earlier (spatial interpolation and spatial transferability) coincide in this extreme scenario, and exploration of E-space by the species is not constrained by movement, nor by unsuitable biotic conditions, so a well-executed sampling scheme will provide a good representation of E-space, and the estimate $\mu(\mathbf{G}_+, \mathbf{E}) = \mathbf{G}_A$ should be close to reality.

Now, assume the slightly more complex case in which an invadable region exists that is distinct from $\mathbf{G}_O$ (see figure 3.5, panel D). Since $\mu(\mathbf{G}_+, \mathbf{E})$ seeks only similarities with the environments in $\mathbf{G}_+$, most algorithms first estimate an environmental subset that may often be manifested in areas outside $\mathbf{G}_O$. Such models will be able to predict beyond a simple spatial envelope of $\mathbf{G}_O$ to provide a spatial transferability prediction. This situation is one in which $\mathbf{M}$ constrains sampling of geography by the species, but perhaps not the species' sampling of E-space. Whether this scenario is realistic or not is little explored. In any case, the niche $\hat{\mathbf{E}}_O$ that the algorithm estimates is an approximation of $\mathbf{E}_O$, and we may suppose that the best possible $\mu(\mathbf{G}_+, \mathbf{E})$, given the data available, will lie between the known occupied distributional area and the abiotically suitable areas: $G_O \subseteq \mu(G_+, E) \subseteq A$, but where exactly will depend on how well the data points in G_+ sample E_P, the particular configuration of the BAM diagram, and other factors to be discussed in later chapters. Uniform sampling across E will shift the position of $\mu(G_+, E)$ toward G_A. Uneven (biased) sampling of E will shift $\mu(G_+, E)$ toward sections of G_O, or perhaps will provide even-poorer estimates of other areas in the BAM diagram. Since, by hypothesis in this configuration (see figure 3.5, panel D), a significant nonaccessible but suitable area exists, the model has the potential to estimate the invadable area G_I if we are able to assume the equivalence of the environmental spaces corresponding to G_P and G_O.

Finally, in the configuration shown in panel A of figure 3.5, environmental representation of G-space is most seriously constrained, by both limited dispersal and unfavorable biotic circumstances. If the environments $\eta(G_O)$ are a representative sample of E_A, and the data also sample those environments well, we expect $\mu(G_+, E)$ to be shifted to the right in the expression $G_+ \subseteq G_O \subseteq \mu(G_+, E) \subseteq A$, providing a good estimation of the potential areas, which is an objective often desired. However, if the environments in G_+ are a biased sample of $\eta(G_O)$, we expect $\eta(G_+, E)$ to be shifted to the left, underestimating even G_O.

In sum, estimation of regions in the BAM diagram is achieved by using algorithms that search for similarities between the environments associated with sets of observed presences (and perhaps absences) G_{data} and the environments associated with the rest of the geographic area of the study at hand. Even assuming exhaustive and uniform sampling across the region by researchers, depending on the configuration of the BAM regions, presences may occur in all habitable regions, or only in some. Constraints of limited dispersal, or of lack of appropriate biotic contexts, may restrict what parts of the environmental universe are available to be occupied by a species. Therefore, any algorithm $\mu(G_+, E)$ that looks for similarities will estimate a region biased toward either G_O or G_A, depending on the specific configuration of the BAM regions, on how uniformly sampled E is, and on the logical structure and parameterization of the algorithm itself. Utilization of some version of the BAM diagram as a methodological framework is indispensable for the proper interpretation of the results of niche modeling.

SUMMARY

In this chapter, we provide an operational context for the theoretical relationships between ecological niches and geographic distributions (Hutchinson's

Duality). Ranges are defined as sets of cells in a grid subdividing the geographic space of interest (**G**). The requirements, biotic relationships, and movements of a species define several types of distributions. Grinnellian niches are defined by the sets of environmental vectors, out of a total space denoted by **E**, occurring in those sets of cells. Hence, we define two spaces that we call G-space, standing for geography, and another called E-space, standing for the environmental conditions represented in **G**. We note the very different topologies of **G** and **E**, and show how ranges in **G** can be defined via properties of spatially explicit population equations. These equations suggest a simple heuristic device, called the "BAM diagram," which provides useful ways of classifying areas and their environments, and thus the results of modeling exercises. We thus discuss fundamental niches in the context of scenopoetic spaces, and the need to define the *existing* fundamental niche, and the biotically reduced niche, analogous to the realized niche. Finally, we explore the implications of different configurations of the BAM framework for estimating niches and distributions from incomplete information as in correlative niche modeling.

PART II
PRACTICE

CHAPTER FOUR

Niches and Distributions
in Practice: Overview

Part I of this book set out a conceptual framework for understanding relationships between niches (in environmental space, or E-space) and spatial distributions (in geographic space, or G-space). This theory forms the base for the next sections, which deal with the practice of modeling ecological niches and estimating geographic distributions (part II) and applications of these methods (part III). Although we cover a wide variety of modeling methods and applications in this set of chapters, the same basic approach is used throughout. This process can be outlined as follows (Hirzel et al. 2002):

1. The study area is conceptualized as a raster map with extent G, composed of grid cells at a specific resolution (grain).
2. The dependent variable is the distribution of the species (G_O, G_P, or G_A), as inferred from occurrence records G_+, sometimes with true absences also being known (presences and absences together referred to as G_{data}).
3. A suite of environmental variables is collated to characterize each cell of the study area in environmental terms; known as E.
4. A function $\mu(G_{data}, E)$ that characterizes the dependent variable in terms of the environmental variables is generated, to indicate the degree to which each cell in G is suitable for the species.

In this section of the book (part II), we describe the process of building ecological niche models, mainly in very practical terms, but based on the conceptual foundation presented in part I. The process includes selecting and obtaining appropriate biological occurrence records and environmental data, choosing and applying modeling algorithms, and assessing the accuracy of the predictions quantitatively. In this chapter, we first set out general principles and definitions necessary for understanding the coming chapters, and then describe—in less formal terms—principal steps to be followed in building niche models.

GENERAL PRINCIPLES

Quite generally, the setup for a prediction problem is as follows. Nature issues a response $\mathbf{Y}$. Two sets of predictor variables, denoted generically by $\mathbf{X}$ and $\mathbf{Z}$, exist that represent conditions that cause effects on $\mathbf{Y}$. The distinction between two types of predictor variables is that $\mathbf{X}$ is observable and easily estimated and visualized (e.g., temperature), while $\mathbf{Z}$ is not (e.g., most biotic interactions). Nature's response is denoted as $\mathbf{Y} = f(\mathbf{X}, \mathbf{Z})$, for some function f. In this book, $\mathbf{Y}$ refers to the presence or absence ($\mathbf{Y} = 1$ or $\mathbf{Y} = 0$, or $\mathbf{Y} = 1, 0$ for simplicity) of a species at a given site or in a given environment, giving rise to what is termed a classification problem. $\mathbf{X} \subset \mathbf{E}$, while $\mathbf{Z}$ refers to other, complex variables that are less easily observed and characterized. As noted already, and as is developed more fully in chapter 5, the very notion of what "presence" and "absence" truly mean is critical, and has many interpretations. That is, in the function f, the variables in $\mathbf{X}$ can change, and $\mathbf{Y}$ may change accordingly. In defining terms, we assume here that this function is given with clear meaning, albeit unknown.

The notion of approximating nature through use of a model is to concede that the most that we can do is to attempt to discover $\mathbf{Y} = f(\mathbf{X})$. Note that the dependence on the additional factor $\mathbf{Z}$ is disregarded completely: we implicitly assume that any effects caused by $\mathbf{Z}$ are minor (see chapter 3). This set of assumptions necessarily introduces a notion of possible randomness, incompleteness, or even outright error, such as variability in $\mathbf{Y}$ even for equal values of $\mathbf{X}$, since the unobserved $\mathbf{Z}$ variables may still cause variation in responses. Models $\hat{f}(\mathbf{X})$ of the true relation $f(\mathbf{X})$ can be obtained by means of algorithms for the purpose of approximating nature's true f. The term "model" is frequently used as a synonym with the words algorithm, prediction, or method, represented in the last chapter by the expression $\mu(\mathbf{G}_{\text{data}}, \mathbf{E})$. Therefore, $f(\mathbf{X})$ is estimated by use of a given algorithm $\mu(\mathbf{G}_{\text{data}}, \mathbf{E})$. The different algorithms summarized in chapter 7 have the common goal of producing functions $\hat{f}(\mathbf{X})$ that can be used to compute a prediction of $\mathbf{Y}$ for a given $\mathbf{X}$. Some methods produce a 0,1 output directly, such that their predictions are of the form $\hat{\mathbf{Y}} = \hat{f}(\mathbf{X})$. Other methods produce a continuous output $\hat{c}(\mathbf{X})$, with the property that larger values indicate greater likelihood of presence (or, more precisely, larger values represent areas more similar to pixels at which the species has been recorded; the degree to which this similarity reflects likelihood of presence depends of the representitiveness of the input data). From such $\hat{c}(\mathbf{X})$ predictions, $\hat{f}(\mathbf{X})$ is defined as $\hat{f}(\mathbf{X}) = 1$ if $\hat{c}(\mathbf{X}) \geq u$ and $\hat{f}(\mathbf{X}) = 0$ if $\hat{c}(\mathbf{X}) < u$, where u is some constant called a "threshold of occurrence" (see chapter 7).

Model Calibration and Evaluation

Model calibration (sometimes called "training") aims to estimate $f(\mathbf{X})$ based on the set of empirical occurrence data $\mathbf{G}_{\text{data}}$. Also available are values of the v variables $\mathbf{X}_1, \mathbf{X}_2, \ldots, \mathbf{X}_n$, which are the environmental values corresponding to each of the n cells in $\mathbf{G}_{\text{data}}$; in symbols, $\{(\mathbf{Y}_i, \mathbf{X}_i)\} = \{(\mathbf{Y}_g, \vec{e_g}) | g \in \mathbf{G}_{\text{data}}\}$. In most applications, these values are environmental values associated with cells where a presence has been recorded $\mathbf{G}_+$, but sometimes true absences exist, and can also be incorporated; also, many applications incorporate environmental information associated with pseudoabsences or background pixels as part of the calibration process (see chapter 7).

Model calibration refers to steps internal to the process that allows an algorithm to form (and in many cases refine) its estimate of f, which in turn may be used to make geographic predictions of distributional areas. Often, an algorithm uses data (both occurrence records and environmental data) in an iterative process (e.g., rule development, internal testing, and rule refinement or selection of weights for variables) to form a model of the species' niche.

Model calibration is a necessary first step, but most of our interest focuses on applying the resulting model for prediction purposes over a *different* set of $\mathbf{X}$'s—say, the values of $\mathbf{Y}$ for an independent set of m occurrence data $\mathbf{X}_1^*, \mathbf{X}_2^*, \ldots, \mathbf{X}_m^*$. This step is where "model evaluation" becomes important (chapter 9). Model evaluation is about examining how successful $\hat{f}$ is in representing observations used for evaluation based on the model derived from the original set of observations used in calibration. A model is deemed useful if the predicted values $\hat{\mathbf{Y}}_i^* = \hat{f}(\mathbf{X}_i^*)$ (where the asterisk indicates that the $\mathbf{X}_i^*$ are additional data) are in some way close to nature, which would be denoted $f(\mathbf{X}_i^*)$. A useful notion is that when evaluating whether $\hat{\mathbf{Y}}$ is "close" to $\mathbf{Y}$, we are implicitly or explicitly assuming what is termed a *loss function*, or evaluation criterion, $L(\mathbf{Y}, \hat{\mathbf{Y}})$. A basic example is "zero-one" loss, described by

$$L(\mathbf{Y}, \hat{\mathbf{Y}}) = \begin{cases} 0 \text{ if } \mathbf{Y} = \hat{\mathbf{Y}} \\ 1 \text{ if } \mathbf{Y} \neq \hat{\mathbf{Y}} \end{cases}. \tag{4.1}$$

No loss is incurred in cases in which prediction and observation agree; otherwise, a loss of 1 is tallied. It should be noted that endorsing "number of correct predictions" as a criterion for what is meant by "close" implicitly favors zero-one loss over other alternative loss functions (see chapter 9). Importantly, we note that zero-one loss is symmetrical, in the sense that omission error and commission error are deemed to bear equal consequences and are assigned equal weights; these concepts are fundamental to model evaluation and are discussed in further detail in chapter 9.

Essentially, all statistical theory on optimality regarding unbiased point estimation is based on the premise that the zero-one loss function is a useful measure of model performance, yet it is not always recognized explicitly that this particular loss function is being employed. Another example of a symmetrical loss function common in statistics is squared error, given by $L(\mathbf{Y}, \hat{\mathbf{Y}}) = (\mathbf{Y} - \hat{\mathbf{Y}})^2$. The key issue here is that any notion of optimality and evaluation is relative to the loss function employed, implicitly or explicitly. No "optimal model" exists; rather, a model can be optimal only relative to a given loss function. Determination of which loss function to adopt is not a mathematical question, but rather a question for the user, to be based on the characteristics of the application and the data at hand.

Once the notion of whether the predicted values $\hat{\mathbf{Y}}_i^* = \hat{f}(\mathbf{X}_i^*)$ are close to $f(\mathbf{X}_i^*)$ is summarized in a loss function, model evaluation is about quantifying $L[f(\mathbf{X}_i^*), \hat{f}(\mathbf{X}_i^*)]$; this mathematical simplification is nonetheless a considerable challenge. On one hand, values of $f(\mathbf{X}_i^*)$ are imprecise or unknown, even if we have data samples that approximate them (see classification of absences in chapter 5). On the other hand, these values contain inherently random factors, as explained earlier. A notion of "expected loss," or average loss over typical values of $\mathbf{X}_i^*$, emerges as an obligatory term quantifying model imprecision, denoted by $E\{L[f(\mathbf{X}_i^*), \hat{f}(\mathbf{X}_i^*)]\}$. To illustrate that this notion of expected loss has been encountered before, albeit perhaps inadvertently, consider zero-one loss, where the expected loss is simply the probability that prediction and observation disagree.

Computing expected loss analytically is not generally possible, so the practical solution is to try to *estimate* expected loss. Suppose that the model $\hat{f}$ has been fitted by using observations $(\mathbf{Y}_1, \mathbf{X}_1), \ldots, (\mathbf{Y}_n, \mathbf{X}_n)$, and suppose that a second dataset $(\mathbf{Y}_1^*, \mathbf{X}_1^*), \ldots, (\mathbf{Y}_m^*, \mathbf{X}_m^*)$ is also available, where the $\mathbf{X}_1^*, \mathbf{X}_2^*, \ldots, \mathbf{X}_m^*$ are representative of the range of values where applying the model is of interest. In this case, one could compute

$$E_{\text{val}} = \frac{1}{m} \sum_{i=1}^{m} L[\mathbf{Y}_i^*, \hat{f}(\mathbf{X}_i^*)], \tag{4.2}$$

and use this equation to approximate expected loss (or average validation error, prediction error, or testing error). This estimate E_{val} is indeed very simple, but it follows from assumptions made regarding the origin of the data. It can be interpreted as "typical" loss incurred in the prediction process.

A closely related set of distinctions is among verification, validation, and our use of the more general term "evaluation." Whereas "verification" refers to assessment of a model's ability to fit to the calibration data (i.e., nonindependent data) and "validation" corresponds to the ability of the model to predict *independent* data (ideally independent spatially, taking into account spatial autocorrelation patterns in the landscape; Araújo and Guisan 2006), the more general term "evaluation" refers collectively to the entire diversity of testing situations: evaluation data that are nonindependent, semi-independent, and fully independent of the calibration data. Typically, two datasets exist or are derived from a single one: the first dataset is referred to as the "calibration dataset" (or "training dataset" in some literature) and the second one as the "evaluation dataset" (or "testing dataset" or "validation dataset"). This distinction between two pools of occurrence data is paramount, but is ignored with surprising frequency. Ideally, the evaluation dataset should consist of observations not included in the calibration dataset. Although such a separation has not always been employed (see chapter 9 and review in Araújo et al. 2005a), we emphasize the critical need for independence of the calibration and evaluation datasets, to allow rigorous assessment of model predictions. When two distinct datasets are not available or possible, it is common to produce two sets artificially by means of random partitioning of a single dataset, which has implications for interpreting the results of the evaluation. This and other schemes such as jackknifing, "cross-validation," and bootstrapping, which generally do *not* achieve full independence of the evaluation dataset, will be detailed in chapter 9. In these cases, estimating E_{val} is not as simple as was described earlier in this chapter.

In the preceding, we could compute an average over calibration data using the first set,

$$E_{ver} = \frac{1}{n} \sum_{i=1}^{n} L[\mathbf{Y}_i, \hat{f}(\mathbf{X}_i)]. \tag{4.3}$$

This measure estimates lack of verification (verification error, calibration error, training error), in contrast to the measure of lack of validation estimated earlier using the independent dataset, the set $\{\mathbf{Y}_i^*, \mathbf{X}_i^*\}$.

It is also important to note that many of the same principles and metrics used for model validation (e.g., log-loss functions and data partitioning) are also used by various model algorithms in model calibration (see chapter 7). In the notation developed here, model calibration amounts to minimization of error E_{ver} by modifying or adapting $\hat{f}$, while keeping the calibration dataset fixed (although subsets of it can be split within the calibration process).

A problem distinct from assessing error (i.e., estimating E_{val}) is the question of examining a prediction for statistical significance. In this case, the investigator quantifies the rates of correct and incorrect predictions given by a model, and compares this rate to those expected under some null hypothesis, which is generally the idea that the predictions are no better than random with respect to the validation data (see chapter 9).

As a final point of introduction, we note that the set of observations $\{(\mathbf{Y}_i, \mathbf{X}_i)\}$ used for calibrating and evaluating niche models must be sampled from conditions characteristic of the space over which predictions are to be made. Because of the two crucial spaces involved here ($\mathbf{G}$ and $\mathbf{E}$), this point may be interpreted in two ways: that $\{(\mathbf{Y}_i, \mathbf{X}_i)\}$ are typical in geographic space or that they are typical in environmental space. We should thus consider the space in which evaluation occurs, since the data may be biased geographically yet unbiased environmentally or possibly vice versa (see chapter 7).

STEPS TO BUILDING NICHE MODELS

More practically, and complementary to the list given at the outset of this chapter, we outline concrete steps that constitute the modeling process (figure 4.1). This set of steps structures the chapters presented in part II of this book.

STEP 1. DATA PREPARATION

The first task in building a niche model is to collate, process, error-check, and format the data that are necessary as input. Two types of data are required: (1) occurrence data documenting known presences (and sometimes absences) of the species (see chapter 5), and (2) raster-format GIS datasets summarizing scenopoetic environmental variables that may (or may not) be involved in delineating the ecological requirements of the species (see chapter 6). With just these two inputs, initial niche models can be developed.

Occurrence records are point localities (defined by x and y coordinates, such as latitude and longitude) that specify what is known about the species' geographic distribution. In most cases, occurrence data comprise records of where the species has been observed to be present (i.e., "presence-only" data). However, in some cases, "records" of places where sampling has occurred but the species has not been documented are also available; when both types of occurrence information are available, the data are termed "presence/absence data." Some modeling methods are able to function based on presence-only data, whereas others can take good advantage of the additional information available in presence/absence data. Still, the advantages and disadvantages of each approach must be considered carefully (see chapter 7). Issues concerning different types of occurrence data, including potential data sources and biases, are discussed in detail in chapter 5.

The environmental datasets characterize variation in scenopoetic variables across the study area. Variables derived from weather-station data (e.g., daily temperature and precipitation), on-ground surveys (e.g., soil characteristics),

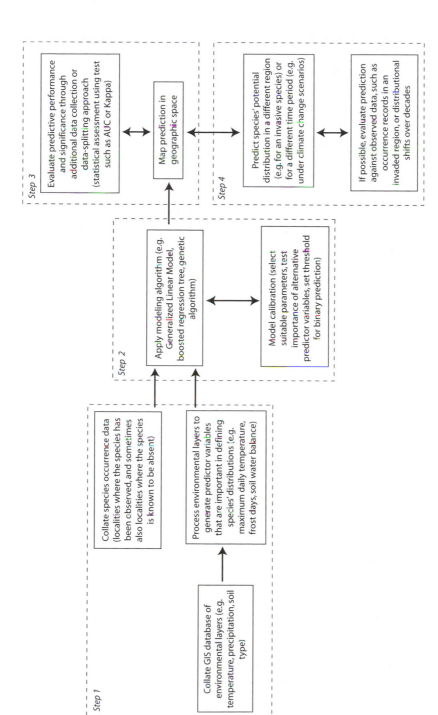

FIGURE 4.1. Summary of steps and challenges in the overall ecological niche modeling process. See text for detailed description.

or remote sensing imagery (e.g., spectral reflectance and cloud cover) are processed into raster GIS layers. Ideally, model inputs are provided that are thought to have a direct physiological role in delimiting the ecological niche of the species. For example, Pearson et al. (2002) collated a suite of seven climatic variables and five soil variables, from which they generated five model-input variables, including maximum annual temperature, minimum temperature over a 20-year period, and soil-moisture availability. The general aim is to develop variables that summarize qualities of the environment that are likely to be relevant to the species' distributional biology, such as availability of water, energy, space, and/or time.

The biological and environmental datasets used in niche modeling are ideally viewed, formatted, and prepared in a geographic information system. GIS facilitates many key operations, including changing geographic projections (all data must be referenced to a common coordinate system, so that occurrence records can be matched to corresponding values of environmental variables), changing spatial resolution, and transforming vector-format occurrence data to raster formats. GIS also provides crucial functionalities for visualizing model results, as well as additional processing of model outputs, such as removing predicted areas that are geographically isolated from known occurrence records by dispersal barriers (see chapter 8). The steps of choosing, obtaining, and formatting environmental data lead to important considerations of data quality, spatial extent, resolution, and richness of environmental dimensions that are discussed in detail in chapter 6.

STEP 2. NICHE MODELING

Having collated occurrence records and environmental variables, the next step is to use a modeling algorithm to characterize the species' ecological niche as a function of the environmental variables. Put simply, the model aims to identify environmental conditions associated with the species' occurrence (and perhaps its absence as well), which is what we have called the estimation, $\hat{f}$. In practice, the function $\hat{f}$ assigns values to points in E-space. Depending on the algorithm employed, the values represent various probabilities, suitabilities, likelihoods, or "degree of membership" to certain sets (see chapter 7). These values and their precise meaning may or may not be apparent, depending on the software used. Although sometimes the software presents only a geographic expression (i.e., a map) of the environmental valuation, GIS methods can extract the actual values assigned to the points in E-space, so model "response surfaces" can be estimated for any algorithm. The task of classifying environments would be relatively straightforward if just one or a few environmental variables were used; in practice, however, we usually seek models that can integrate many predictor variables, since species likely respond to combinations of multiple environmental factors. The task is therefore to build a model of the species' ecological niche in a multidimensional E-space. In chapter 7, we review modeling algorithms that have been applied to this task, and classify models according to their functioning and data requirements.

At this stage of the modeling process, one task is calibrating the model to ensure that the algorithm provides optimal results. Depending on the particular method used, various decisions are necessary. For example, if applying the Maxent algorithm (Phillips et al. 2006), it is necessary to choose a suitable regularization parameter; in using the Genetic Algorithm for Rule-set Prediction (GARP; Stockwell and Peters 1999), a convergence criterion must be set; and generalized additive models (Hastie and Tibshirani 1990, Guisan et al. 2002) require the degrees of freedom to be selected. More generally, the relative importance of alternative environmental predictor variables may be explored at this stage, to select variables for inclusion in the final model. Another consideration is how best to select an appropriate threshold of occurrence for converting continuous model output (e.g., probability of occurrence) to a binary prediction of "present" versus "absent." These topics are covered in detail in chapter 7.

Here, it is particularly important to understand conceptually how the model is assembled in E-space and then projected back into G-space. Several factors, including the degree of equilibrium between environmental conditions and species' occurrences (Araújo and Pearson 2005) and the adequacy of sampling of the species' distribution, will affect the degree to which the model is able to predict elements of the Hutchinsonian Duality of niche (i.e., $\mathbf{E}_A$, $\mathbf{E}_O$, $\mathbf{E}_I$, or $\mathbf{E}_P$) and biotope (i.e., $\mathbf{G}_A$, $\mathbf{G}_O$, $\mathbf{G}_I$, or $\mathbf{G}_P$). Chapter 7 provides a discussion of the potential for models to predict different elements of the niche and distribution, and shows that the choice of a "best" modeling approach depends on the aims of the modeling exercise.

STEP 3. MODEL PROJECTION AND EVALUATION

The next tasks are to map ("project") the prediction in G-space (or possibly E-space) and to evaluate how well the model predicts independent data. To map the prediction, algorithms display in G-space one of several possible values: as we will see in chapter 7, these output values may be "suitability" of an environment for the species; probability of occurrence of the species $P(\mathbf{Y} = 1 | X = g)$; probability of finding cells similar to those already visited $P(X = g | \mathbf{Y} = 1)$; or envelope quantities describing membership in a set. As mentioned earlier, depending on the biological realities or assumptions of the particular case (i.e., the configuration of the BAM diagram for that species × landscape × resolution combination; see figure 3.5), the resulting map summarizes environmental suitability across the landscape (i.e., an estimate of $\mathbf{G}_A$), which may or may not correspond closely to the occupied distributional area of the species ($\mathbf{G}_O$). Additional refinements to and analyses of model outputs at this stage can be used to improve predictions of the occupied distributional area (i.e., modifying or processing $\hat{\mathbf{G}}_A$ to estimate $\mathbf{G}_O$ more closely); for example, areas predicted as suitable, but isolated from observed occurrence records by dispersal barriers (such as a river or mountain range) may be removed (Peterson et al. 2002b), and the influences of finer-scale habitat features (James et al. 1984, Thuiller et al. 2004a, Heikkinen et al. 2007) or interacting species (Anderson et al. 2002b, Araújo and Luoto 2007) on species' distributions may be incorporated to improve model results. Methods for estimating geographic distributions from ecological niches are treated in chapter 8.

Finally, before model predictions can be interpreted or used for any of various applications, we must evaluate the predictive performance and significance of the model. Ideally, data used for evaluation of models would be collected independently of the calibration data, specifically for the purpose of evaluating the model; however, it is common to use a data-splitting approach to generate subsets for calibrating the model and subsets for evaluating it (see the preceding discussion and chapter 9). Several alternative statistical tests have been applied to evaluating niche models, including tests based on either presence-only or presence/absence evaluation data and tests based on binary predictions (i.e., presence versus absence) or continuous suitability (or probability, in some cases) surfaces (Fielding and Bell 1997). We will discuss these statistical tests, along with important conceptual difficulties associated with evaluating model performance and significance, in chapter 9.

STEP 4. MODEL TRANSFERABILITY

In some niche modeling applications, it is necessary to transfer modeled niche conditions to predict environmental suitability across a new region or for a different time period. For example, predicting the potential for spread of alien invasive species requires that a model calibrated using occurrence records for one range area be applied to a different region to identify suitable areas (see chapter 13). Likewise, predictions of potential distributional shifts under climate change require that the model be used to predict under scenarios of future or past climatic conditions (see chapter 12). Many potential applications exist for these predictions, including addressing questions in ecology, evolutionary biology, and conservation planning, which we describe in chapters 10 to 15 (part III).

Predicting into new regions and under alternative climate scenarios raises important difficulties, such as the problem of "extrapolating" beyond the range of environmental conditions over which the model was calibrated (see chapter 7). An important distinction should be made here. "Extrapolating," or rather transferring, models in G-space means identifying regions of the world with environments similar to those modeled for a certain species in the calibration region. This question may be perfectly valid and interesting (see the discussion of spatial interpolation versus spatial transferability earlier). In contrast, extrapolating *in E-space* means predicting beyond the ranges of environmental variables used for the calibration. This procedure is risky, and can be justified only under certain assumptions, such as linearity or monotonicity of responses (see chapter 7).

We discuss many problems and caveats as they are encountered in part III, but an important final step—where possible—is to evaluate predictive performance using occurrence records from the new region or different time period, providing a further validation that is certainly more independent and rigorous. For example, occurrence records from a new distributional area for an invasive species may be used to test a prediction for the region based on the native range (Peterson 2003a, Thuiller et al. 2005b). Similarly, distributions recorded across different time periods may be used to test predictions of range shifts over time (Araújo et al. 2005a, Martínez-Meyer and Peterson 2006). These tests utilize approaches and statistics that are discussed in detail in chapter 9.

CHAPTER FIVE

Species' Occurrence Data

Although most of biological diversity is poorly known (Wilson 1988, Erwin 1991), one commonality among species is that something is generally known about where they occur on Earth. That is, an integral part of every scientific description of a species is information about the geographic provenance of the available type specimen material (e.g., for animals, Article 76, ICZN 1999). Therefore, all known species should have at least one geographic occurrence locality available. In many cases, of course, the situation is better than just the type locality, with data from specimens or observations documenting many additional occurrences also available.

This book focuses on the process of transforming such primary occurrence data into a synthetic understanding of the geographic and ecological conditions under which a species occurs. As noted previously, an alternative set of methods for characterizing ecological niches takes a mechanistic approach to the challenge (Porter et al. 2002, Natori and Porter 2007). However, our focus is on correlative models based on occurrence data, since such models can take advantage of the near-universal availability of some form of occurrence data, are easily implemented for large numbers of species, and hence can have quite broad applicability.

TYPES OF OCCURRENCE DATA

Although the concept of occurrence data seems quite straightforward—dots on maps showing places where the species has been found—the details can be much more complicated. Starting with the very basics, occurrence data for use in ecological niche modeling would ideally be drawn from the accessible areas that are environmentally suitable for the species to maintain long-term stable, source populations (Pulliam 2000, Araújo and Guisan 2006)—that is, G_O in the discussions in chapter 3. Absence data would ideally also be carefully considered—in this case, these points would come from areas that have been accessible to the species (M), and where the environment is *unsuitable* for the species to maintain populations, or the area outside of $\mathbf{G}_P$ (or in some cases the area outside $\mathbf{G}_A$), although in practice they will frequently be drawn from the area outside of $\mathbf{G}_O$ (see chapter 5). With occurrence data and absence information of this nature (i.e., presences from $\mathbf{G}_O$ and absences from outside of $\mathbf{G}_A$ but within $\mathbf{M}$ and $\mathbf{B}$), niche modeling would be relatively straightforward. All, however, is not as simple as it might seem, or as one might wish, as the numerous considerations of biases and problems in this and succeeding chapters will illustrate.

What Makes an Occurrence Record?

In reality, a long chain of events connects the suitability of a site to the existence of a data record documenting the species' presence or absence at that site. Among these factors are the following:

1. The area may be unsuitable, and for that reason the species is not present.
2. The area may be unsuitable, but the species is present (at least occasionally) owing to dispersal from suitable areas.
3. The area may be suitable, but the species has never been able to disperse to it.
4. The suitable area was at one point occupied by the species, but the species has since been extirpated from the area. (Note that this idea can include temporal variation in presence of a species at a site, which may, in some cases, be important.)
5. The suitable area may be occupied by the species, but no researcher has ever visited the place to sample.
6. The suitable area may be occupied and may have been visited and sampled by researchers, but they did not detect the species.
7. The suitable area may be occupied and may have been visited and sampled, and the species may also have been detected, but the record is not among those available to the researcher.
8. The suitable area may be occupied and may have been visited and sampled, the species detected, and a record is available to the researcher.

This set of factors can be considered quite usefully in the form of a probability tree diagram, in which nodes are arranged sequentially to represent factors, and branches denote the possible ways each factor can result (figure 5.1). These factors can be divided into at least two groups: biological factors (i.e., mobility, abiotic suitability, and biotic suitability, as discussed in the context of the BAM diagram in chapter 3) and factors related to exploration, detection, and data (e.g., the species must have been identified correctly). For a presence record to occur at a given site, all of the biological factors and all of the exploration-

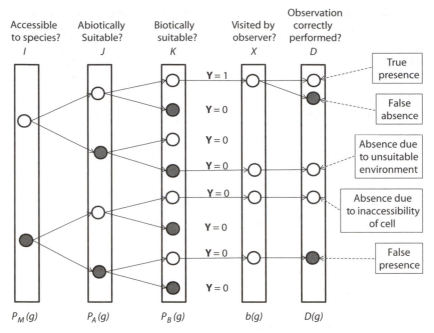

FIGURE 5.1. Probabilistic events leading to presence and absence observations, including erroneous results. Each bar represents a choice, with open circles representing "yes," and gray circles representing "no." A presence observation requires not only that the species is present ($\mathbf{Y} = 1$) in a cell, which is the result of the three biological processes represented in the first three columns, but also that the cell was visited by observers [with probability $b(g)$], and that the observation was performed properly (e.g., no identification errors, no typographical errors) with probability $D(g)$. Several opportunities exist for mistakes to be committed. We highlight an example of false absence resulting from faulty observation, two absences due to radically different biological causes, and a false presence resulting from faulty observation or recording.

and detection-related factors must be fulfilled simultaneously. This schema is represented sequentially in figure 5.1, assuming simple, yes-no alternatives for each biological factor, which we will explore more quantitatively.

The tree is a graphical representation of a random multistep schema explaining presences and absences of species at nodes (denoted $\mathbf{Y} = 1, 0$), and their detection by an observer (denoted $D = 1, 0$). Five tiers (at least) of random outcomes must be considered. The first through third tiers are based on the ideas of the BAM diagram. Three binary random variables I, J, K are used to represent access to a site by the species, abiotic suitability, and biotic suitability, respectively. The event $\{\mathbf{Y} = 1\}$ is equivalent to $\{I = 1\} \cap \{J = 1\} \cap \{K = 1\}$. In other words, $\mathbf{Y} = 1$ signals the existence of a source population in a given cell. Sink populations are found in regions where $I = 1$, but either $J = 0$ or $K = 0$. We use the following succinct notation for conditional probabilities that label specific segments in the probability tree:

$$P_M(g) = P(I = 1|X = g),$$
$$P_A(g) = P(J = 1|I = 1, X = g), \text{ and}$$
$$P_B(g) = P(K = 1|J = 1, I = 1, X = g). \quad (5.1)$$

Here, the sums of $P_M(g)$, $P_A(g)$, $P_B(g)$ over all g do not necessarily add to one (that is, they are not necessarily density functions over g). For example, $P_M(g) = 1$ for *all* g is possible, denoting the situation when the region is fully accessible to the species. The number $P_A(g)$, as developed in chapter 3, depends on the Grinnellian environmental variables $\eta(g)$; one could really write $P_A[\eta(g)]$, but we maintain $P_A(g)$ as shorthand. The probability $P_B(g)$ is an Eltonian set of factors, and $P_M(g)$ depends on rates of mobility, time spans, and the geographic configuration of $\mathbf{G}$.

As we saw before, $P_A(g)$ can be obtained mechanistically, although this estimate may require substantial effort (Kearney and Porter 2004, Crozier and Dwyer 2006, Buckley 2008). $P_B(g)$ is unknown for virtually all species (Araújo and Guisan 2006, Soberón 2007). Given the huge disparities in scale between the variables defining $P_B(g)$ and $P_A(g)$, it may be possible to regard $P_B(g)$ as constant, meaning that the Eltonian environment is roughly similar over the entire region, or perhaps just a random variable that is uncorrelated spatially, which is the Eltonian Noise Hypothesis (see chapter 3). $P_M(g)$ may be completely known, may be postulated on biogeographic or ecological grounds, or may be constant, meaning that all sites g are equally accessible to the species.

In the fourth tier, we ask whether a given cell has been accessed by an observer or not. The event $\{X = g\}$ represents the fact that cell g has been visited, and its probability is denoted by $P(X = g) = b(g)$. Variation in $b(g)$, biological sampling, accounts for sampling bias. The last tier of nodes focuses on detection of the species once the collector has examined a site and the species is in reality present at the site. We use the notation $D(g) = P(D = 1|Y = 1, X = g)$ for the conditional probability that the species is detected. Note that $D = 0$ can occur even if $Y = 1$ (a false negative detection), with conditional probability $1 - D(g)$. $D = 1$ can occur even if $Y = 0$ (a false positive detection). Each last-tier segment of the tree beginning at $Y = 0$ may actually have a specific probability value for a false detection. For example, a false positive may be more probable if biotic conditions are suitable than if they are unsuitable. This general case for detection is quite complex, so some simplifications will be made. The most benign of the detection configurations is the "perfect detection" setup, where $D(g) = 1$ and all seven conditional probabilities in the tree $P(D = 1|Y = 0,$

$X = g$) are set to zero. In general, $D(g)$ depends on the methods and experience of the collector, as well as on g.

The notion of "statistical independence" means that conditional probabilities obey a special structure spread out over all segments in the tree. For example, abiotic suitability may be independent of access (a generally reasonable assumption, since abiotic suitability can be seen as a physical property of a site that can be defined regardless of access by a species; marine situations, in which currents may determine both access and suitability, may present exceptions), then

$$P_A(g) = P(J = 1|I = 1, X = g) = P(J = 1|I = 0, X = g) \qquad (5.2)$$

[that is, the value $P_A(g)$ would also apply to the single segment connecting nodes $I = 0$ and $J = 1$]. If biotic suitability is independent of abiotic suitability and accessibility, then

$$\begin{aligned} P_B(g) &= P(K = 1|J = 1, I = 1, X = g) = \\ &P(K = 1|J = 0, I = 1, X = g) = \\ &P(K = 1|J = 0, I = 0, X = g). \end{aligned} \qquad (5.3)$$

The multiplication rule in probability theory states that the probability at any single branch as a whole is obtained by successive multiplication of the conditional probabilities of component segments. For example, the probability of the uppermost branch in our tree (labeled $\mathbf{Y} = 1, D = 1$) is given by

$$\begin{aligned} &P(X = g, I = 1, J = 1, K = 1, D = 1) = \\ &P(X = g)P(I = 1|X = g)P(J = 1|I = 1, X = g) \times \\ &P(K = 1|J = 1, I = 1, X = g)P(D = 1|K = 1, J = 1, I = 1, X = g) = \\ &b(g)P_M(g)P_A(g)P_B(g)D(g). \end{aligned} \qquad (5.4)$$

This last compact notation, because of the interpretation stated earlier, is really a product of conditional probabilities, and it does not imply that we are always assuming independence of collector visitability, species access, abiotic suitability, biotic suitability, and detection. Independence is only a special case, represented by structure over other branches in the tree, as explained earlier.

The probability tree keeps track of the possible ways in which a single unit of sampling effort can result in an observed presence or absence record. If a collector sets out on an arbitrary, single data-collection effort, the probability that he or she records an observed presence ($O = 1$) is given by

$$P(O = 1) = \sum_g P(X = g)P(\mathbf{Y} = 1|X = g)P(D = 1|\mathbf{Y} = 1, X = g) = \\ \sum_x b(g)P(\mathbf{Y} = 1|X = g)D(g). \qquad (5.5)$$

In the special "perfect detection" case, $\{O = 1\}$ is equivalent to $\{\mathbf{Y} = 1\}$. But in general, because a biological presence is not equivalent to an observation, it is important to distinguish these two concepts.

In our tree, biological presence of the species at site g is explained by favorable biotic and abiotic conditions, and the species' dispersal abilities. By the rule of successive multiplication in probability trees, the conditional probability of a presence at site g is given by

$$P(\mathbf{Y} = 1 | X = g) = P_B(g)P_A(g)P_M(g). \qquad (5.6)$$

Equation (5.6) relates what may be called a statistical representation of probability of presence to a more ecological representation, based on causal factors. This equation excludes collector activity, and summarizes how uncertainty regarding a biological presence is a combined effect of three separate factors that in turn depend on the environment of site g.

The simple diagram in figure 5.1 also serves as a bookkeeping device for several types of errors present in species' occurrence data. The second box from the top on the right side of figure 5.1 represents a case in which the species is present and the site was visited, but no detection occurred (a false absence in G_{data}, or akin to omission error for a prediction; chapter 9). Similarly, false positives ($\approx$ commission errors) can occur (the last box in figure 5.1) when a site is in reality unsuitable for a species, but the site is visited and the species' presence is recorded, either owing to errors in geographic referencing or errors in taxonomic identifications (see chapter 9 and latter portions of this chapter for more discussion of these errors).

Figure 5.1 shows that the probability distribution of presence records (observed data) depends on the combined effect of all five components. Although interest always lies in inferring the biological aspects of the tree, observed data are nonetheless produced as a function of the combined effects of all factors. It is important to understand that occurrence data inevitably contain information that may mask the true objects of interest, so results must be interpreted critically. Under some special circumstances, distributions of observed data truly depend only on the biological quantity of interest, such as a region that is very well sampled, so that $b(g) = D(g) = 1$ for all g in figure 5.1.

Overall, then, we can see that occurrence data are not simple documentations of species' presences and absences, but rather the result of complex filtering by very diverse processes. This complexity must be considered early in the design of each study, as a means of controlling the quality of the data input into the modeling algorithm and avoiding the classic "garbage in, garbage out" situation. Without such precautions, niche modeling studies run a clear risk of modeling the "niche" of the data collector, the sampling, the accessibility, or any of several other nonbiological factors.

TYPES OF OCCURRENCE DATA

On a more practical note, we can consider the sources of occurrence data available to a researcher. We define primary occurrence data (or frequently, for brevity, just "occurrence data") as records that place a particular species in a particular place at a particular time. As such, primary occurrence data can take many forms. Labels associated with specimens in the research collection of a natural history museum or herbarium provide a rich resource of primary occurrence data, as they connect an identified (or at least identifiable) specimen with information on its provenance in time and space. Another vast resource of primary occurrence data is that of observational information, including visual sightings, auditory detection, records of tracks, and other similar techniques for establishing a species' presence without collecting and preserving the individual. For some taxonomic groups for which visual or auditory identification is feasible (e.g., birds, butterflies, large mammals), large observational datasets have been developed (e.g., BBS 2008, SSIC 2009), which can offer enormous quantities of occurrence information. Finally, diverse technologies are under exploration for automated and remotely sensed species identification: for example, automated identification of vocalizations (Chesmore 2004, Chesmore and Ohya 2007), or identification of tree species from remotely sensed information (Castro-Esau et al. 2004, Clark et al. 2005, Papeş et al. 2010).

Not all occurrence data are equally useful. Note that primary occurrence data place the species in question at a point in space, down to the limits of the precision and accuracy of its georeferencing (coordinates of latitude and longitude, or other system; see the following). As such, primary occurrence data are useful in theory in analyses conducted at any spatial resolution as coarse as or coarser than the spatial precision and accuracy of the record itself. Also, some primary occurrence data are documented by evidence that can be revisited and reidentified if necessary—these records are termed "vouchered" primary occurrence data. Specimens in natural history museums and herbaria are the best examples here, although photographic and recorded auditory records may also be considered to be vouchered to some extent. Although nonvouchered observational data can certainly be of value for ecological niche modeling, vouchering adds an additional level of quality control and future utility that is not possible when a data record cannot be revisited in view of new evidence or analyses.

We distinguish carefully between primary occurrence data and what we can term "secondary occurrence data," which represent summaries or syntheses of primary occurrence data. Although "synthesis" sounds attractive, secondary

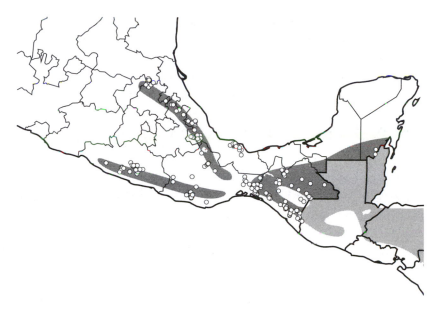

FIGURE 5.2. Example of the lack of precision and detail in polygon-based, second-ary sources of distributional information. Shown is the example of the Emerald Toucanet (*Aulacorhynchus prasinus*); polygons (shaded gray) are from Ridgely et al. (2005), with primary occurrence data for the Mexican portion of the distribu-tion from the *Atlas of Mexican Bird Distributions* (Navarro-Sigüenza et al. 2006), each point corresponding to a vouchered occurrence record of the species.

occurrence data often include subjective elements that can detract seriously from their utility. Common sources of secondary occurrence data include range maps digitized from field guides (e.g., Howell and Webb 1995), digital extent-of-occurrence GIS datasets (e.g., Ridgely et al. 2005), and biodiversity atlases (e.g., Hall and Moreau 1970, McGowan and Corwin 2008). In each case, some level of subjective interpretation is involved in the production of the secondary product. Typically, the range summary of secondary occurrence data is overly simplified (see example in figure 5.2).

Generally, secondary products suffer from a broad-stroke approach (large grain/coarse spatial resolution, sometimes as crude as 1:40,000,000), and for this reason, they may omit areas of known occurrence or include areas where the species does not occur (Hurlbert and Jetz 2007). Although they can be quite useful as range summaries, and are receiving increasing use in macroecologi-cal analyses (Ceballos and Ehrlich 2006, Davies et al. 2006, Orme et al. 2006), secondary summaries are certainly not generally recommended for use in mod-eling species' niches and potential distributions, and results will require careful interpretation.

Species Characteristics

Clearly, even if sampling considerations (see the following) do not complicate
the picture, species vary in the geographic and ecological characteristics of
their distributions and occurrences. A useful framework for thinking about these
points is the "seven types of rarity" concept of Rabinowitz et al. (1986). This
framework separates species into a $2 \times 2 \times 2$ matrix structured by overall local
population size, size of the occupied distributional area, and level of habitat
specificity. One of the cells in this matrix (large population, broad geographic
range, low level of habitat specificity) corresponds to common species, but the
species in the other seven categories are "rare," in at least one dimension. These
three basic ways in which species can be rare can explain many of the ways in
which their occurrence data vary as well.

In this sense, species' distributional characteristics can be visualized by the
spatial patterns in their occurrence data at different spatial scales (figure 5.3).
Species may be broadly distributed or narrowly distributed geographically. For
example, the most broadly distributed species approach cosmopolitan distributions (e.g., House Mouse *Mus musculus*, Osprey *Pandion haliaetus*, Barn Owl
Tyto alba), whereas the narrowest distributions of species may constitute a
single mountain peak or a single cave or spring. At a different scale, species—
whether broadly or narrowly distributed globally—may be broadly distributed
in terms of habitats across a particular landscape. Furthermore, within appropriate regions and habitats, a species may be present in high or low numbers
(Ricklefs 2004). Species in different portions of this conceptual matrix may offer
very different challenges for developing ecological niche models.

Sampling Complications

Bias in geographic space.   Sampling of biodiversity is complex, with numerous factors clouding any straightforward relationship between true distributions
of species and the occurrence data that are derived from them and document
them (see figure 5.1). A major motivating factor in developing methods for
modeling ecological niches and predicting potential distributions is precisely
the fact that primary occurrence data often provide a distorted representation
of true distributional patterns, even when large amounts of such data are available. As such, detailed consideration of the vagaries of biological sampling and
their effects on primary occurrence data is worthwhile.

A first consideration in this regard is that of detectability—quite simply,
some species are more easily detected than others. In its simplest form, detectability is associated with how apparent members of the species are, and as
such can be seen as a random reduction in detection probability from unity to

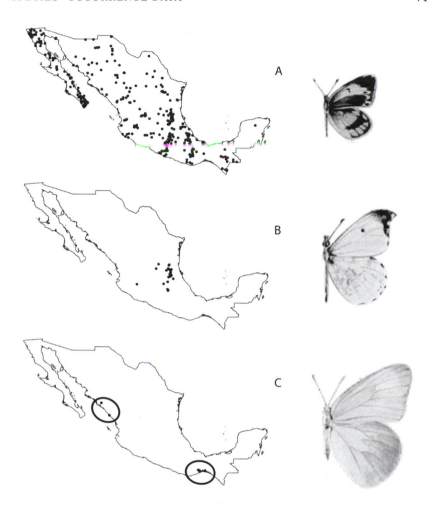

FIGURE 5.3. Maps showing three sets of occurrence data for three butterfly species across Mexico. (A) Available presence data for *Nathalis iole*, an abundant and widespread butterfly. (B) Available presence data for *Paramidea limonea*, a species that is quite geographically restricted but that is locally abundant within that area. Finally, in (C), available occurrence data are shown for *Prestonia clarki*, a geographically widespread species with only a small number of low-density populations across western Mexico (occurrences are highlighted within ellipses). Data from Llorente-Bousquets et al. (1997).

something below unity in any site across the range (MacKenzie et al. 2002). This sort of detectability effect on biodiversity data acts simply to rarefy the content of biodiversity databases for each species, but should not introduce systematic biases for any given species. Detectability may, however, vary for a single species among regions and habitat types, which can indeed bias the information on a particular species toward or against certain regions (Buckland et al. 2005). In this latter case, the primary occurrence data available for a particular species may misrepresent its ecological tolerances, and thus bias any estimate of its ecological niche.

Another common source of misrepresentation of species' distributions in primary occurrence data is that of bias in the distribution of sampling effort. This effect is commonly manifested in the form of "road bias": the pattern of most sampling taking place near roadways (Kadmon et al. 2004), although rivers, areas of high human population density, and other access points may also serve to create such biases (Heyer et al. 1999, Lim et al. 2002). (We suspect that roads cut across very different environments more frequently than will rivers, so road-biased sampling may not be as problematic as other sources of geographic bias; see the next section.) Sampling bias may be most common, extreme, and likely in short-term or single-investigator efforts, but major, pervasive biases exist even in major biodiversity compilations accumulated over centuries of effort (Reddy and Dávalos 2003).

Finally, uneven sampling may result simply by historical accident, particularly when data were not collected for the purpose of a particular study, but rather were assembled post hoc from other sources containing data deriving from previous field efforts. Because of the presence of a particular institution or investigator in a given area, or the location of a popular field station, a concentration of records may accumulate, providing a skewed representation of the species' distribution (see figure 5.4). This sort of effect can develop thanks to some countries mounting major survey efforts, the existence of state-level monographic treatments, mandates by particular agencies, intense sampling in regions surrounding universities or field stations or other study sites, increased sampling in heavily populated areas (e.g., through "citizen science"), and around type localities as researchers often return there to sample topotypes.

Bias in environmental space. Sampling that is biased in geographic space may or may not be biased in environmental dimensions as well. For example, samples clustered along a road that cuts across a large mountain range are obviously biased spatially, but may sample environments very well—but other situations may not be as fortunate. Countering environmental bias in sampling is difficult when working in an ecological niche modeling framework, because

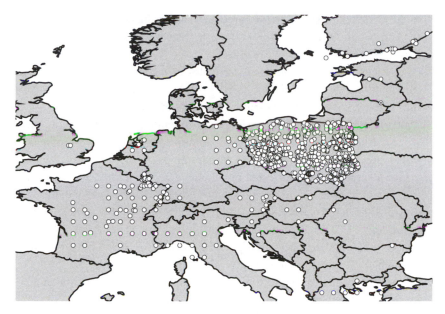

FIGURE 5.4. Example of the varied problems often found in occurrence data derived from heterogeneous sources. Shown are known occurrences of the European butterfly *Lycaena dispar* derived from queries to data resources mediated by the Global Biodiversity Information Facility. Discernible are biases in survey effort (e.g., compare Poland and Luxembourg with Germany), as well as secondary data likely derived from coarse-resolution atlases across much of the species' range (see, e.g., points aligned in rows in northern Italy).

nonrandom associations between occurrence records and environmental variables are precisely how we characterize ecological niches. That is to say, the constraints that the ecological niche places on the presence and absence of a species across a landscape are in essence themselves a bias (in the sense of a deviation from random) in representation of the species across environmental dimensions. As such, to characterize sampling bias in ecological dimensions and potentially rectify its effects, additional information regarding the nature of the sampling per se becomes necessary (Zaniewski et al. 2002, Phillips et al. 2009).

Unfortunately, if sampling bias in ecological dimensions does exist, it can have rather pervasive adverse effects on the results of niche modeling (see chapter 7). For example, perhaps in a particular case higher elevations in a landscape were not sampled as much because of increasing difficulty of access, producing an inverse correlation of sampling probability with elevation, which in turn is inversely correlated with temperature and other features of climate.

A niche model calibrated using occurrence data resulting from such sampling will likely predict low suitability in those high-elevation regions, because no occurrence records from such regions were available (Nogués-Bravo et al. 2008a). Clearly, any environmental correlates of sampling bias limit the predictive ability of the niche models based on that sampling, especially for techniques that use background or pseudoabsence sampling (chapter 7).

Appropriate means of characterizing sampling bias and distinguishing sampling bias from niche-based ecological limitation would include examination of either direct information regarding the sampling scheme or information that provides a surrogate of that sampling (Zaniewski et al. 2002, Phillips et al. 2009). Direct information regarding sampling effort may be available in some cases for single, large planned survey efforts, such as the North American Breeding Bird Survey (BBS 2008) or the Smithsonian Venezuelan Project that surveyed mammals and their ectoparasites (Handley 1976). However, reconstructing sampling effort per se from biodiversity databases including the results of many surveys by different collectors would require careful consultation of field notes in museum archives; even if such information exists, the process of compilation will often prove unfeasible.

An alternative to direct compilation of sampling effort for the species of interest is to use primary occurrence data for a broader sample of species as a surrogate (Phillips et al. 2009). To provide a valid surrogate (or index of sampling effort), such data should correspond to species that are sampled in the same manner as the species of interest would be sampled. Such a "target group" (*sensu* Anderson 2003) is not necessarily simply a taxonomic group. For example, surveys for small nonvolant mammals in the northern Neotropics often capture marsupials, shrews, and rodents. Even though these three groups do not represent a monophyletic lineage, they would constitute an appropriate target group when estimating sampling effort for any individual species of those groups. Data for a well-chosen target group provide information regarding which environments have been sampled, and in particular, environments that are well-sampled but from which the species has not been documented (Heyer et al. 1999, Anderson 2003). Data regarding species in the target group must also correspond to the same data sources used for the species of interest, so that the actual sampling effort across the landscape is captured. Such data may be easy to obtain when online databases are involved, but will typically be difficult to assemble if direct consultation of specimens in museums and herbaria is necessary.

Given appropriate data regarding sampling effort or a surrogate of it, testing for sampling bias in ecological dimensions becomes feasible. Randomization tests offer a powerful approach to solving these challenges. In this approach, the observed distribution of sampling points is compared with multiple, repeated, random samples of the same number of samples or sampling points from the entire study region to assess whether biases with respect to the environmental spectrum are present in the biological sampling. For example, Kadmon et al. (2004) characterized sampling of woody plants in Israel, showing that, while a strong road bias exists in geographic space, biases in E-space are relatively subtle. Biases in environmental dimensions were manifested principally in terms of precipitation, while sampling of temperature regimes was fairly representative. In a complementary approach, Funk and Richardson (2002) provided a visualization of environmental variation across Guyana, and demonstrated several intriguing gaps in the coverage of environmental variation across the country by existing biodiversity sampling.

CHARACTERISTICS OF ABSENCE DATA

Careful consideration of absence data is necessary in order for niche modeling based on this information to be rigorous and successful, and not confused by introduction of additional biases. The probability tree of types of occurrence data presented in figure 5.1 gives an example of the complex set of possibilities that can lead to real or apparent "absence" data. Absence data or something akin, however, are required by several modeling approaches that use both presence data and some sort of absence, pseudoabsence, or background data for calibrating models (Hirzel et al. 2002, Elith et al. 2006; see chapter 7).

Three basic approaches are available for assembling and using occurrence data for modeling: (1) use presence-only data; (2) use presence and absence data, where available; and (3) use presence data and a sample of background or "pseudoabsence" data, in lieu of absence data (see chapter 7 for discussion of algorithms that follow each of these strategies). In the latter option, number of grid pixels is resampled by some means, typically at random, from the study area as a whole or from areas where the species has not been detected (Graham et al. 2004a). Although the terms "pseudoabsence" and "background" are sometimes used indiscriminately, background sampling from the study area as a whole can be used to characterize the environmental conditions present across the study region (i.e., creating a sample from the "background" that potentially includes sites where the species' presence has been observed). In contrast, pseudoabsence sampling entails selecting from areas or sites where the species has not been detected (or sometimes only from the subset of nondetection sites where sampling has occurred). Pseudoabsence sampling thus holds the intent of mimicking absence data; however, such data typically suffer from

inclusion of pixels from areas that the species does or could occupy, contrary
to the intent of this approach. In any case, neither approach can be considered
to yield true absence data; accordingly, resulting analyses must be interpreted
differently from those based on real absence data (if such data were to exist).

At this point, it is necessary to return to the considerations presented in
chapter 3 regarding why species are present where they are present and why
they are absent where they are absent. The limitations to a species' movement
referred to as $\mathbf{M}$ are of particular importance, as they constrain species to occur
in only a subset of otherwise suitable sites. That is to say, areas exist that are
suitable for a species to maintain populations, but that are not inhabited because of dispersal limitations; we term such regions $\mathbf{G}_I$. Species' invasions offer
a clear demonstration of the existence of the area $\mathbf{G}_I$ (Peterson 2003a).

However, if background or pseudoabsence data are generated via resampling from undocumented sectors of a broad landscape, some of the resulting
sites will very likely derive from areas *suitable* for the species but outside of $\mathbf{M}$
(i.e., in $\mathbf{G}_I$), causing problems in the modeling process (see chapter 7). Even if
"real" absence data are available—that is, if one has information on sites where
the species was not present during intense sampling (MacKenzie et al. 2002)—
one is never sure that those sites are genuinely not habitable by the species, or
rather may not hold the species simply because they lie outside of the species'
dispersal range, or the species was not present at the time of sampling owing
to temporal differences in occupancy, or it was present in other unsampled parts
of the cell (see figure 5.1; Soberón and Peterson 2005).

More specifically, we can refer to the BAM diagram from chapter 3 to identify precisely what sort of data would be ideal for in niche-modeling exercises.
As has been discussed already, presence data are ideally sampled from $\mathbf{G}_O$ (i.e.,
source populations), whereas absence data would be selected from regions
outside of $\mathbf{G}_P$ (although differentiating $\mathbf{G}_P^C$ from $\mathbf{G}_O^C$, where the superscript C
indicates the complement, is a serious challenge), and within $\mathbf{M}$ and $\mathbf{B}$. (We
note that distinguishing source and sink populations is important, but quite challenging—possibilities include hypothesizing sink populations as those occurring only in environments located close to source populations, but no clear
methodology has as-yet been developed.) Careful consideration of these points
is key in niche modeling exercises, as models can otherwise be seriously
flawed, corresponding rather to aspects of species distribution modeling (i.e.,
estimating $\mathbf{G}_O$; see the sections treating probability of occurrence and study
region in chapter 7). Given these concerns, a general conclusion is that it is best
to avoid using absence data in developing niche models, and similar complications exist for background and pseudoabsence data (Anderson and Raza 2010).

OCCURRENCE DATA CONTENT AND AVAILABILITY

Recent decades have seen dramatic increases and improvements in the availability of biodiversity data. What was fairly recently a challenge of harvesting data from analog formats (e.g., specimen labels, card files, data sheets) has now become convenient and efficient, and great quantities of data can be accessed quickly. In the excitement of this "new age," however, it is important to bear in mind constantly the need to consider the provenance, quality, and possible biases that the data may carry.

IDEAL INFORMATION CONTENT

Data content (i.e., which fields are included), database design, and data integration are rapidly expanding fields of inquiry (Soberón 1999, Stein and Wieczorek 2004, Guralnick et al. 2006). As such, we do not attempt a comprehensive review herein, but rather touch briefly on a few important points. Although primary occurrence data are in essence quite simple (i.e., documenting the occurrence of a species at a particular place at a particular time), more information regarding the record is usually needed.

Data that lack documentary information can be perilously imprecise or misleading, not for what they say, but for what they imply. For example, older natural history specimens may have very general locality information, for example "Brazil, Rio de Janeiro." While this locality description sounds simple, a careful user would consider whether this locality descriptor refers to Rio de Janeiro the city, or Rio de Janeiro the state. Also, Rio de Janeiro in 2011 is considerably more extensive than it was in 1880. Finally, one must consider how far outside of the limits of the city of Rio de Janeiro the specimen could have been collected and still be labeled as "Rio de Janeiro" by the original collectors.

The solution to this quandary, and many other confusions regarding occurrence data, is to attach critical metadata to the primary occurrence data record. Specifically, metadata should include information regarding the location of the voucher specimen (if one exists), the means of identification and person making it, the species concept underlying the identification provided, the source of the geographic reference (e.g., GPS reading, map, gazetteer, etc.), and the datum and spatial precision (maximum error) of the quantitative geographic reference (e.g., geographic coordinates; Wieczorek et al. 2004). Additional helpful metadata might inform the user as to how the records were accumulated (e.g., sampling strategy, focal taxa for sampling, accumulation curves for sites sampled, etc.), which might be useful in designing strategies for assembling absence information. Several data architectures and schemas have been developed to attempt to capture a broad and consistent picture of information associated with primary occurrence records (Graham et al. 2004a).

Certainly, in ecological niche modeling, geographic coordinates are key, as they are the link between known occurrences of species and environmental characteristics of occurrence sites. Hence, considerable effort has been invested in development of georeferencing tools and information schemas that preserve maximal amounts of information from the original collecting event. For example, the MaNIS georeferencing protocol (Wieczorek et al. 2004) provides a useful framework, in which the point to which the geographic reference corresponds is associated with a radius of possible error or uncertainty, based on several criteria. This protocol is now mostly automated, with metadata created in formats that can be attached seamlessly to the original record. In the future, the error "radius" could be refined to reflect the fact that uncertainty is not always uniform in its directionality, so that uncertainty in georeferencing would be represented as a probability density cloud around the point. The key benefit to this overall approach is that once uncertainty has been assessed quantitatively, the researcher can filter available occurrence data so as to use only those occurrences that are georeferenced with sufficient precision and accuracy for the modeling exercise at hand. As such, georeferencing and error assessment constitute critical steps in the process of niche model development.

Availability of Occurrence Data

One of the most challenging bottlenecks in ecological niche modeling is that of obtaining useful occurrence data, but that situation is changing rapidly. Just a few years ago, occurrence data had to be accumulated via specific queries to information "owners," via visiting stand-alone Internet-based databases to develop queries, or by visiting natural history museums or other repositories of vouchered primary occurrence data. Beginning in the late 1990s, however, efforts began in several regions to offer efficient, integrated access via distributed biodiversity databases (i.e., biodiversity databases linked via a common access portal on the Internet; Soberón 1999). This concept has been successful, and early technologies have given way to DiGIR and TAPIR, both highly effective information-transfer protocols that have been developed specifically for the transmission of biodiversity information. Most prominently, the Global Biodiversity Information Facility (GBIF) offers a portal to >270 million primary species' occurrence data records.

The availability of Internet-based digital primary occurrence records has augmented rapidly through time. From near-nil in the early 1990s, these data resources have increased by about eight orders of magnitude to the present

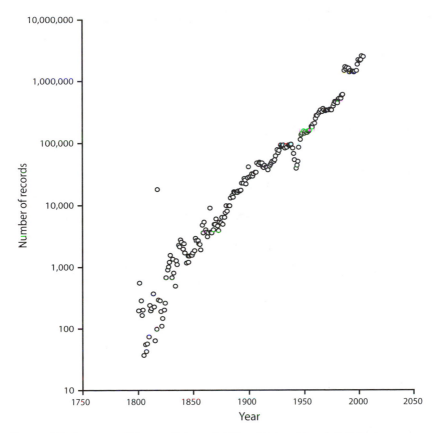

FIGURE 5.5. Summary of the growth in availability of Internet-based digital primary occurrence data records provided by the Global Biodiversity Information Facility data portal. Note that the number of records presented refers to records originally collected *in that year*, and as such the log-linear increase shown in this figure is impressive. Redrawn from Soberón and Peterson (2008).

hundreds of millions. Indeed, Soberón and Peterson (2008) recently analyzed ~85 million records from data served via the GBIF portal, and detected an intriguing trend based on year of original collection: a linear temporal increase of biodiversity information on a semilogarithmic scale. That is, the number of records available *per year* increased approximately 10-fold every 50 to 55 years (figure 5.5), which bodes well for the continued growth of biodiversity information networks. Of course, an important question is whether this trend will continue into the future. It may in the short term through digitization and integration of existing large datasets, in the medium term through new large-

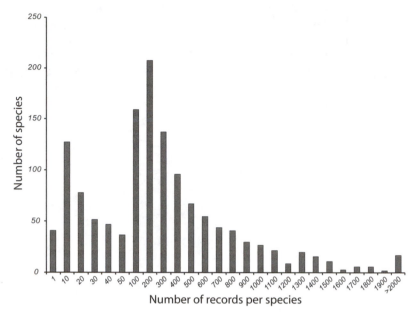

FIGURE 5.6. Example of the uneven nature of biodiversity information across species. This graphic summarizes >400,000 records of Mexican birds (Navarro-Sigüenza et al. 2006), showing frequencies of representation for each of >1300 species occurring in the country. Notice that the intervals on the horizontal axis change to provide detail on the left end of the spectrum.

scale observational and collection efforts, and potentially in the long term via applications of new remotely sensed data and remote identification of individual species.

Naturally, the usual problems of large datasets apply to these new resources. Many species are characterized by only a single record or very few records (figure 5.6); error rates are far from negligible; many records lack identifications and even more identifications are surely incorrect; and furthermore, the great bulk of records are yet to be georeferenced. All of these complications together imply significant work before data are usable in niche modeling exercises (Soberón et al. 2002, Chapman 2005). However, the investment involved in such efforts is near-trivial when compared with the prospect of repeating the biological sampling that led to the wealth of information currently held in natural history museums, especially because the temporal dimension of sampling cannot be recreated and because so many habitats have been altered drastically by humans.

SUMMARY

Data documenting occurrences of species across geography exist—in some form or another—for almost all known species. The series of events leading to the existence of those particular data, however, is less clear, and the diverse and complex factors associated with occurrence data need to be considered in order to produce a view of the challenges behind niche modeling. These factors can be summarized in the form of a probability tree, which illustrates how presence data should generally be reliable indicators of presence (not always, though), whereas absence data spring from many and varied phenomena. Occurrence data—mostly in the form of data documenting presences of species—can be discussed as primary (i.e., deriving directly from observations of species) versus secondary (i.e., processed or interpreted from primary data), and must be pondered in terms of possible biases in geographic and environmental spaces. Absence data, if to be used at all, must be considered carefully with respect to the probability tree and to the BAM diagram.

CHAPTER SIX

Environmental Data

Ecological niche models are built from two sources of input data: (1) the known occurrences of the species of interest discussed in chapter 5, and (2) environmental predictors in the form of raster-format GIS layers. Whereas the quality of and biases in occurrence data have seen considerable documentation and discussion (e.g., Soberón et al. 2000; see chapter 5), the nature, quality, and biases of environmental datasets have seldom been considered in detail in niche modeling analyses, despite the key role that they play in the process of calibrating models (Peterson and Nakazawa 2008). In this chapter, we discuss conceptual and applied aspects of environmental data, in the context of building and interpreting ecological niche models.

SPECIES-ENVIRONMENT RELATIONSHIPS

In Hutchinson's conceptualization of the ecological niche, discussed in detail in chapters 2 and 3, persistence and abundance of populations of species are determined by suites of scenopoetic and bionomic variables within an n-dimensional hypervolume of environmental space (Hutchinson 1957). As this concept is critical for ecological niche modeling, we must consider carefully how its details, ambiguities, and difficulties should guide us in implementation.

In the first place, it is clear that the type and number of variables comprising the dimensions of the ecological niche vary from one species to another (Leibold 1995, Pulliam 2000). For example, bats are highly sensitive to low temperatures, while felids are more sensitive to vegetation structure (Kitchener 1991, Patten 2004). Moreover, even within one of these broader groups, a particular species may respond to one set of variables, while another responds to other features of the environment. Finally, the relative importance of particular environmental variables for a species may vary according to the geographic and biotic contexts.

Environmental variables have been classified in various ways, depending on their relationships with, and influences on, geographic distributions of species.

While Hutchinson distinguished scenopoetic from bionomic dimensions of environments (Hutchinson 1978), Austin (2002) proposed two different break-downs, as follows:

1. *Idealized variables*—This categorization focuses on the degree to which variables have direct physiological effects on organisms, which Austin (2002) subdivided into:
   * *Indirect*—Variables with no causal physiological effects on individuals, but that have a correlation with species' occurrences because of correlations with other factors. Examples would include latitude and elevation.
   * *Direct*—Variables that affect organisms physiologically, but that are not consumed by them. Equivalent to Hutchinson's scenopoetic variables; examples would include temperature.
   * *Resource*—Variables that are consumed by or affected by organisms. Equivalent to Hutchinson's bionomic variables; examples would include food resources, presence of predators or parasites, or light in shade-limited plants.
2. *Distal/proximal variables*—Here, Austin (2002) sorted variables by the degree of causality of species' responses to environmental factors (see figure 6.1). He divided variables into:
   * *Proximal*—Organisms respond directly to such variables; an example might be freeze durations that affect survival of cacti in northern latitudes directly.
   * *Distal*—Responses of organisms to these variables are not direct, but rather via multiple additional causal links; an example would be annual mean temperature, which manifests as a causal factor only via the freeze durations just mentioned earlier.

It should be borne in mind that these categories are not mutually exclusive of one another—a variable may be proximal or distal and direct or resource at the same time.

Beyond this basic classification, different environmental factors operate at different spatial and temporal scales. As a consequence, their relative importance in defining a species' distribution and abundance can be highly scale-dependent (Mackey and Lindenmayer 2001, Pearson and Dawson 2003, Soberón 2007). In the early twentieth century, Joseph Grinnell proposed that species' distributions are hierarchically structured in space, the most inclusive classes consisting of climatic variables (particularly temperature and humidity) as the main drivers at coarse resolutions, whereas availability of food and

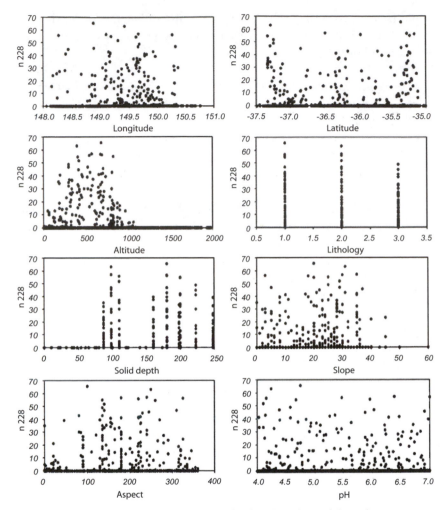

Figure 6.1. Example summary of responses of a virtual species to eight environmental characteristics. Adapted from Austin et al. (2006).

refuges are the most important factors at finer resolutions (Grinnell 1917). More recently, it has become generally accepted that different controlling factors typically operate at certain corresponding "scale domains" (Pearson and Dawson 2003): again, macroclimatic variables influence distributions at coarser scales, whereas landscape features (e.g., vegetation cover) act at meso-scales, and specific habitat features and biotic interactions have strongest influences at local scales (Mackey and Lindenmayer 2001, Pearson and Dawson 2003; figure 6.2).

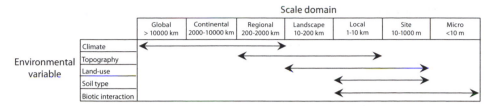

FIGURE 6.2. Schematic example of how different suites of environmental factors may affect distributions of species across a range of spatial scales. Scale domain indicates ranges of extents over which particular suites of ecological factors are surmised to act most strongly. Adapted from Pearson and Dawson (2003).

ENVIRONMENTAL DATA FOR ECOLOGICAL NICHE MODELING

A key question is thus which and how many variables are needed for rigorous modeling of a species' niche? The answers clearly depend on the scale at which the question is addressed, the knowledge available regarding the autecology of the species in question, the complexity of the ecological niche of the species, and the availability of high-quality data. To be able to understand ecological and geographic distributions at the finest scales, a combination of scenopoetic and bionomic variables would be necessary, covering features of microclimate, habitat structure, resource supply, competitor density, fine-scale solar radiation, and other fine-grained parameters (Soberón 2007). At intermediate scales, land cover, aspects of terrain, and vegetation phenology are likely to become relevant. Finally, at coarser scales (regional to global), the signature of scenopoetic variables is generally expected to be dominant (Mackey and Lindenmayer 2001, Pearson and Dawson 2004).

Thanks to the growth of spatial modeling techniques and remote-sensing systems, high-quality digital environmental data are increasingly available. Such data are being generated at diverse spatial extents and resolutions, from global to local, and with pixel sizes ranging from $<10^0$ to 10^5 or more meters (Quattrochi and Goodchild 1997). However, it should be kept in mind that most environmental dimensions summarized by such easily available datasets correspond to scenopoetic variables, such as climate, topography, and land cover.

Indirect variables are useful for niche modeling only to the extent that they correlate consistently with direct factors (Austin 2002). For example, elevation (an indirect variable) may provide a good surrogate for temperature (a direct variable) across certain extents and latitudinal ranges, but may be misleading across broader areas. This mismatch occurs because correlations between elevation and direct variables such as minimum temperature differ across space (especially across latitude): species do not respond directly to elevation, but rather to changes in temperature, oxygen availability, and air pressure that are themselves affected by elevation. Hence, a species living at high elevation in a low-latitude region may be restricted to lower elevations in higher-latitude areas. Some indirect variables combine different types of direct or resource variables, thus effectively reducing dimensionality in the analysis (Guisan and Zimmermann 2000). A good example is the Normalized Difference Vegetation Index (NDVI), derived from the red and infrared bands of remote sensing systems, that summarizes concentration of chlorophyll (Myneni et al. 1995); NDVI has been used successfully in niche modeling applications (Egbert et al. 2002, Anderson et al. 2006). Mechanistic models may play a potential role in clarifying which variables are direct and which are indirect.

Another factor affecting decisions regarding which variables to include in modeling is whether the goal is to reconstruct occupied ($\mathbf{G}_O$) or potential ($\mathbf{G}_P$) distributional areas of species (Peterson 2006c), and over what time period. Some environmental variables are more static than others, in the sense that they change only slowly (e.g., topography and climate), while others are more dynamic (e.g., land cover), especially given the pervasive influence of humans upon the environment. When the goal of a study is to approximate potential distribution patterns over several or many decades, inclusion of static predictors is convenient, while when the study question centers on the current distribution of a species over a very short timescale of one or a few years, dynamic and more proximal variables become necessary (Anderson and Martínez-Meyer 2004, Peterson et al. 2005a). Ecological niche modeling (ENM) has to date been applied mostly in global to regional analyses, and to a lesser extent in meso- and microscale studies, but not reflecting a limitation of the approach. Rather, as a correlative technique, ENM is not scale-sensitive, except that at some point bionomic habitat considerations may come to dominate distributional processes.

In practice, different strategies have been developed for selecting variables for inclusion in niche-modeling studies. Some researchers prefer to use a few preselected, relatively uncorrelated variables that correspond directly to known physiological rules, such as temperature and water stress (Huntley et al. 1995, Pearson et al. 2002, Huntley et al. 2004, Baselga and Araújo 2009). At the other end of the spectrum, when no previous knowledge exists regarding key factors, large datasets (i.e., 10 to 10^3 variables)—some highly correlated—have been used (Stockwell 2006), albeit not without strong criticism (Peterson 2007b).

Both extremes have their drawbacks. When few variables are used, the models risk missing important factors thus resulting in an undercharacterization of the niches; in this case, niche models are likely to produce overly broad potential distributional areas (Barry and Elith 2006). On the other hand, over-dimensionality (i.e., excessive numbers of variables) can lead to overfit models or loss of degrees of freedom in regression-based approaches, and geographic distributions may be underrepresented, particularly when small numbers of occurrence records are used (Peterson and Nakazawa 2008; chapter 7). Collinearity among these often highly correlated suites of variables can impede application of some statistically based model algorithms (Guisan and Zimmermann 2000). Correlated variables may cause additional problems, especially in terms of producing overly dimensional (i.e., highly complex) niche models, which can complicate model development and confound efforts to characterize niche models in ecological terms.

As a consequence, to avoid these problems, several schemes have been proposed for exploring variable interrelationships and reducing dimensionality prior to model development (Sweeney et al. 2006). Some investigators have used correlation analysis to identify least-correlated sets of variables (Baselga and Araújo 2009), and others have rotated (transformed) original variables to produce new, uncorrelated (orthogonal) axes via principal components analysis or other ordination techniques, and then selected the most important of these dimensions as input for modeling (Manel et al. 2001, Hirzel et al. 2002, Peterson et al. 2007c). Finally, particularly among investigators using nonstatistical, evolutionary computing techniques, it is common to include large numbers of variables, and simply to allow the algorithm to "sort it out" (i.e., determine which variables hold the most predictive information for the species of interest) while controlling complexity via different means in different algorithms (see chapter 7). These issues of dimensionality and data transformation and how they affect niche model quality are little-explored, so further research is needed to improve current practices in this area, particularly as regards incorporation of prior knowledge about organisms, their natural history, and their physiology.

ENVIRONMENTAL DATA IN PRACTICE

The conceptual aspects mentioned earlier regarding environmental data for niche modeling offer general guidance for designing and choosing such datasets. However, a number of more practical considerations must also be weighed. The following section details some of these points.

DATA PREPARATION

Regardless of the number and type of environmental variables chosen, all variables need to meet certain cartographic standards to match the characteristics of the occurrence data. First, in current systems, all environmental datasets must be in a raster GIS format—that is, cell- or pixel-based files, in which each cell is assigned a value of the corresponding environmental variable. Data in other formats (e.g., point-based measurements such as weather stations, vector GIS polygon-based data layers for soil or vegetation classes) must be converted to raster grids, and any cross-scale issues managed (e.g., the assumption of uniformity across entire grid cells, coarse spatial resolution of polygon data). Hence, the environmental dataset is composed of a series of raster files (= "layers") representing the variables to be included in the analysis. Second, both the raster files and the corresponding occurrence data must be in the same cartographic system (e.g., geographic projection, datum, coordinate system, units). Finally, for many analytical platforms, the raster files must match spatially, i.e., they must have the same spatial extent (numbers of rows and columns), geographic position (coordinates of origin), and resolution (cell size or grain).

In studies involving transferring the ecological niche model $\hat{f}(\mathbf{X})$ onto alternative spatial or temporal geographic scenarios, additional environmental datasets are necessary. For example, in studies of biological invasions (chapter 13) and climate change (chapter 12), sets of environmental layers corresponding to the other place or the other time period, respectively, must be assembled. These datasets must contain the same variables as the ones used in calibration, but not necessarily the same spatial extent or resolution.

DATA QUALITY

As in any other quantitative analysis, the quality of input data is fundamental. However, researchers all-too-commonly ignore known or possible deficiencies and biases inherent in the geographic data they use in developing models (Barry and Elith 2006). In part, these lapses may result from the fact that most users are not involved in generating environmental data products, but rather, use what is available to them. Modelers—before beginning analyses—should examine the metadata and ancillary files accompanying the datasets that they plan to use to learn about the sources, methodologies, and known limitations. In this way, it is often possible to ascertain whether a raster grid is useful or not in a particular analytical challenge.

Similarly, it is not uncommon to have access to more than one source for the same type of variable. By knowing the source and type of primary data used in generating such maps, one can decide which is more reliable, or one can identify means of combining multiple sources, provided that the information that they include is compatible. In this sense, the metadata accompanying geospatial datasets can be as important as the actual *data* in assuring the proper use of the information.

SPATIAL EXTENT

Spatial extent refers to the surface covered by the area of analysis (for example, one spatial extent might encompass all of South America, while another, narrower extent may encompass only Colombia). In general, the spatial extent of the analysis will match the extent of the phenomenon under study. For example, it would not make much sense to choose all of South America as an extent of analysis for model calibration if the goal of the study is to identify and map potential dispersal routes of a bird species endemic to a narrow elevational range in the Eastern Andes of Colombia. Rather, given considerations discussed in chapter 7, considering the natural history and dispersal abilities of the species in question (i.e., particularly **M** in the BAM diagram, and in some cases **B** as well), an appropriately restricted extent should be chosen.

Indeed, choice of a particular extent of analysis has direct implications for model characteristics and quality. For modeling algorithms that use absence, "pseudoabsence," or "background" data (see chapter 5) data as inputs, such as Maxent (Phillips et al. 2006), GARP (Stockwell and Peters 1999), and many multivariate statistical approaches (Elith et al. 2006), an analysis over too broad a spatial extent may lead to inclusion of absence, pseudoabsence, or background data from areas that a species fails to inhabit for reasons other than its scenopoetic existing fundamental niche (E_A). For example, such regions may lie outside the dispersal abilities (at present or historically) of the species. Absence, pseudoabsence, or background data from such regions are irrelevant or even misleading to the model (Anderson and Raza 2010; see chapter 7 for further discussion on this point).

Spatial extent also has direct implications for the results of essentially all approaches to evaluation of such models (see chapter 9). Evaluation methods are highly sensitive to the percentage of the overall study region in which the species is predicted to be present (Jiménez-Valverde et al. 2008). This point is developed in much greater detail in chapter 9.

RESOLUTION IN SPACE AND TIME

Resolution refers to the size of the subdivisions—in time or in space—that are applied to the datasets under consideration. In this book, unless otherwise noted, we refer to spatial resolution (or "grain"): the size of the cells or pixels into which the extent of analysis is subdivided in the raster maps (figure 6.3). As with spatial extent, the spatial resolution of analysis should match the spatial

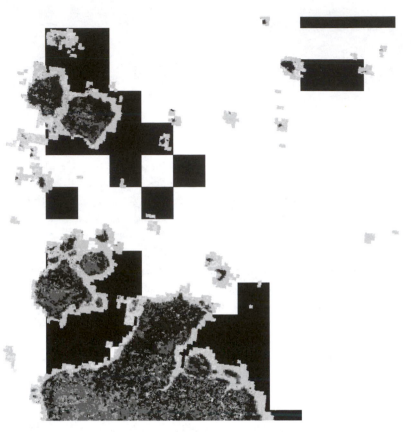

FIGURE 6.3. Illustration of different spatial resolutions in geospatial datasets. Shown are MODIS satellite data from the region of the northern tip of Australia summarizing photosynthetic mass at 250 m resolution in gray scale, on top of WorldClim climatic data at 17 km resolution in the background, in black. Note that even the definition of islands as separate depends on the pixel resolution.

resolution of the biological phenomenon under study. For example, a pixel size of 1 km^2 may be an appropriate spatial resolution for analyzing distribution patterns of terrestrial mammals, but it is far too coarse for analyzing the distribution of a soil nematode, which may be responding to environmental signals on much-finer spatial scales (see chapter 3).

Spatial resolution also impacts the sample sizes of occurrence data that are available. At very coarse resolutions, nearby occurrence records will often fall within the same large pixels. Because most niche modeling algorithms and applications reduce all occurrences in the same pixel to a single data record (so as to give presence data as opposed to something akin to abundances, and also in some cases reduce the artifacts of unequal sampling intensity across geography), the number of unique records available for analysis thus may be reduced substantially at coarse resolutions, but is maximized at the finest resolutions.

Frequently, when environmental data are obtained from diverse sources, spatial resolution will vary among maps. Most niche modeling algorithms require uniform cell sizes across environmental datasets, so resampling (changing cell size) procedures are often necessary. Reducing large pixels to sets of small pixels (downscaling) by simply subdividing larger cells is common, but introduces an appearance of fine resolution in the dataset (false precision) that must be interpreted with caution. When downscaling is necessary, more-advanced spatial techniques (e.g., kriging) may provide more realistic results than simple nearest-neighbor methods; however, the latter retain the *information* from the original data source, even though the resolution has been increased. On the other hand, when data are coarsened in resolution (upscaling), fine-grained information may be lost because small pixels are averaged or otherwise summarized to assign values to larger pixels.

Temporal resolution—the time span that a particular environmental parameter covers—is another important issue for consideration that is linked to the distinction between scenopoetic and bionomic variables. Some variables, particularly scenopoetic ones, have such slow dynamics that they can be considered effectively static over years to many decades or more—for example, topography does change, but not markedly within the ecological time spans important to the ecological and geographic distributions of species. Other variables are much more dynamic, influencing species' distributions in the middle and short terms, such as plant phenology or circadian cycles. In essence, ecological niche models are static, in the sense that correlations between species' occurrences and variables are made in a single temporal snapshot, over which time period we assume that the species' niche does not evolve. Hence, to capture a biogeographic or ecological phenomenon of interest adequately, occurrence and environmental data must coincide temporally.

Temporal correspondence between occurrence and environmental datasets also becomes important for studies through evolutionary history. For example, fossil records may be used to estimate paleodistributions of species for biogeographic and evolutionary analyses. In such cases, the time periods of both the fossils and the climatic reconstructions on which the models are based must match. Currently, however, such matching has only been possible at coarse temporal resolutions, such as 10^3 to 10^4 years, given the low density of fossil data and the coarse time slices available for paleoclimatic data (Martínez-Meyer et al. 2004a). Finer matching of occurrence records and climatic data (e.g., by decade) has been accomplished for studies of species over the past century (Araújo et al. 2005a, Nakazawa et al. 2007), and indeed month-by-month matching has been achieved in one study of insect distributional dynamics (Peterson et al. 2005a).

TYPES OF ENVIRONMENTAL DATA

Environmental data for niche modeling come in many different formats and resolutions, and summarize many different environmental parameters. The basis for characterizing environmental landscapes depends on how fundamental environmental factors (e.g., water, energy, time, and space) are summarized in measurable correlates (e.g., precipitation and humidity, solar radiation and prey abundance, growing days and season length, and patch size, respectively), and how these factors in turn relate to intrinsic growth rates (see equation 3.2) of populations of the species in question, or at least to their presence in G_{data}. In the following sections, we present a brief review of several relevant characteristics of widely used environmental datasets, and how these characteristics likely influence niche model performance.

Source. Geospatial environmental data used in niche modeling are derived from diverse sources and are generated via diverse methodologies. A first categorization of environmental datasets is thus based on the origin of the information. Some environmental surfaces have been generated from data collected from ground-based sampling stations (e.g., weather stations) and then interpolated spatially (often with environmental covariates like elevation) to provide information for the whole region. A typical implementation of this approach is the generation of climatic datasets for the present and recent past (New et al. 1997, Hijmans et al. 2005). These climatic datasets are the most widely used environmental-data resources in niche modeling, and yet at times can be somewhat "empty" with regard to original information content: e.g., a raster climate layer at 1 km^2 resolution across the huge Congo Basin in central Africa may in truth be based on data from only a few widely spaced climate stations.

Environmental raster maps may also be generated using biophysical modeling, as in the case of the General Circulation Models (GCMs) used for producing future (IPCC 2007) and past [e.g., Last Glacial Maximum (LGM); Shin et al. 2003] climatic scenarios. Thanks to enormous efforts by the Intergovernmental Panel on Climate Change (IPCC), future climate scenarios are now well-coordinated, and resulting data are openly and readily available (IPCC 2009). Numerous GCMs are now available for the whole world at coarse resolutions (i.e., >1°); Regional Circulation Models (RCMs) are now also increasingly available for several parts of the world with improved resolution (i.e., 0.1 to 0.3°). When RCMs are not available, or available spatial resolution is not sufficient for the purposes of a given study for whatever reason, techniques for downscaling environmental data now have been explored in some detail (Wilby and Wigley 1997). Because of the complex process of generating models, scenarios built and modified in this way may present high levels of uncertainty that can be propagated into niche models (Beaumont et al. 2007).

A third important data source is remotely sensed imagery. Several environmental dimensions can be measured via remote sensors, including elevation and derivatives (e.g., slope, orientation, etc.), land cover features (at levels of processing ranging from raw reflectance data to vegetation indices to fully interpreted land-cover classifications), and ocean-surface features. Remotely sensed datasets have been used as predictor variables in diverse niche modeling applications, with interesting results (e.g., Anderson et al. 2006, Bradley and Fleishman 2008), proving particularly convenient and information-rich when modeling distributions with short-term temporal resolutions. Here, we stress the importance of temporal matching between occurrence data and satellite-derived predictors. For instance, remote sensing products such as vegetation indices are quite suitable for use with occurrence data obtained for a species in the same time period as when the images were collected, but should be applied with caution when using natural history museum data collected over a broader time span (e.g., Raxworthy et al. 2003, Graham et al. 2004b). These problems result from the fact that many areas where the species was originally collected may now have been transformed into different land cover types, thus potentially leading to erroneous predictions. Similarly, the spatial resolution of occurrence data and environmental data must match (i.e., they should be on the order of the maximum estimated error or uncertainty between the two datasets) to avoid problems with error propagation (see chapter 5).

Measurements and units. Environmental variables can be represented in raster maps in several ways. The most common systems are "categorical" or "nominal," in which elements are described with names or numbers with no quantitative value or order among categories (e.g., soil classes); "ordinal," in which elements are sorted in a meaningful order, but no quantification is assigned to the categories, and no information on the numeric difference between categories is provided (e.g., "low," "medium," and "high" values for an index to degree of habitat conservation); "interval," or quantitative measurements in which values are grouped into categories (e.g., 0 to 10 mm, 10 to 20 mm, 20 to 30 mm, etc.); and "continuous," in which precise measurements are reported along a continuous gradient, without categorization or other grouping (for instance, most temperature and precipitation data). In niche modeling, the continuous variable representations are both the most commonly used and also generally the most informative. Categorical data are frequently not handled appropriately (e.g., in GARP) or not at all (e.g., in BIOCLIM). A few programs have implemented features that both recognize and manipulate categorical data correctly (e.g., Maxent). These issues are treated in greater detail in chapter 7.

Diverse environmental variables are measured in different units and measurement systems. Some niche modeling algorithms, like Ecological Niche Factor Analysis (ENFA; Hirzel et al. 2002), require standardization or normalization of variables before analysis, while others (e.g., desktopGARP) incorporate standardization within their processing. Other approaches, such as Maxent (Phillips et al. 2006), are able to manage variables in native units. Most modeling algorithms are based on real-number or integer variables, but a few variables are expressed in alternative measurement systems. For example, directional exposure ("aspect") is often expressed in radial units, or in terms of northing and easting. It is quite important to assure that the modeling algorithm is able to manage such measurement systems appropriately in its rule development.

For convenience in GIS processing, real-number representations can be cumbersome, whereas integer representations are much more convenient. As a result, investigators frequently multiply decimal values by 10, 100, or 1000, and then convert to integer representation to simplify the data (e.g., Hijmans et al. 2005).

ANCILLARY DATA

Niche modeling has been challenged by the need to integrate into analyses additional factors that influence distributions of species beyond simple scenopoetic and bionomic variables; we refer to such factors as "ancillary data." For example, geographic barriers and historical events play important roles in shaping occupied distributional areas of species G_O, and data pertaining to these factors can also be included in the latter stages of niche modeling exercises (i.e., after calibration; Soberón and Peterson 2005). Another example is the use of land-use summaries or vegetation classifications to refine distributional estimates calibrated based on coarser-resolution variables (Sánchez-Cordero et al. 2005).

In general, for the challenge of modifying raw model predictions (e.g., of G_P) into more accurate representations of occupied distributional areas of species (G_O), data layers summarizing environmental factors associated with dispersal potential become of prime importance (see chapter 8 for in-depth treatment of these issues). For example, investigators have used maps of biogeographic regions as hypotheses of regions accessible to species over biogeographic time periods (Anderson and Martínez-Meyer 2004); others have explored the structure of spatial autocorrelation in species' occurrences (Smith 1994, Araújo and Williams 2000, Keitt et al. 2002, Segurado and Araújo 2004) to provide this sort of information. However, much work needs to be done in this area.

Assuming that spatially referenced variables can provide information about distributions, the researcher may ask why not include these factors as covariates in niche models, just like the scenopoetic or bionomic variables. Examples that have been used in the literature include latitude and longitude, and distance to coast, all of which are explicitly spatially referenced. Concerns exist on conceptual and statistical grounds. The conceptual concern is that historical and other dispersal-based factors that limit and shape *actual* geographic distributions (i.e., the occupied distributional area G_O) operate in the geographic domain, whereas niche modeling reconstructs ecological niches in the *environmental* domain. This separation of spaces makes many inferences possible that would otherwise be obfuscated; by inserting history into model calibration, the researcher would lose the predictive power of niche models, i.e., the ability to estimate G_I. The statistical concern is that the spatial structure of species occurrences is determined both by autocorrelation among environmental variables and autocorrelation in biological factors driving distributions (e.g. dispersal, speciation, population extinctions, etc.); distinguishing the two sources of spatial structure in species occurrences is a challenge in itself (Segurado et al. 2006). In addition, when spatial covariates are included within models, it is common to find that they explain the greatest proportion of the variation in the data, thus limiting the useful explanatory power of scenopoetic or bionomic variables (Araújo and Williams 2000, Segurado et al. 2006, Dormann et al. 2007), which can be a serious problem when models are used for explanation, as indicated earlier. An alternative procedure, which we favor, at least in concept, is to separate the process of modeling niches from that of modeling spatial processes determining species' distributions. For an overview of approaches to postprocessing of distributional predictions (model outputs), see chapter 8.

SUMMARY

Environmental datasets that are of potential utility in ecological niche modeling are numerous and come from very diverse sources. These datasets must be pondered in terms of the information that they may offer to the modeling process. Direct, indirect, and resource-related variables can be distinguished with respect to whether they have relevance to the physiology of the organism in question, and by whether they are consumed or affected by it; somewhat distinctly, distal and proximal variables can be distinguished with regard to how immediate their effects on the organism are. In more practical terms, ideas behind variable selection for niche modeling are outlined, and challenges of preparing data, checking data quality, and assessing data characteristics (e.g., extent, resolution, source) are reviewed.

CHAPTER SEVEN

Modeling Ecological Niches

In the preceding two chapters, we discussed the biological occurrence (chapter 5) and environmental (chapter 6) data necessary for developing ecological niche models. In this chapter, we focus on how to use these data to create models that characterize species' ecological niches in E-space (which can then be applied to and visualized in regions of $\mathbf{G}$). The task is to characterize every cell within a region in terms of quantitative values related to probability of occurrence (or group membership), as a function of the environmental conditions presented in that cell. In the terminology presented in chapter 4, we aim to create a model, $\hat{f}$: a function constructed by means of data analysis, $\mu(\mathbf{G}_{data}, \mathbf{E})$, for the purpose of approximating the true relationship (i.e., the niche) in the form of the function f linking the environment and species occurrences (at least in $\mathbf{G}_O$ and hopefully also in $\mathbf{G}_I$).

An important consideration here is the use for which one is developing the model (Peterson 2006c). That is to say, a very basic initial consideration is whether the model is intended for the purpose of prediction or explanation (Araújo and Guisan 2006). In the specific case of niche models, a prediction-oriented model might use methods and data that produce optimal predictions in geography, yet provide little in the way of interpretable environmental information regarding the specific qualities of the niche being modeled. An explanation-oriented model, on the other hand, might emphasize characterizing the niche in useful and understandable terms, even though it may not necessarily produce the best predictions across geography. However, some models may fulfill both realms well. These choices are quite fundamental, and should be borne in mind in reading the sections that follow.

We first describe general principles regarding how modeling algorithms $\mu(\mathbf{G}_{data}, \mathbf{E})$ approach the task of finding some model $\hat{f}$ (several possible probabilities, suitability functions, and so on exist that may be of interest) about the true, but unknown, function. We then describe some commonly used algorithms. We do not aim to give detailed descriptions of individual algorithms, but rather to provide an overview of the different *types* of approaches that are used (e.g., climatic envelopes, general linear models, machine learning algorithms), and then to classify algorithms according to their requirements in terms of occurrence data. We proceed to describe important considerations about model calibration and definition of thresholds for converting continuous or ordinal suitability values into binary predictions of presence versus absence. Finally, we discuss differences among alternative modeling methods and the difficulties associated with selecting a general "best" approach.

WHAT IS BEING ESTIMATED?

An important consideration that should perhaps precede all others in this chapter is what is the "meaning" of the function f that is being estimated by the algorithms. Ideally, the model would produce a response variable that would relate directly to some quantity of biological interest, but what variable? Many publications presenting modeling algorithms have indicated that they estimate "probability of occurrence, given the environment" (e.g., Keating and Cherry 2004), which generally requires the existence of true absence data and an unbiased sampling scheme; others estimate degree of resemblance to the environment in the sample points (~ suitability; Hirzel et al. 2002, Farber and Kadmon 2003), or simply membership (or not) in a well-defined set (Busby 1991, Carpenter et al. 1993). Still others (e.g., Maxent) estimate very different quantities that can be transformed to yield probabilities of presence under certain assumptions related to the BAM diagram (Phillips and Dudík 2008).

In fact, it has been argued that, strictly speaking, probabilities of occurrence cannot be estimated without rigorous comparisons of presence and absence data (Ward et al. 2009), and that most niche modeling applications using presence-only data can at best estimate indices of relative suitability (Ferrier et al. 2002). The issue of what different algorithms do is central to all phases of niche modeling, and particularly to model evaluation. To clarify this important point, we will use arguments based on the BAM-related probability tree, developed in chapter 5. Equation 5.6 relates probability of presence to probabilities of different conditions being fulfilled; this equation, by application of Bayes's rule, gives:

$$P(\mathbf{Y} = 1 | X = g) = P_M(g)P_A(g)P_B(g) = \frac{P(\mathbf{Y} = 1)}{b(g)} P(X = g | \mathbf{Y} = 1). \quad (7.1)$$

In words, equation 7.1 relates probabilities related to the circles in the BAM diagram (a mechanistic or process-based approach) to correlative approaches requiring absence information [i.e., $P(\mathbf{Y} = 1 | X = g)$] and others that can operate with background or pseudo-absence data [$P(X = g|Y = 1)$]. Different objects of ecological interest can now be described in terms of this equation, and some confusions can be clarified. First, we discuss the difference between ecological niche modeling and species distribution modeling. A multiplicity of interests and objectives has caused considerable confusion regarding the activities encircled in those terms (Peterson 2006c). Under the niche modeling perspective, interest focuses on the set of environmental variables $\eta(g)$ at the sites g where the combined effect $P_{AB}(g) = P_A(g)P_B(g)$ is large; niche modeling also focuses on environments occurring in cells having large $P_A(g)$ (i.e., the abiotic contribution to the distribution). On the other hand, distribution modeling focuses interest on large values of $P_{BAM}(g) = P_M(g)P_A(g)P_B(g)$, which characterizes G_O. The two approaches often overlap in their aims and methods, but too often their specific objectives are left implicit (i.e., inadvertently hidden to the reader). It is not uncommon in the literature to apply methods that estimate different parts of equation 7.1 as if they were equivalent, and to use the two names interchangeably or carelessly. In what follows, we clarify this situation.

THE PROBABILITY OF OCCURRENCE OF A SPECIES, CONDITIONAL TO A SITE

A quantity of paramount importance is the probability of a species being present in a given cell: $P(Y = 1|X = g)$. If true absence data are legitimately available and are drawn from all parts of G (i.e., from all areas inside and outside of A, B, and M, then $P(Y = 1|X = g)$ is obtained directly by modeling procedures such as GLM, GAM, regression trees, and others (Thuiller et al. 2003, Keating and Cherry 2004, Pearce and Boyce 2006, Phillips et al. 2006, Phillips and Dudík 2008). $P(Y = 1|X = g)$ obtained directly represents a purely correlative result, without any hypothesis about the mechanisms that generate a presence record for a species. In fact, it is perfectly feasible to obtain fairly well-calibrated correlative models by fitting the preceding probability to geographic coordinates as covariates (Bahn and McGill 2007). However, equation 7.1 suggests the equivalence of $P(Y = 1|X = g)$ with $P_{BAM}(g) = P_M(g)P_A(g)P_B(g)$. Therefore, if true absences were available and the sampling points were randomly taken from throughout G, one is estimating the probability that the three conditions in the BAM diagram (defining G_O) are met, and the corresponding environments represent the occupied niche space $\eta(G_O) = E_O$. If, however, absence points are drawn from areas outside G_P but within G_M (and also within G_B, if the Eltonian Noise Hypothesis is not true), one can estimate the probability that the conditions B and A (or just A) are met (see also Anderson and Raza 2010).

Probability of Visiting a Site Conditional
to Presence of the Species

A second important quantity is the conditional probability $P(X = g|Y = 1)$, which in words means the conditional probability that the observer visits site g given that it holds populations of the species (Phillips and Dudík 2008). This probability is important because it is obtainable when true absences are lacking, by means of using background data. This probability, as can be seen from equation 7.1, is proportional to the probability of presence only if the visitation probability $b(g)$ is constant:

$$P(X = g|Y = 1) = \frac{P(Y = 1|X = g)b(g)}{P(Y = 1)}. \tag{7.2}$$

Again, as in the previous section, the conditional probability of presence being modeled depends on from which regions of **G** the absence (or background or pseudoabsence) data derive. Although constant $P(X = g) = b(g)$ represents an (often unstated) assumption of niche modeling, it is seldom constant or true in reality. Serious sampling biases exist in the great majority of occurrence datasets, where sites near to roads and other access points are visited with very high probabilities, and large inaccessible regions may have very low probabilities of having been sampled (Bojórquez-Tapia et al. 1995, Soberón et al. 2000, Graham et al. 2004b; see chapter 5). $P(X = g|Y = 1)$ can be estimated by a variety of methods (for example, Maxent or regression methods using background or pseudoabsences; see the following). However, since data are really about *observed* presences, rather than actual presences, the problem becomes still more complicated: Phillips et al. (2009) showed that, when biases in sampling and detection are taken into account, $P(X = g|Y = 1)$ relates monotonically to $P(Y = 1|X = g)$ only if the sampling bias of the background data is equal to the bias of the occurrence data.

Direct Estimates of Subsets of Niche Space

It is important to notice that several methods do not provide estimates of probabilities, but rather are procedures designed to construct certain sets, including particularly the so-called envelope methods (Pearce and Boyce 2006). We discuss this possibility by proposing the following formal identity:

$$P_{BAM}(g) = P[g \in \mathbf{M} \wedge \eta(g) \in \eta(\mathbf{A}) \wedge g \in \mathbf{B}] \tag{7.3}$$

where $\eta(g) = \vec{e}_g$ represents the vector of scenopoetic variables in cell g (see chapter 3), and $\eta(\mathbf{A})$ is the set of environments in the abiotically suitable region, or the scenopoetic existing fundamental niche, and **B** is the set of areas with favorable biotic environments. Basically, equation 7.3 is a translation of the probability tree term to an expression in terms of membership of the cell to a certain geographic set (the accessibility set **M**), and of its environment to the environmental set of abiotic (**A**) and biotic (**B**) niches (the latter frequently cannot be expressed in terms of scenopoetic variables). Neglecting the **B** term, the right side of equation 7.3 can be transformed to

$$P(g \in \mathbf{M} \wedge \eta(g) \in \eta(\mathbf{A})) = P[g \in \mathbf{M}|\eta(g) \in \eta(\mathbf{A})]P[\eta(g) \in \eta(\mathbf{A})]. (7.4)$$

Different envelope or "profile" methods obtain subsets $\{\eta(\mathbf{A})\}$ of environmental vectors directly, by using the observed environmental values of presences to classify the rest of the set. Envelope methods like BIOCLIM (Nix 1986), Support Vector Machines (SVMs; Guo et al. 2005), or HABITAT (Walker and Cocks 1991), which will be discussed later, simply use the outlying observations in environmental space to enclose a larger subset of environments. BIOCLIM uses a hyper-rectangle, SVMs use hyperspheres, and HABITAT uses a convex polygon. Distance methods, like the Mahalanobis distance approach, use the entire set of observations in E-space to classify every environmental vector in terms of distance to some aspect of the known observations. These methods generate, for every element in **E**, an index, normally interpreted as an index of habitat suitability for the species (e.g., they may be indices of environmental similarity to the centroid of the observations). In general, these methods provide direct estimations of $\{\eta(\mathbf{A})\}$ (albeit with a number of accompanying assumptions).

MODELING ALGORITHMS

We use the term "modeling algorithm" to refer to the procedure, rule, or mathematical function used to estimate the species' ecological niche as a function of a suite of environmental variables; we denote this function as the model $\mu(\mathbf{G}_{\text{data}}, \mathbf{E}) = \hat{f}$. The algorithm is, in some ways, the "core" of the model, but it should be remembered that the modeling algorithm is just one part of the broader modeling process: other factors, including selection of the reference regions **G** and **M** (see the following and chapter 3), choice of environmental variables (see chapter 6), characteristics of the occurrence data (see chapter 5), approaches to model calibration (this chapter), and choice of decision thresholds (this chapter), are key elements of the process that may be varied regardless of the algorithm being used.

As we said earlier, diverse algorithms have been applied to the task of modeling ecological niches and delineating suitable geographic distributional areas ($\mathbf{G}_P$ and $\mathbf{G}_O$). We do not attempt here to provide a detailed review or exhaustive list of approaches. Rather, we aim to provide an overview of the general principles used by different algorithms, and to provide practical advice for selecting suitable methods; numerous approaches are used commonly for modeling niches and distributions, and new methods (or variants on existing methods) are published every year. Published studies have given names or acronyms to various algorithms and/or methods—for example, "Maxent" refers to an implementation of the maximum entropy principle (Phillips et al. 2006); others include GLM, GAM, GARP, ANN, etc. In several cases, algorithms have been implemented in user-friendly software packages that are freely available.

Some important differences exist between modeling algorithms, including: (1) the sorts of biological data that the method requires (i.e., presences only, presences and true absences, presences and background data, and presences and pseudoabsence data); (2) the underlying methodological approach (e.g., algorithms may be based on regression methods, classification procedures, machine-learning procedures, or Bayesian statistics); (3) the form of output (e.g., continuous predictions versus binary or ordinal predictions); (4) the capacity to generate highly complex versus relatively simpler modeled response surfaces with respect to particular environmental variables; and (5) the ability to incorporate categorical environmental variables (see chapter 6). Hence, algorithms may be classified in many different ways. In the following paragraphs, we classify models by their biological input data requirements, which is a very practical consideration when selecting a method. Requirements of different algorithms with respect to type of biological input data are as follows:

1. *Presence-only approaches.* Some approaches rely solely on presence records, without need for reference to other samples or any other information drawn from the study area.
2. *Presence/absence approaches.* Some algorithms function by contrasting sites at which the species has been detected with sites where the species has been documented as absent, and therefore require both presence and absence data (see chapter 5). It is important to note that, in principle, any presence/absence algorithm can be implemented using background or pseudoabsence data (see the following).
3. *Presence/background approaches.* These approaches assess how the environment where the species is known to occur relates to the environment across the entire study area (the "background"). Therefore, such approaches use presence records along with environmental data drawn from the whole study area, or at least a very large sample drawn from across the study area, potentially including the known occurrence localities.

4. *Presence/pseudoabsence approaches*. Some approaches sample "pseudo-absences" from the study area, aiming to compare known occurrence localities with a set of localities having some probability of constituting presence localities that is below unity. The subtle difference from the background approach earlier is that pseudoabsence data are sampled only from sites at which the species is not known to occur (but note that this distinction has not been applied consistently in the literature). We do note, however, that unless a very broadly distributed species is very well sampled, background- and pseudoabsence-based results will not differ dramatically.

For the sake of completeness, we note that additional approaches have been developed that also incorporate information on distributions of other species, sometimes termed "community modeling" (Ferrier et al. 2004, Elith et al. 2006, Baselga and Araújo 2009), although these methods are not treated in detail in this book. This idea should not be confused with the use of interactions with a possible or known host-plant species as a biotic covariate in models (Araújo and Luoto 2007) or information from a target group to estimate sampling bias or for tests of artifactual absences (Anderson 2003). We note that a critical feature is what kind of absence data are used in analyses, as is detailed in the next several sections; we describe examples of commonly used algorithms that fall within each of the four categories of biological input occurrence data. Our descriptions are intended only to provide a broad overview of the approaches taken by different algorithms, so we encourage readers to consult the materials referenced for more in-depth discussion of particular algorithms.

PRESENCE-ONLY METHODS

Perhaps the simplest approach to modeling niches is by fitting "environmental envelopes" (or climatic envelopes), identifying shapes in multidimensional E-space that enclose environments associated with known occurrence localities. One of the earliest methods developed for this purpose is BIOCLIM (Busby 1991), which calculates a box-like minimal rectilinear envelope (figure 7.1A). BIOCLIM envelopes thus enclose within a rectangular (in a multidimensional space) envelope values in the range between the minimum and maximum values occupied by a species across all environmental variables (after Nix 1986). This box-like envelope is often estimated at different percentiles so as to incorporate different proportions of the observed occurrence localities (for example, the multidimensional envelope may be calculated so as to include 90% of occurrence localities, excluding the farthest outliers; see, e.g., Lindenmayer et al. 1991). Although transparent and straightforward, it suffers from a number of

A

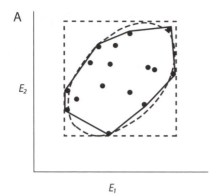

B

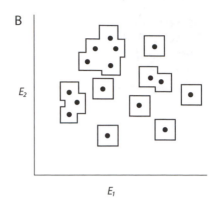

C

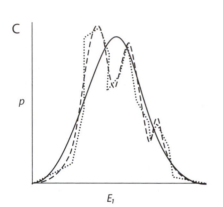

D

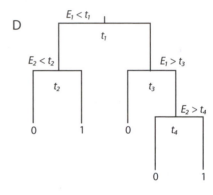

E

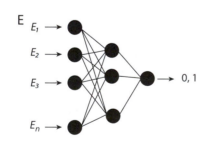

F

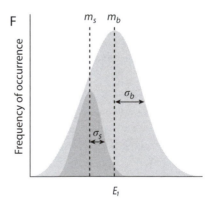

drawbacks, notably the inability to model interactions among variables and the inability to fit more complex "shapes" in ecological space.

Similar is the HABITAT approach (Walker and Cocks 1991), which fits environmental envelopes using minimum convex polygons or polyhedra, as illustrated in figure 7.1A. HABITAT thus encloses occurrence records more tightly within E-space than does BIOCLIM, allowing for reconstruction of interactions among variables, at least to some degree. If the environmental variables of the observations are uncorrelated, the two methods yield similar results (Guisan and Zimmermann 2000). Any notable obliquity in the arrangement of observations in E-space, however, will lead to overprediction by the rectangular box of BIOCLIM relative to HABITAT reconstructions.

Another set of methods based on presence-only records uses distances in E-space to fit models. One such distance measure is the Mahalanobis distance, which can be used to measure the difference in multidimensional E-space between the set of values for each site and the mean values for each environmental variable, across all occurrence localities. Thus, the closer that a site is to the mean (or the closest) of observations in E-space, the smaller the Mahalanobis distance and the closer to optimal the site is considered (Rotenberry et al. 2006).

FIGURE 7.1. Different approaches to calibrating ecological niche models in environmental dimensions (e.g., E_1 and E_2). Dots represent occurrence records. (A) BIOCLIM (dashed rectangle), HABITAT (solid line = minimum convex polygon; Walker and Cocks 1991), and Mahalanobis distance-based models (dashed ellipse; Farber and Kadmon 2003). (B) A DOMAIN model, based on proximity in E-space to sites of recorded presence, where proximity is measured by the Gower metric (Carpenter et al. 1993); solid lines delimit regions of E-space classed as "within" the envelope, based on a set threshold. (C) Example response curves generated by GLM (solid line), GAM (dashed line), and MARS (dotted line) methods (Elith et al. 2008). (D) An example classification tree, in which two environmental variables E_1 and E_2 are divided at split points t_1, t_2, etc.; a decision tree then classifies suitability as positive (1) or negative (0) along each branch (after Elith et al. 2008). (E) An example artificial neural network composed of "neurons" (shown as black circles) connected by weights (connecting lines); environmental variables E_1, E_2, etc. are input into the network and a machine-learning algorithm is used to "learn" a response (0, 1) at another point in the network by adjusting the weights (after Tarassenko 1998). (F) Illustration of Ecological Niche Factor Analysis (ENFA), comparing a focal species' distribution with respect to an environmental dimension E_1 (s, dark gray) against the background of the study region (b, light gray); the distribution of the focal species may differ from the background with respect to the means ($m_s \neq m_b$; marginality), and/or with respect to standard deviations ($\sigma_s \neq \sigma_b$; specialization; Hirzel et al. 2002).

An underlying assumption, therefore, is that deviations from mean observed occurrence conditions are associated with lower niche "suitability." Thus, unlike the rectilinear approach, which relies on extremes of a distribution in E-space, Mahalanobis distance-based approaches measure distance to the mean of the observed distribution. This approach allows for oblique positioning of an elliptical envelope in E-space (see figure 7.1A), and is less sensitive to outliers because it is based on all occurrence localities, rather than on peripheral localities only (Farber and Kadmon 2003). Mahalanobis distances also account for collinearity among variables in calculating the variance-covariance matrix. Several improvements on the original approaches based on the Mahalanobis distance measure have been proposed (Rotenberry et al. 2006, Calenge et al. 2008). An alternative distance measure developed by Gower (1971) has been applied in the DOMAIN approach (Carpenter et al. 1993; see figure 7.1B).

The so-called Support Vector Machines (SVMs) are envelope-based classification methods that can be used with presence-only data in niche modeling applications (Guo et al. 2005). With this type of data, an SVM fits a hypersphere that minimally encloses known presences in E-space. SVMs can fit complex shapes (Tax and Duin 1999), but for the presence-only case they are envelope techniques perhaps most analogous to Mahalanobis distance methods.

All of the methods mentioned here assign numerical values to different environments that are the elements of the set $\{\eta(g)\}$. These values express the similarity to observed environments in different ways. The user then, by explicitly or implicitly assigning a threshold, separates the elements of $\{\eta(g)\}$ into those predicted as belonging to $\eta(\mathbf{A}) = \mathbf{E}_A$ and those outside of $\eta(\mathbf{A}) = \mathbf{E}_A$. These methods identify sets of environments, or of geographic cells, without making reference to any probabilities.

PRESENCE/ABSENCE METHODS

As discussed earlier, when both presence and true absence records are available, diverse approaches may be applied to estimate $P(\mathbf{Y} = 1|X = g)$ directly, and thus, in theory, and given representative sampling of $\mathbf{G}$ (see the preceding discussion), $P_{BAM}(g) = P_M(g)P_A(g)P_B(g)$. Numerous such methods are based on regression analysis, a general statistical approach for relating a response (dependent) variable to values of predictor (independent) variables. Generalized linear models (GLMs) are extensions of basic least-squares regression methods that are well suited for modeling ecological relationships because they can fit a flexible range of error structures. GLMs are based on an assumed relationship (the "link function") between the mean of the response variable and a linear combination of the explanatory variables. This relationship defines the correspondence between the species' occurrence and individual environmental variables (see the following), and the error distribution can be selected from among numerous alternative distribution types, including normal, binomial, Poisson, or negative binomial distributions (see figure 7.1C; Guisan et al. 2002). Generalized additive models (GAMs) are semiparametric extensions of GLMs: like GLMs, GAMs also use link functions, but fewer assumptions are made regarding the form of the function, so it is possible to fit highly complex, nonlinear and nonmonotonic relationships (see figure 7.1C; Guisan et al. 2002). A related set of approaches includes Multivariate Adaptive Regression Splines (MARS), which are nonparametric and can fit response curves with levels of complexity similar to GAMs (Elith and Leathwick 2007). MARS fit linear segments to the data, meaning that MARS link functions consist of series of connected straight line segments, in contrast to the smooth curves of GAMs (see figure 7.1C).

Another approach for modeling with presence/absence data is to use classification and regression trees (CARTs), which provide a fairly intuitive way to classify locations in multidimensional E-space. Classification trees are constructed by repeated partitions of data into pairs of mutually exclusive groups, each as internally homogeneous as possible. In cases in which the response variable is categorical (e.g., presence/absence), homogeneity is assessed by minimizing the diversity of categories within each group; in this case, the method is referred to as a classification tree. When the response variable is numeric, homogeneity is assessed using methods related to regression analysis, including measuring sums of squares associated with group means, and the method is termed a regression tree (De'ath and Fabricius 2000). This splitting procedure is applied to each group separately, and repeated until a tree is "grown." Usually, a large tree is grown, and then is "pruned" back to achieve a desired level of simplicity or complexity (see figure 7.1D; for details see De'ath and Fabricius 2000).

Another suite of modeling approaches within this category are some (but not all) evolutionary-computing algorithms. While statistical approaches (e.g., GLM) assume the shape of a response curve, and estimate parameters for this model from the data, evolutionary-computing approaches do not assume a particular response curve, but rather "learn" the relationship between the response variable and its predictors (Breiman 2001). Such ideas have been applied to develop boosted regression trees (BRTs), which aim to improve performance of CARTs by fitting many models and then combining them to create predictions. BRTs build on trees fitted previously by focusing sequentially on the observed records that are hardest to predict (Elith et al. 2008).

Other machine-learning approaches that have been applied to the challenge of modeling ecological niches with presence/absence data include artificial neural networks (ANNs) and genetic algorithms (GAs). ANNs are computational systems inspired by the structure and operation of biological neural systems that "learn" species' responses to environmental predictor variables by repeatedly passing training data through a network of artificial "neurons." By adjusting internal structures of the neural network after each iteration, ANNs estimate a response in one part of the network based on inputs (environmental variables) at a different point in the network (see figure 7.1E; Hilbert and Ostendorf 2001, Pearson et al. 2002).

GAs are also computational tools that are inspired by biological phenomena. The "genetic" part comes from the idea that rules are in the form of linear strings (i.e., linear combinations of variables or Boolean statements) that are "evolved" similar to how biological chromosomes evolve: point mutations, insertions, deletions, and crossing-over events. At each step in the evolutionary process, the "fitness" of the resulting rule set is evaluated based on an internal subsetting of the calibration dataset, changing rule characteristics randomly, and then selecting an optimal set of rules by testing results internally against known cases (e.g., McClean et al. 2005). A GA method that has seen considerable use in niche modeling applications (GARP) will be described in more detail later, as it uses pseudoabsence data in place of absence data.

Finally, SVMs (see the preceding discussion) can work with presence/absence data. In this situation, SVMs use an algorithm that searches an optimal separating hyperplane capable of discriminating between presences and absences in E-space (Guo et al. 2005). SVMs do not estimate probabilities, but rather build sets on the basis of a discriminating function.

It is worth stressing that when (1) the objective is to model $\mathbf{G}_O$, (2) true absence data are available, and (3) sampling is unbiased and drawn from throughout $\mathbf{G}$, then presence/absence methods allow estimation of the occupied area $\mathbf{G}_O$ and its associated niche $\eta(\mathbf{G}_O)$ directly. However, frequently, the goal of modeling is to estimate $\mathbf{G}_A$ or $\mathbf{G}_P$, in which case use of true absence data will likely prevent the algorithm from doing so (see chapter 3), unless sampling of absences is restricted to areas outside $\mathbf{G}_P$ but within $\mathbf{G}_M$ and possibly also within $\mathbf{G}_B$ (see also Bahn and McGill 2007, Anderson and Raza 2010). As discussed in chapter 5, however, true absence data are complicated—not only can absence per se be difficult to establish, but also absence data for niche model development should be drawn only from relevant areas.

PRESENCE/BACKGROUND METHODS

Methods that utilize background data are distinct from presence/absence methods in that they do not require absence records, and are distinct from presence-only methods in that they incorporate information on environmental variation across the study area (the "background") in model development. As such, presence/background methods may frequently have better discriminatory power than presence-only methods.

One widely used presence/background method is based on the principle of maximum entropy, implemented in Maxent (Phillips et al. 2006). Maxent formalizes the maximum entropy principle that estimated probability distributions should agree with what is known (or inferred from the environmental conditions where the species has been observed), but should avoid assumptions not supported by the data. The approach is thus to find the probability distribution of maximum entropy—that which is most spread out, or closest to uniform—subject to constraints imposed by the information available regarding the observed occurrence records and environmental conditions across the study area (Phillips et al. 2006). Maxent estimates $P(X = g | Y = 1)$, which constitutes a density (i.e., if summed over all cells, it adds to one). However, the different outputs provided by the Maxent software are transformations from these raw probabilities, including a logistic output that estimates the probability of suitable environmental conditions (or probability of occurrence, if **B**, **A**, and **M** all coincide within **G**; Phillips and Dudík 2008). Maxent is able to fit extremely complex response curves, so we might expect that the challenges—as with other approaches that can fit highly complex models—are to provide input data that meet the assumptions of modeling (Phillips 2008) and to avoid overfitting (Peterson et al. 2007c; details are provided in the remainder of this chapter).

Another presence/background method is Ecological Niche Factor Analysis (ENFA), implemented in a software package called Biomapper (Hirzel et al. 2002), which compares the species' distribution in E-space with conditions across the study area via two factors: "marginality" and "specialization." Marginality quantifies how the mean of the occurrence records differs from the mean of the entire study area along each environmental axis, while specialization quantifies how the variance of the occurrence records relates to the variance across the study area (see figure 7.1F). To characterize marginality and specialization in multidimensional E-space, ENFA uses factor analysis (an ordination technique related to principal components analysis), to transform predictor variables into uncorrelated factors, specifically focusing on marginality on the first axis and specialization on the second. Then, the Biomapper software finds the distance to the median centroid in the space of niche factors, and assigns suitability scores to each pixel according to its distance to presence points in the transformed environmental space.

Statistical approaches like GAM and GLM can also be used when true absence data are lacking, by means of substituting either background or pseudo-absence sampling for absence data (Elith et al. 2006). Note, however, that when background samples or pseudoabsences are used in lieu of true absences, the estimation of $P(\mathbf{Y} = 1|X = g)$ requires the use of a correction term in what is known as "case-control sampling" (Pearce and Boyce 2006, Ward et al. 2009). The essence of the complication lies in estimation of the occurrence probability $P(\mathbf{Y} = 1)$, which with true absences and random sampling is estimated simply by the number of occurrences divided by the total sample size:

$$P(\mathbf{Y} = 1) = \frac{\text{Number of presences}}{[\text{Number of presences} + \text{Number of absences}]} \quad (7.5)$$

but that otherwise is more complex, as is explored in the remainder of this chapter.

Presence/Pseudoabsence Methods

Another, and indeed similar, method of dealing with the lack of true absence data is to resort to sampling "pseudoabsence" localities (see the preceding discussion and chapter 5). Here, "absence" information is resampled from the broader study area from sites lacking presence records for the species; however, these so-called pseudoabsence data are not always chosen with care, and thereby may not provide the appropriate contrasts necessary for rigorous model calibration. One widely applied method that utilizes pseudoabsence sampling is the GA called GARP (Stockwell and Peters 1999). Internally, the GARP algorithm itself requires absence data, but to allow use in the common cases where true absences are not available, the GARP software generates pseudoabsence records automatically from the study area. GARP combines three variants on climate envelopes with simple GLM methods in a GA framework that should— one would expect—produce a solution that is always as good as or better than the models that any component algorithm would yield. Because GARP does not yield deterministic solutions, multiple (100 to 10,000) runs of the algorithm are typically developed, which are then filtered based on error characteristics (Anderson et al. 2003), and summed to produce a consensus ordinal prediction.

Other Techniques and Multimethod Models

Numerous additional approaches have been published that do not fit so comfortably into the preceding classification scheme. For example, Bayesian approaches have permitted estimation of detection probabilities (Latimer et al. 2006), in some cases incorporating the occurrence data probability tree shown in chapter 4 explicitly (Argáez et al. 2005). Similar methodological adaptations may be possible for background sampling, but the theoretical implications need to be taken into account in each case.

In addition, some approaches produce multiple model outputs by producing multiple iterations of portions of the modeling process (e.g., multiple models with different algorithms, multiple iterations of algorithms with random incorporation of different types of rules), and then combining results to achieve single predictions (Araújo and New 2007). For example, GARP uses several envelope and regression methods that compete with each other in the GA framework (Stockwell and Peters 1999). Another example is the BIOMOD framework (Thuiller et al. 2009), which runs several algorithms (GLM, GAM, ANN, CART, MARS, multiple discriminant analysis, random forest, generalized boosting trees, and surface-range envelopes that are analogous to BIOCLIM).

When running multiple methods, it is necessary to identify a single result from among the different outputs. In the case of BIOMOD, one approach is to calculate evaluation statistics (see chapter 9) for each individual method, using as the final result the one that performs best in the evaluation tests (Thuiller et al. 2003, Thuiller et al. 2009). Alternatively, results from multiple-method approaches may be summarized by identifying areas of agreement among models (Araújo et al. 2005c, Araújo et al. 2006): for example, summary results may include areas identified as "all models predict" or "any model predicts" (e.g., Waltari et al. 2007). Another possibility is to create an "ensemble" of results from different methods, alternative parameterizations of the same method, or multiple iterations of stochastic methods (Anderson et al. 2003), to generate a suitability value (Araújo and New 2007).

A risk with combining multiple methods is that good results may be made *worse* by dilution with poor results. Possible solutions include utilization of only models shown to be robust for the particular application under consideration, combination of results from different methods only after poor results have been removed based on examination of evaluation statistics, or combination of model outputs via some objective weighting scheme, such as weighting based on evaluation statistics (Marmion et al. 2009). A second risk is that since, as we have argued earlier, different methods estimate rather different mathematical objects (different probabilities, different transformations, membership to sets, and so on) that may identify different portions of the BAM diagram ($\mathbf{G}_O$, $\mathbf{G}_O \cup \mathbf{G}_I$, $\mathbf{A}$, or their respective environments), mixing the results of different methods may create difficult problems of interpretation. Although these considerations may not matter much if one is interested primarily in predictive ability, if the aim of the modeling exercise is in explanation, to understand the ecological (niche) side of a biological problem, it may be preferable to use a more limited set of models with greater consistency in their internal workings.

IMPLEMENTATION

Ecological niche modeling is a fast-evolving area of research, with new algorithms and new implementations of niche models being published frequently.
Consequently, we provide here only a general overview of the main types of
implementations that are currently available. In some cases, software tools to
implement methods must be requested directly from the authors, usually simply because time and resources have not permitted development of versions
suitable for open distribution. However, in many cases, tools for implementing
these methods are freely available to interested users via the Internet. Most
techniques use software for Windows-based personal computers, but a few can
also be used with Macintosh and other platforms (e.g., Maxent).

    Many methods can be implemented using general-purpose statistical soft-
ware. For example, Thuiller et al. (2009) provided a platform for ensemble
forecasting of species' distributions using libraries available in the open-source
software environment R, and Elith et al. (2008) provided R code and a tutorial
for building niche models using BRTs. Other authors have worked in different
computing environments: for example, Latimer et al. (2006) provided code for
building Bayesian niche models in the WinBUGS software package.

    Other methods require specialized software. For example, Maxent, GARP,
and BIOMAPPER are all available freely in stand-alone packages. Other ini-
tiatives are developing software environments for implementing different
methods, such as openModeller (Muñoz et al. 2009). Niche modeling methods have also been implemented within GIS software platforms—for example, BIOCLIM and DOMAIN models are both implemented in DIVA-GIS
(Hijmans et al. 2001), and several methods can be run within the commercial
GIS packages IDRISI and ArcGIS (e.g., BIOCLIM). Software for computer-
intensive ensemble niche modeling has also been developed (BIOENSEM-
BLES; Diniz-Filho et al. 2009), in which production of comprehensive simulations across initial conditions, model classes, model parameterizations, and
boundary conditions is automated in a high-capacity distributed (grid) processing environment.

MODEL CALIBRATION

In chapter 4, we defined model calibration as referring to the process of adjusting model parameters until differences between model predictions and observations meet predefined, algorithm-specific criteria. Different algorithms
calibrate models in different ways, and diverse parameters and constants may

require adjusting. For example, the internal structure of an ANN must be optimized (Pearson et al. 2002), CART and BRT models should be pruned to an optimal level of complexity (Elith et al. 2008), and an appropriate "regularization" parameter must be set in Maxent (Phillips et al. 2006). We emphasize one important point: although theoretical and empirical research may have led to suggestion of default settings by the researchers developing the software (e.g., Anderson et al. 2003, Phillips et al. 2006, Phillips and Dudík 2008), it is generally poor practice to use default settings provided by software without justification, testing, and exploration of these values for a particular application. It is not our goal here to describe calibration methods for particular methods—for this information, readers should consult detailed literature on individual methods and experiment with the effects of varying particular parameters. Rather, we outline three general principles of importance during model calibration: data splitting, variable selection, and threshold selection.

DATA SPLITTING

A key goal of model calibration is to construct a model that fits well to the known data, but that does not overfit in ways such that its predictive ability is low when presented with independent data (see chapter 4). Many of the algorithms discussed earlier are able to fit complex response curves, if allowed. A common goal for these methods is to select a degree of complexity that is optimal for balancing E_{ver} and E_{val}, often achieved through an automated process of data splitting *within* the calibration dataset and internal to the algorithm's processing. If an algorithm does not carry out these steps automatically, a researcher may need to "tune" settings via species-specific experiments.

Note that this subdivision is in addition to that which divides the data into calibration and evaluation sets. For example, suppose we have 500 observed species occurrence records. We may first randomly set aside 100 of these records to use for model evaluation (see chapter 9); then, of the remaining 400 records, we might randomly hold out another 100 records against which the performance of the model will be tested for overfitting during the calibration process. These internal evaluations do not constitute assessments of prediction of independent data (validation). As such, they have the potential to assess overfitting to *noise* but not to any biases present in the overall calibration dataset, since such biases will also be present in any subset of the calibration data.

To test for overfitting to noise, several modeling methods assess model performance against the held-out data at regular intervals during model calibration (e.g., each generation in the evolution of a GARP model). If the model becomes too complex and overfit to the data being used to fit the model, then model performance will be poor on the internally held-out data. This approach

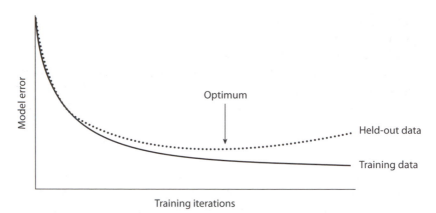

FIGURE 7.2. Illustration of data splitting to identify optimal performance in an artificial neural network model. One portion of the available data is used for model calibration (= training), and the other is held out for identifying when the model becomes overfit. Model performance is tested against the held-out data at regular intervals during training (e.g., every 100 iterations), and the iteration at which error is lowest on the held-out data is selected as the optimal model.

is exemplified in use of ANNs, whereby the network is calibrated on one subset of the data and then tested against the held-out data at regular intervals (measured by the number of learning iterations) during the training process (Pearson et al. 2002). Although the model will perform better and better on the training data (i.e., $\hat{f}$ will become increasingly complex, as the number of learning iterations increases), performance on the held-out data will begin to decrease once overfitting begins (i.e., when model complexity becomes excessive). The stage at which the model performs best on the held-out data is usually interpreted as the $\hat{f}$ with optimal complexity (figure 7.2). Other algorithms employ different strategies for avoiding overfitting, including discarding rules with low performance on held-out data in GAs, the regularization parameter in Maxent, and others.

Data to be held out to avoid overfitting are generally selected from the calibration dataset randomly in a one-time split; for example, Pearson et al. (2002) held out 50% of the calibration data to assess overfitting with ANNs, and GARP holds out 50% of calibration data for assessing rule predictivity (Stockwell and Peters 1999). However, more advanced data-splitting approaches can offer advantages. For example, with k-fold cross validation, calibration data are split into k roughly equal sized subsets ($k > 2$), and each part is held out successively while the other $k - 1$ parts are used for model building (Elith et al. 2008). An important advantage of this approach, particularly when sample sizes of available occurrence records are small-to-moderate, is that all calibration data are used at some point in the process to fit models. Other data splitting methods (e.g., bootstrapping and jackknifing) are used more commonly in model evaluation, and are therefore described in detail in chapter 9. However, similar approaches may be used during model calibration to minimize overfitting.

VARIABLE SELECTION

Different types of environmental predictor variables have been discussed in detail in chapter 6. Decisions regarding which variables to use should be based on biological reasoning, including selecting specific variables known or suspected to have a physiological role in limiting a species' distribution. For example, the role of hard, long-term freezes in limiting the distribution of saguaro cactus (*Cereus giganteus*) is well known (Drezner and Garrity 2003), so inclusion of variables related to time below freezing temperatures would be particularly relevant in a model of that species. More generally, however, such information is often not known, or not known completely, so it can be useful to explore which variables, and how many variables, to include in the model during the calibration process.

Inclusion of large numbers of environmental variables (i.e., 20 to 50) enables more complex models to be built and provides more information on which a model can be based. However, inclusion of fewer variables (e.g., Green et al. 2008 used only three variables in calibrating models) may be sufficient for some applications, and can help to avoid overfitting by limiting model complexity. Numbers of variables that should be used will also depend on the modeling method and on numbers of occurrence records available, with fewer variables warranted when fewer occurrence records are available (Fielding and Bell 1997). Some approaches operate such that no limits exist on numbers of variables that can be input into the model, regardless of the number of known occurrence records; however, clearly, as large numbers of variables are considered, the potential for overfitting rises dramatically (Peterson 2007b). To reduce overfitting in such cases, an algorithm should include some process of balancing model complexity against predictivity (e.g., the Akaike Information Criterion). For example, Maxent (Phillips et al. 2006) implements a form of regularization that can exclude variables from the final model (although they are considered in the calibration process), thus eliminating the requirement for prior variable selection. On the other hand, parametric statistical approaches (e.g., GLM, GAM) have rather firm requirements regarding the balance between numbers of occurrence records and numbers of parameters to be estimated, which in turn depends on the number of environmental variables included in the model.

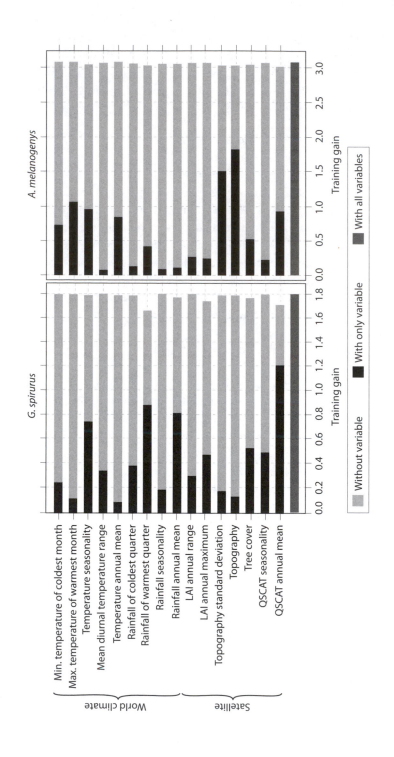

A. melanogenys

G. spirurus

Training gain

Min. temperature of coldest month
Max. temperature of warmest month
Temperature seasonality
Mean diurnal temperature range
Temperature annual mean
Rainfall of coldest quarter
Rainfall of warmest quarter
Rainfall seasonality
Rainfall annual mean
LAI annual range
LAI annual maximum
Topography standard deviation
Topography
Tree cover
QSCAT seasonality
QSCAT annual mean

World climate

Satellite

Without variable   With only variable   With all variables

A generic approach for exploring the importance of different environmental variables is jackknifing of the variables included in model calibration (not to be confused with the jackknife evaluation method described in chapter 9, from Pearson et al. 2007, which varies the occurrence records). In the calibration case, multiple models are built, each time excluding set numbers of variables (Peterson and Cohoon 1999). Model performance is assessed using one of the evaluation measures described in chapter 9 for each replicate model, and the importance of different variables estimated based on comparisons of performance with and without each variable (i.e., model performance will decrease markedly when variables with important unique contributions are excluded). Since environmental variables often are correlated with one other (chapter 6), interpretation of jackknife results must be undertaken with caution: jackknifing by excluding single variables ($n - 1$ jackknife) informs about the *unique* contribution of that variable—if other variables are highly correlated, then the test may not detect the relevance of such variables, making other "$n - d$" jackknife manipulations (where d is the number of variables deleted from each iteration; $d > 1$) more informative (Shao and Wu 1989, Efron and Tibshirani 1993). This procedure may also include calibrating models with each variable individually to estimate the single contribution of that variable (figure 7.3; Peterson and Cohoon 1999). Automated jackknifing of environmental data is included in some modeling programs, including desktopGARP and Maxent.

Finally, as intimated earlier, the overall *number* of variables, even if they are independent, can produce overfitting. In this sense, maintaining **E** simple, ecologically meaningful, and not too highly dimensional becomes critical. Procedures for reduction of numbers of variables, such as principal components analysis or other ordination techniques, can prove useful for correlated variables (see chapter 6). These steps offer the double benefit of reducing numbers of variables *and* producing uncorrelated variables that will cause fewer problems

---

FIGURE 7.3. Illustration of jackknifing as an approach for exploring the importance of different environmental variables in model calibration. Multiple models are built (in this case, using Maxent), each time excluding set numbers of variables to observe the effect of their exclusion from the model calibration process. In this illustration, results of two jackknifing analyses are shown for each of two South American bird species (*Glyphorynchus spirurus, Adelomyia melanogenys*). The dark bar for each variable shows its contribution (in terms of training gain) in single-variable analyses; the lighter-gray bar shows the training gain of models *excluding* that variable; the bar at the bottom shows the training gain of a model including all variables. Note that variable importance would be maximal if its single-variable-model gain were high and omit-one-model gain were low (each relative to experiments with other variables). From Buermann et al. (2008).

for modeling algorithms sensitive to nonindependent environmental variables. The cost associated with this step, however, is that the intuitive interpretability of the resulting (rotated) axes may be less than that of the original variables.

SETTING THRESHOLDS

Many modeling algorithms produce a continuous surface (e.g., probability values or relative suitability scores) as output, rather than binary predictions of presence and absence. It is often useful to convert continuous or ordinal model predictions to binary ones by choosing a threshold value (known commonly as a "threshold of occurrence," denoted by u in chapter 4), at or above which the environment is predicted as suitable for the species (i.e., within $\mathbf{G}_P$ or $\mathbf{G}_A$, depending on the particular absence, background, or pseudoabsence data). For instance, thresholding may be necessary if one is combining models of multiple species to yield predictions of species richness (e.g., Graham and Hijmans 2006; see chapter 12), or if the goal is to identify specific sites to be surveyed for species discovery (e.g., Pearson et al. 2007; see chapter 11) or selection of important areas for conservation action (e.g., Araújo and Williams 2000; see chapter 12). Selecting an appropriate threshold affects results considerably: as the threshold increases (i.e., imposing a more restrictive condition on suitability), the proportion of $\mathbf{E}$ estimated as being suitable for the species decreases, as does the proportional area predicted as suitable in $\mathbf{G}$.

Choice of threshold is therefore determined in large part by the proposed application of the model (Peterson 2006c). For example, if the aim is to identify areas within which disturbance may impact a species negatively (e.g., as part of an environmental impact assessment), then a low threshold may be chosen so as to identify a larger area of potentially suitable (and therefore, potentially vulnerable) habitat. In contrast, if the goal is to identify areas for conservation of important populations or potential reintroduction sites for an endangered species, then a relatively high threshold would be more appropriate, since this reduces the risk of selecting unsuitable sites; at least for some algorithms, such a step may identify the *best* sites.

Numerous methods have been employed for selecting thresholds for niche models (table 7.1). The simplest approach is by selecting an arbitrary value (e.g., 0.5, for outputs varying from 0 to 1), but this approach is very subjective, may lack ecological reasoning, assumes a fixed threshold when different species might require different thresholds and, not surprisingly, has been shown to perform poorly (Liu et al. 2005, Freeman and Moisen 2008). More objective methods, which can be applied to any algorithm producing nonbinary output, are frequently based on criteria applied during model calibration (i.e., based on the calibration dataset). As such, we include this topic in this chapter, even

TABLE 7.1. Some published methods for setting thresholds of occurrence, to convert continuous or ordinal model output to binary predictions of "present" and "absent."

Method	Definition	Occurrence data type[b]	Example reference(s)
Fixed value	An arbitrary fixed value (e.g., probability = 0.5).	None needed	Manel et al. 1999; Robertson et al. 2004
Least training presence	The lowest predicted value corresponding to an occurrence record.	Presence-only	Pearson et al. 2007; Phillips et al. 2006
Fixed sensitivity[a]	The threshold at which an arbitrary fixed sensitivity is reached (e.g., 0.95, meaning that 95% of calibration occurrence localities will be included in the prediction).	Presence-only	Pearson et al. 2004
Sensitivity-specificity[a] equality	The threshold at which sensitivity and specificity are equal.	Presence/absence	Pearson et al. 2004
Sensitivity-specificity sum maximization	The sum of sensitivity and specificity is maximized.	Presence/absence	Manel et al. 2001
Maximize Kappa[a]	The threshold at which Cohen's Kappa statistic is maximized.	Presence/absence	Huntley et al. 1995
Average probability/ suitability	The mean value across model output.	None needed	Cramer 2003
Equal prevalence	Species' prevalence (the proportion of presences relative to the number of sites) is maintained the same in the prediction as in the calibration data.	Presence only	Cramer 2003

Note that the least training presence threshold approach is a particular implementation of fixed sensitivity.

[a] Sensitivity is the proportion of presences correctly predicted; specificity is the proportion of absences correctly predicted; Kappa is a measure of model performance that reflects correct prediction of both presence and absence records. These indices are described in detail in chapter 9.

[b] Species occurrence data required to set the threshold. Note that methods suitable for presence-only data can also be applied to presence/absence data, but not vice versa. Based in part on Liu et al. (2005).

though some of the evaluation statistics used in setting thresholds (e.g., Kappa) are described in detail in chapter 9; the reader may need to refer to that chapter for a more complete understanding of the methods. Options for selecting thresholds are limited by types of data that are available (i.e., presence-only versus presence/absence), and we discuss these separately below.

Presence/absence thresholding.   When using presence/absence data, a simple approach is to balance numbers of observed presences and absences that are correctly predicted (although all of the cautions mentioned earlier regarding absence data in niche modeling apply here as well). Thus, the threshold is adjusted so that the number of observed presences incorrectly predicted (i.e., that fall below the threshold) is balanced against the number of observed absences that are incorrectly predicted (i.e., that fall at or above the threshold). Methods based on such data usually assess model performance across the range of possible thresholds (e.g., for a model that predicts probability of occurrence, performance could be assessed at thresholds at increments of 0.01 between 0 and 1), selecting the threshold that maximizes performance according to the quantitative measure employed (e.g., number of records incorrectly predicted).

It is common to calculate the Kappa statistic (which has high values when both presences and absences are predicted correctly; see chapter 9) to assess performance based on the calibration data (or, in some cases, an internally held-out subset of the calibration data) and then to select the threshold at which the statistic is maximized (commonly termed "maximizing Kappa"; figure 7.4). A similar alternative is to select the threshold at which the proportion of presences correctly predicted and the proportion of absences correctly predicted coincide (more formally, this approach means balancing sensitivity and specificity; see chapter 9 and figure 7.4). We note, however, that the cautions expressed earlier regarding what quantity is estimated by each method may also apply to this situation, and (most critically) also that all of these methods assume a symmetrical loss function (i.e., they give equal weight to omission and commission errors).

Liu et al. (2005) tested 12 methods for setting thresholds using presence/absence data for two European plant species. Based on four evaluation statistics (sensitivity, specificity, Kappa, and overall accuracy; see chapter 9), they concluded that the best methods were maximizing the sum of sensitivity and specificity, using the average probability/suitability score, and setting equal proportional area predicted (termed *prevalence*) between calibration data and the prediction. In a similar analysis using presence/absence records for 13 tree species in western North America, Freeman and Moisen (2008) also found support for thresholds based on equal prevalence between the calibration data and the prediction, and for those based on maximizing Kappa.

Presence-only thresholding.   Clearly, approaches using presence/absence data depend rather critically on the absence information, so it is important to consider what can be done in the frequent case in which such data are lacking. Because background and pseudoabsence samples may suffer from inclusion of

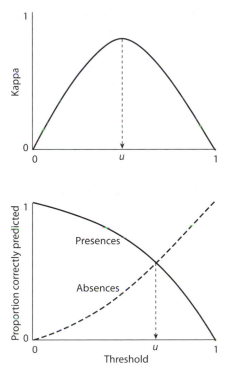

FIGURE 7.4. Illustration of approaches to selecting a threshold of occurrence u based on presence/absence data. Top: The Kappa statistic, which includes a combination of information on omission and commission errors (see chapter 9), is calculated for each possible threshold (in this case, shown across the range 0 to 1), and the threshold at which Kappa is maximized is chosen. Bottom: Proportions correctly predicted for presence and absence data are calculated over all thresholds, and the point at which they cross indicates the optimal threshold. Note that both of these approaches weight omission and commission errors equally, and as such assume a symmetric loss function.

grid pixels falling within G_P, leading to apparent commission error (Anderson et al. 2003), strictly speaking, methods that integrate any estimate of commission cannot be used when true absence data are lacking, at least without modifying the loss statement and interpretation (see chapter 4). Fortunately, a few methods *can* be applied to presence-only data. A common approach is to use the lowest suitability value associated with a calibration presence record, termed the "least training presence threshold" by Pearson et al. (2007). This approach assumes that presences of the species are restricted to sites at least as suitable as those at which the species has been observed so far, and that sites at or above that threshold are indeed suitable for the species. The approach therefore identifies a minimum area in **G** within which the species can occur, while ensuring that no known presence records are incorrectly predicted. This approach can be either overly restrictive (e.g., when few occurrence records are available) or overly broad (e.g., when some records are incorrectly identified or georeferenced, derived from sink populations, or associated with factors favorable to suitability but not included in the environmental dataset). Because of the possibility of overly broad predictions, an alternative is to choose a threshold such that a certain percentage of presence records are included (e.g., 95% of presences included in the prediction). This method is less sensitive to outliers than the lowest presence threshold, but errors of omission accumulate as a consequence (i.e., some presences are incorrectly predicted; Pearson et al. 2007). These approaches have the advantage of weighting errors of omission and commission, and thereby fit well with our understanding of the relative importance of presence versus absence data.

In addition to these thresholding rules based on omission of calibration data (that can be implemented with many modeling approaches), some techniques provide theoretical expectations by which thresholds can be chosen. For example, with the cumulative output from Maxent (which ranges from 0 to 100), under certain assumptions, the value of the prediction equals the expected omission rate for independent samples of occurrence records for the species; hence, the evaluation omission rate obtained by applying a given threshold, e.g., 10 out of 100 can be compared to the theoretical expectation (Phillips et al. 2006). Similarly, the logistic output of Maxent ranges from 0 to 1 and provides the probability of suitable conditions, or probability of occurrence if both model calibration and model evaluation occur in a study region where **B** and **M** do not affect the species' distribution. Here, similar to presence/absence thresholding with GAM/GLM, etc. (techniques that produce estimates of probability of occurrence), a threshold can be set (0.5) that should, in theory, minimize both omission and commission rates. These probabilistic interpretations of Maxent outputs, however, have not gone without criticism in light of overfitting when model assumptions are violated, as will frequently be the case (Peterson et al. 2008a).

Finally, we note that presence-only approaches to setting thresholds may be justified even in cases when absence records are available. Indeed, it may be argued that maximizing numbers of correctly predicted observed presences is more important than is minimizing numbers of incorrectly predicted absences. This consideration relates to the concept of "false absences" discussed in chapter 5: if some absences are likely to be recorded in environments that are nonetheless suitable (owing to detection probabilities <1, environmental heterogeneity within coarse-resolution grid cells, or uncertain estimation of **M** as the area of analysis), it is inappropriate to give them the same weight as observed presences. This issue is central to the theory behind model evaluation (see discussion of apparent commission error, chapter 9), but it is also relevant to selecting a threshold of occurrence. This issue may be addressed by modifying the loss function (see chapter 4): for example, the "best subsets" implementation of the GARP algorithm allocates unequal weights to commission and omission errors (Anderson et al. 2003); similarly, recent proposals for modified ROC evaluation statistics weight different errors differentially (Peterson et al. 2008a).

MODEL COMPLEXITY AND OVERFITTING

Model complexity refers to the degree of flexibility in the function $\hat{f}$ to conform to calibration data, which minimizes E_{ver}. When models show close fit to calibration data, but are less able to predict independent or even semi-independent evaluation data, that is, E_{ver} is small but E_{val} (validation error; see chapter 4) is large, this situation is formally termed "overfitting," providing a quantitative definition of a term already used at several points in this book.

It is generally true (Hastie et al. 2001) that too little complexity induces both E_{ver} and E_{val} to be large, whereas more complexity successively produces smaller E_{ver}, and initially also reduces E_{val}. However, increasing complexity still further causes E_{val} to increase again (see figure 7.2), which is precisely the lack of generality caused by overfitting models. Overfitting occurs because models are overparameterized (for example, in the case of niche modeling, too many variables are included in the model or too strong a weight is assigned to them). This condition is not caused by the data, but rather by the methods used to produce the model. Many modeling methods do not allow explicit control of the degree of model complexity, except for the number of variables; others can be calibrated to yield optimum degrees of complexity (see the section "Data Splitting" earlier). Because almost all uses of niche models require some degree of predictive ability (Peterson 2006c), eliminating, minimizing, or at least being fully aware of overfitting is of great importance.

A useful way to think about model complexity and overfitting is as regards response curves (figure 7.5). Response curves describe probabilities of presence for a species across the range of values of an environmental variable (or in some cases, for combinations of variables, such as the product of two variables). Here, species occurrence is the dependent variable, and the environmental variable is the independent variable. The task for the modeling algorithm is thus to estimate response curves for each of the environmental variables (of course,

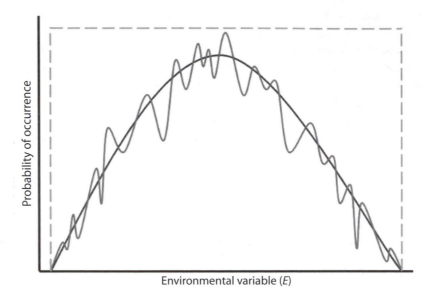

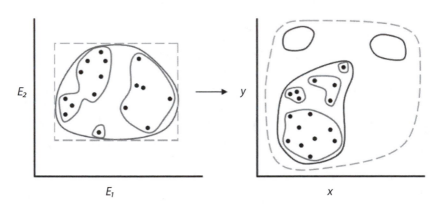

FIGURE 7.5. Illustrations of model complexity and its relationship to environmental dimensions and geography. Top: Visualization of three response curves along a single environmental dimension—dashed line = overly simple response curve; dark solid line = probable biologically realistic response curve; light solid line = overly complex response curve. Bottom: The same three response curves (overly simple, biologically realistic, overly complex) visualized in a two-dimensional E-space and G-space, showing their likely tendencies to generalize excessively or to specify too much.

taking into account any interactions between variables, as necessary). Figure 7.5 contrasts three hypothetical response curves: one that offers a biologically realistic situation, another that is a highly complex model and probably unlikely to represent a realistic response, and finally an extremely simple response curve also unlikely to be biologically realistic or to have much predictive power.

Niche models generally combine response curves across multiple predictor variables in some way, which can range from very simple to complex. Figure 7.5 illustrates models with differing degrees of complexity in two-dimensional E-space (models are presented as binary predictions for simplicity). A more complex model can fit more complex niche shapes, whereas a simpler model fits less complex shapes but may be more likely to include parts of E-space that are not within the species' true ecological niche. A third model provides an intermediate level of complexity. When projected into G-space, the more complex model predicts a smaller area that fits more closely to known occurrences, while the simpler model predicts a broader area, thus including more areas not known to be occupied.

This trade-off between model complexity and simplicity is a key part of model calibration. If the model is overly complex, it is likely to make predictions that fit too closely to known occurrences and that have poor predictive ability for unsampled cells; in contrast, a model that is too simple may not capture the true ecological complexity, and may therefore underfit the niche, and show poor performance, but generally in the direction of predicting too broad of an area. We discuss these issues in more detail later in this chapter, and we describe how to test for over- and underfitting in chapter 9.

STUDY REGION EXTENT AND RESOLUTION REVISITED

As mentioned several times earlier in this book, selection of the appropriate region for analysis represents a critical step in the modeling process. A species' ecological niche may appear highly specialized at a global extent but it may be much less specialized within regions close to where it occurs. For example, although a species might appear restricted completely to tallgrass prairie when viewed at broad extents, a closer view may reveal that it occurs broadly into nearby land cover types. Changing the extent (i.e., the geographic limits) of the study area G can thus have a substantial impact on model predictions (Hirzel et al. 2002).

With presence/absence, presence/background, and presence/pseudoabsence modeling techniques (for presence-only techniques, these concerns do not apply), the region for model calibration should not include areas where the species is absent for reasons *other than* conditions unsuitable in terms of the environmental variables used in modeling (Anderson and Raza 2010). Specifically, model calibration should be focused over the region within **M** in the BAM diagram, as absence (or background or pseudoabsence) information in this area will be maximally meaningful and not misleading (see detailed treatment in Barve et al. 2011). In fact, ideally, the comparisons in model calibration would also avoid drawing pixels from areas from which the species is absent for biotic reasons, although these areas may be yet more difficult to identify.

Violation of these conditions hinders successful calibration of $\hat{f}$ because "absence" information potentially corresponding to *suitable* environments can provide false negative signals. For example, the model may recognize spurious environmental differences between a region that a species actually inhabits versus another region that it could inhabit but does not because of a geographic barrier (i.e., outside **M**). This problem does not represent overfitting to noise, but rather to a bias: incomplete representation of environments in the occurrence dataset, namely the lack of positive occurrence records from all regions holding suitable conditions, an explicit or implicit assumption of many, if not all, presence/background modeling techniques.

Clearly, detailed information regarding factors producing such biases will generally be difficult to obtain or estimate. However, explicit statement of assumptions regarding **M** constitutes a concrete step forward, outlining why the chosen study region was delimited as it was. Some situations may call for hypotheses of **M** based on the current dispersal capabilities of the species, whereas others may attempt to estimate likely past distributions (e.g., during the Pleistocene). (It should be noted that **M** must also be adjusted by the spatial distribution of sampling, as was clear in figure 5.1: this distribution constrains the spatial possibilities of occurrence similarly to the action of **M**.) Future research is necessary to develop operational guidelines for selecting study regions based on these ideas (Anderson and Raza 2010, Barve et al. 2011).

MODEL EXTRAPOLATION AND TRANSFERABILITY

We distinguish carefully between extrapolation and transferability. Extrapolation refers to use of a model to predict into areas presenting environmental values beyond the environmental range of the area on which the model was calibrated (Williams and Jackson 2007). For example, suppose that a niche model was calibrated using occurrence records sampled from a region with temperatures ranging 10–20°C. If the model is used to predict suitable conditions across a different region (or under a different climate scenario), where

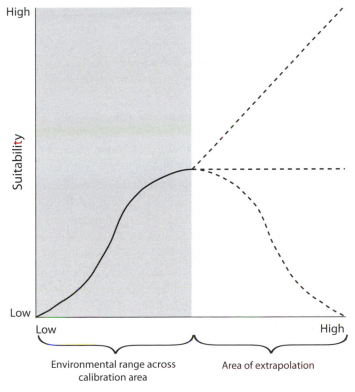

FIGURE 7.6. Hypothetical example illustrating the perils of extrapolation of niche model suitability predictions beyond the environmental range of the calibration data (in gray). In the area of extrapolation (i.e., environmental conditions not represented in the calibration region **G**), the trend characterized in the calibration range could continue rising, level off, or decline, but no information relevant to this point is available. Such situations are referred to as "truncated response curves."

temperatures reach 25°C, then the model must extrapolate. Because the model has no prior information regarding the probability of the species being present at 25°C, the prediction into such conditions is likely to be highly uncertain (Thuiller et al. 2004c, Pearson et al. 2006). Examples of the perils of extrapolation are illustrated in figure 7.6. Given the truncated response curves often experienced in current datasets, extrapolation should be approached with much caution, since the correlative niche-modeling approach does not necessarily identify mechanistic, process-based relationships between species' occurrence and the environment, which would be more likely to hold true under novel circumstances (Kearney and Porter 2004). It is thus essential to understand the assumptions made when extrapolation occurs: extrapolation is particularly risky in situations in which response curves are high or increasing where truncated (Anderson and Raza 2010), such as will frequently be the case when range limits are imposed externally (e.g., a coastline for a terrestrial species) and not related to **A**.

One approach that has been explored to tackle this problem is to calibrate models based on study regions that incorporate the full range of environments of interest. For example, in a study of climate change impacts on plants in Britain, Pearson et al. (2002) calibrated niche models at the extent of all of Europe to incorporate a broad range of environmental conditions, and then projected niches onto future climate scenarios only in the British Isles. This approach minimized the risk that, when applied to future climate change scenarios, the model would be used to extrapolate (in **E**) outside the environmental range of the calibration data. This approach shows promise, but is subject to the limitations described in the previous section on selecting the study region.

Transferability is in and of itself a simpler and more tractable challenge—it refers to the idea of applying a model developed on one landscape to another landscape or to another time period in the same area (Araújo and Rahbek 2006, Randin et al. 2006, Peterson et al. 2007c). Here, the Hutchinsonian Duality of linked spaces that we have discussed in chapters 2 and 3 becomes key: the model developed in **E** can be applied to any space **G** that is characterized in the same dimensions (environmental variables). Problems can arise if the two **G** spaces differ in their associated **E** spaces, which could lead to truncated response curves, and the need for extrapolation, as mentioned earlier. The key challenge in achieving high transferability is really one of developing models that are not overfit in the calibration region and that treat extrapolation with care when necessary.

DIFFERENCES AMONG METHODS AND SELECTION OF "BEST" MODELS

We have seen that several similarities exist between ecological niche modeling and species distribution modeling, as well as many differences. In other words, the nature of the f being modeled may correspond to different areas in the BAM diagram, and/or their corresponding environments. Put much more simply, in our view, species distribution modeling will always involve ecological niche modeling, but must also take on additional challenges, such as incorporating spatial processes of dispersal and its limitation.

We have also discussed a broad diversity of alternative modeling methods in this chapter, which can produce rather contrasting forms of $\hat{f}$ functions, described by several different probabilities, suitability indices, or memberships to sets. In addition, different kinds of data are used to perform these analyses. It is important, therefore, to consider the degree to which these different methods, with different objectives, and based on different data types, yield different results.

Numerous studies have demonstrated that different modeling approaches have the potential to yield substantially different predictions (Loiselle et al. 2003, Thuiller 2003, Brotons et al. 2004, Segurado and Araújo 2004, Thuiller 2004, Araújo et al. 2005b, Elith et al. 2006, Pearson et al. 2006). For instance, Pearson et al. (2006) modeled plants of the family Proteaceae in the Cape Region of South Africa, and found pronounced differences among model predictions regarding change in range size under future climate scenarios, with predicted changes differing in both direction and magnitude among algorithms (e.g., from 92% reduction to 322% range increase for a single species). In another study, Loiselle et al. (2003) demonstrated distinct results when alternative methods were used to identify priority sites for conservation prioritization.

A particularly extensive comparison of methods was provided by Elith et al. (2006), who compared 16 modeling methods using 226 species across six regions of the world. For each species, two sets of occurrence data were collated: (1) presence-only records from unplanned surveys or incidental records (i.e., the sort of data typically obtained from data associated with natural history museum and herbarium specimens); and (2) presence/absence records derived from planned surveys of the same landscapes. The former data were used for model calibration, while the latter were used for model evaluation, ostensibly to provide an optimal evaluation of model performance. Given the lack of absence records for model calibration, methods requiring some form of absence information were implemented using presence/background or presence/pseudoabsence data. Elith et al. (2006) concluded that the "best" models were those able to fit highly complex responses, such as the machine-learning methods Maxent and BRT. Subsequent publications have detailed "experimental" manipulations that were designed to clarify details of robustness to sample size, positional error, grid resolution, etc. (Graham et al. 2007, Guisan et al. 2007, Wisz et al. 2008); results again were interpreted as revealing clear differences in performance among methods (figure 7.7). However, subsequent publications have pointed out weaknesses of these studies, particularly regarding techniques for model validation, which favored certain model types over others for artifactual reasons, and not based on real differences in performance (see chapter 9; e.g., Peterson et al. 2008a).

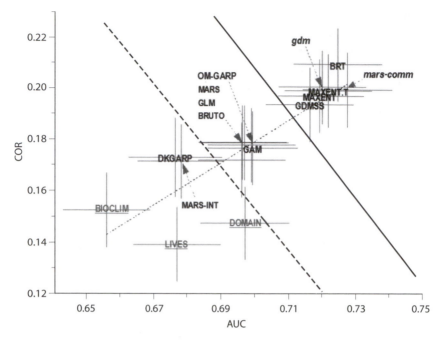

FIGURE 7.7. Summary of results of comparisons of niche modeling methods by Elith et al. (2006), who compared 16 modeling methods using 226 species across six regions of the world. Model performance is summarized via two measures (correlations between predictions and observations, COR, and the area under the curve of the receiver operating characteristic plot, AUC). In the lower-left portion of the figure are poorly performing methods, and in the upper-right portion of the figure are better performers. See text for detailed discussion of the implications of this figure, which is derived from Elith et al. (2006).

Given that different modeling methods can give different results, selection of an appropriate method or methods is crucial, if difficult. Identifying methods that are generically "best" is problematic since the approach used to evaluate model quality should depend on the aim of the modeling (Peterson 2006c). For example, Elith et al. (2006) evaluated the ability of methods to predict species' occupied areas (G_O) via statistical tests that reward models for classifying both presences and absences correctly, when different weights should be accorded to each. What is more, since the evaluation data were drawn from the same geographic area as the calibration data, and were therefore spatially auto-correlated, models able to fit highly complex response curves could fit closely to calibration data and were thus more likely to yield particularly good evaluation statistics. The spatial autocorrelation between calibration and evaluation data certainly compromises the independence of the two datasets (see chapter 9).

In contrast, Pearson et al. (2007) assessed predictive performance of different modeling methods based on their ability to predict only presence records, arguing that the purpose of the modeling was to identify abiotically suitable areas ($\mathbf{G}_A$, and hence also $\mathbf{E}_A$). In this case, use of absence data in assessing which model is "best" is inappropriate, because sites classified as absent may in fact be environmentally suitable but outside of $\mathbf{M}$ and/or $\mathbf{B}$, and thus in truth part of $\mathbf{G}_A$ and $\mathbf{E}_A$. We will return to this discussion in chapter 9, when describing the merits and limitations of different evaluation strategies. The important point is that identifying a best method is in no way straightforward, and that comparative evaluations published to date have results that are far from definitive.

CHARACTERIZING ECOLOGICAL NICHES

We have focused on ecological niches that can be defined as subsets of an environmental space that is defined by noninteractive (unlinked) variables, and we have described several approaches to estimating ecological niches based on known occurrences of species across landscapes. These "niche models," however, are generally defined in multivariate spaces that are complex and irregular (see figure 2.2). This complexity obfuscates efforts to visualize niche models effectively. Therefore, it is necessary to describe strategies for characterizing ecological niches in ways that not only allow visualization, but also permit comparisons, definition of quantitative measures, etc.

We emphasize that Grinnellian niches are defined in relatively coarse-resolution scenopoetic environmental dimensions, so we avoid much of the complexity and intractability of bionomic variables and interactions with other species, which have long plagued efforts to characterize Eltonian niches. In fact, one of the most powerful advantages of Grinnellian niches is that their straightforward and operational definition in terms of subsets of multivariate spaces allows a number of well-known quantitative methods to be used to explore and analyze them. Grinnellian niches can be characterized by their position, size, and shape (Jackson and Overpeck 2000), as we develop in the next several paragraphs.

Much of the early efforts to characterize niches were developed in the 1970s, beginning with attempts to find measures of "niche width" or "niche breadth" for resource or habitat utilization. Most of these early papers belong to the concept of niche that we have called "Eltonian," which is focused on community-ecology problems and mostly uses consumable resources as niche variables. Nonetheless, many of the original ideas can be applied directly to a Grinnellian framework.

In the early literature, niche measures were mostly measures of spread or variance along single axes (Levins 1968, Colwell and Futuyma 1971, Cody 1974), but soon the ideas were generalized to several variables, sometimes in an entirely theoretical way (Yoshiyama and Roughgarden 1977). However, a few pioneering papers (Green 1971, James 1971) began characterizing niches using multivariate techniques on realistic environmental datasets. These first steps were followed later by many more, examples of which are Dueser and Shuggart (1979), Rotenberry and Wiens (1980), Carnes and Slade (1982), Austin (1985), and Austin et al. (1990). Most modern literature on niche characterization is rooted firmly in the multivariate mathematics of ordination methods (Legendre and Legendre 1998), which is the philosophy we will follow here.

Several problems arise when trying to characterize niches in an E-space of v variables. Two technical challenges immediately become apparent: (1) many of the variables are correlated, and (2) their units and ranges are likely to be distinct. It has been customary since the beginnings of multivariate niche characterization (Green 1971, Carnes and Slade 1982) to deal with these challenges via standardization of variables, followed by dimensionality reduction by means of factor analysis or principal component analysis (PCA). In z-standardization, each environmental value has the mean over all combinations subtracted, and the resulting quantity is then divided by the standard deviation, producing standard normal variables with mean of zero and variance of unity. This procedure has the advantage of scaling all variables to a comparable range of values, but can distort the fact that some variables are by nature distributed more narrowly than others (Pielou 1984, Legendre and Legendre 1998).

The challenge of removing redundancy among environmental variables is also nontrivial. Although some have argued that niche exploration should be performed in spaces of thousands of dimensions (Stockwell 2006), as discussed in chapter 6, most of these "dimensions" are highly collinear, so the true number of environmental dimensions in which niches can be characterized is effectively much smaller (Peterson 2007b). Two general classes of approaches are available to remedy this problem: (1) factor analysis or other dimensional-reduction manipulations can be used to produce combinations of original variables that summarize environmental variation along fewer and simpler axes; or (2) correlation analyses can be used to detect highly redundant variables and eliminate repetitive ones. Once again, each approach may have advantages and disadvantages. The dimensional reduction approach loses the ease of interpretability of the original raw environmental dimensions, and reduces the predictive ability of models because the synthetic variables generated with the process of dimensionality reduction (e.g., by generating factors or components that are linear combinations of sets of covarying variables) may not maintain the same correlation (or covariation) structure in other times or regions. The variable reduction approach, in contrast, requires careful thought in selection of variables so as not to lose interpretability.

Selecting one method or another of standardization, and deciding whether to reduce the dimensionality of the space, depends to some degree on the specific problem, and we will not elaborate further on this topic. For simplicity, in what follows we assume that the niche variables have been z-standardized, and that principal component analysis was performed on the resulting data.

We begin then with a matrix $\mathbf{Z}$ of v columns (v equals the number of principal components retained) and n rows (one for each of the cells in the geographic grid; we assume that every geographic cell has been uniquely characterized environmentally) that implicitly or explicitly define $\mathbf{M}$. In other words, the rows of $\mathbf{Z}$ are the elements of $\eta(\mathbf{M})$, effectively the pool of environments that the species has experienced and may or may not find suitable. The modeling algorithm then provides the subset of $\eta(\mathbf{M})$ that is hypothesized to be suitable for the species to maintain populations: $\hat{\mathbf{N}} = \eta(\mathbf{G}_p)$, assuming here and throughout this section a configuration of the BAM diagram in which $\mathbf{B}$ does not reduce $\mathbf{G}_p$—i.e., that the Eltonian Noise Hypothesis is true. This niche corresponds to a certain subset of rows of $\mathbf{Z}$, and we need ways to characterize it.

Let $\hat{\mathbf{N}}$ be an estimated niche (figure 7.8), obtained by any of the methods described earlier in this chapter. The simplest way of characterizing $\hat{\mathbf{N}}$ is by the number of its elements (its cardinality), denoted by bars $|\hat{\mathbf{N}}|$, and that we call its size. An appealing feature of this measure is that if the geographic cells are uniquely described by their scenopoetic variables, then $|\hat{\mathbf{N}}| = |\eta^{-1}(\hat{\mathbf{N}})|$. In words, the geographic projection of the niche equals its size. On the other hand, this measure does not tell us anything about the position or shape of $\hat{\mathbf{N}}$ in E-space, or about its shape.

A very direct way of expressing the position of $\hat{\mathbf{N}}$ in the v-dimensional E-space is by the centroid of its elements. In figure 7.9, we display several typical estimated Grinnellian niches. It is important to notice their very irregular shapes (Austin et al. 1990); as a result, centroids of niches may not be very informative. For example, a generalist and a specialist may have identical centroids but vastly different niches, and this phenomenon would not be detectable via centroid-based measures. Thus, measures of shapes of niches are also needed. Models of the entire surface of response of populations to environmental variables are possible (Austin et al. 1990).

These niche characterizations can be very precise, but are complicated and difficult to compare. In a simpler approach, it is possible to measure the variance along each of the environmental axes, as might be produced by a principal components analysis or discriminant function analysis of niche conditions

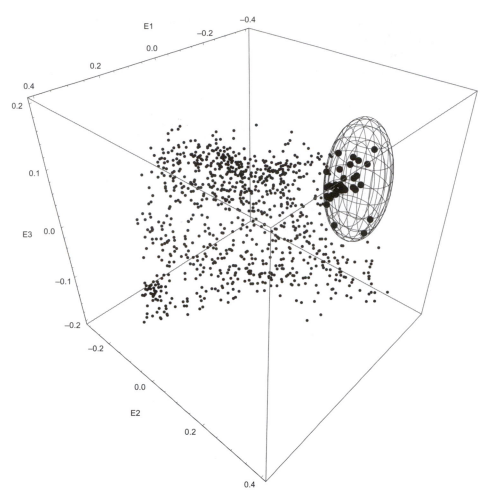

FIGURE 7.8. Illustration of environmental combinations identified as belonging to the niche of a hypothetical species (shown as larger symbols in black), as compared with the remaining combinations present in the broader study area (smaller symbols), depicted in three standardized environmental dimensions. The ellipse shows the niche of the species.

(Colwell and Futuyma 1971, Green 1971, Carnes and Slade 1982), and then take their average. This measure summarizes how spread out the niche is in the dimensions of E-space. It should be remembered that the average of the variances of any multidimensional cloud of points remains the same after a rigid rotation around its centroid (Legendre and Legendre 1998), which means that the average variance is a consistent measure of the shape of the niche cloud,

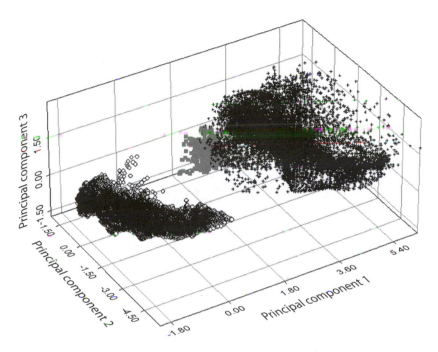

FIGURE 7.9. Illustration of estimated niches for three species of orioles (*Icterus* spp.). The environmental space consists of the first three principal components of the climatic space of the Western Hemisphere (spatial resolution 5′). Black crosses = *Icterus chrysocephalus*; gray squares = *I. auratus*; open circles = *I. galbula*.

not affected by the choice of reference frame (as in comparisons of raw versus principal components-rotated niche clouds).

Given the matrix **Z**, the number of elements of a niche, its centroid, and the average of its variances are all absolute measures, relative to **Z** only in the sense that it defines the environmental space of discourse. However, one needs to define the distribution of habitable environments for a species with respect to the availability of environments across the landscape of interest. Careful definition of this landscape as equivalent to **M**, the area accessible to the species via dispersal, is important (Basille et al. 2008). It has long been recognized that defining a universe of reference in niche measurements is a fundamental challenge (Colwell and Futuyma 1971). The reasoning needed to define **M** is obviously species-dependent, and constitutes a frontier area in the modeling of distributions (treated in detail in Barve et al. 2011). However, many mathematical techniques related to ecological niche modeling require specification of a reference environmental space, which serves to underline the importance of facing this difficult challenge. This is the case, for example, for Ecological Niche Factor Analysis (Hirzel et al. 2002); this technique, using factor analysis methods, describes the niche by a marginality factor, which is the distance of the centroid of the niche to the global centroid (the global centroid represents the average conditions in the dispersal-available region), and then by calculating successive specialization factors orthogonal to the marginality vector (a vector joining the global centroid to the centroid of the niche) that describe the ratios of variance of niche dimensions to the global variance in successively lower-dimensional spaces. This method can be used both to predict areas of suitable habitat and most importantly to answer the question "what does the organism look for?" (Basille et al. 2008); it characterizes the niche in terms explicitly relative to the environments that are regarded as available, that is, $\mathbf{M}$.

In a related methodology that also requires explicit definition of an availability zone, Doledec et al. (2000) characterizes a niche by its marginality, measuring the maximum departure of the centroid of the niche from the global centroid, and by a decomposition of the total "inertia" of the niche $\hat{\mathbf{N}}$, in two "tolerances" that represent niche breadth in two orthogonal subspaces of $\mathbf{Z}$. Inertias are measures of how apart and how scattered points in the niche are respective to the global centroid. This method has been applied to address the question of niche separation of a suite of species (Doledec et al. 2000).

It is necessary to mention briefly how characterization methods can be applied to comparison of several niches. In principle, it would be possible to compare two or more niches by any of a number of multivariate techniques, like discriminant function analysis or multiple analysis of variance (Green 1971, Dueser and Shuggart 1979), but the very non-normal and spatially autocorrelated nature of most estimated niches suggests the need to use nonparametric techniques, like versions of MANOVA (Anderson 2001), inertia analysis (Doledec et al. 2000, Dray et al. 2003, Broennimann et al. 2007), or ad hoc methods that calculate significance by randomization (Warren et al. 2008). Given that estimated niches may be composed of thousands of data points, estimation of significant differences via randomization may be extremely computationally intensive, and not every existing method or software package can cope with realistic datasets.

We see then that characterization and statistical treatment of Grinnellian niches has roots in relatively old literature, and that a number of methodologies capable of dealing with the different challenges of this task are available. The main problems still remaining are conceptual (e.g., how to define the area $\mathbf{M}$), methodological (which techniques are better suited to what measures), and technical (developing software tools adequate, for example, to compare niches containing large numbers of cells). In this section, we have only scratched the surface—we have offered general considerations that we know will be important, but none of the methods presented has as yet seen detailed sensitivity analysis, and much remains to be learned in this area.

SUMMARY

This chapter provides an overview of challenges inherent in estimating ecological niches and corresponding distributions from data documenting known occurrences of species. The first question, of course, is what quantity is being estimated—is it the probability of occurrence, probability of suitability, or something else? Algorithms for developing these estimates may require presence data only, or may also need data documenting absences or at least data from the background to provide a contrast with the presence data. Model calibration is a difficult process and requires careful consideration of strategies for data splitting, variable selection, thresholding approaches, and other considerations. A particularly important consideration is that of model complexity, and how best to avoid overfitting models; a related consideration is transferability of models, and how best to avoid genuine extrapolation in E-space and the risks that it entails. The chapter concludes with a discussion of how to characterize ecological niches—which are generally manifested in numerous dimensions—and as such present serious challenges of visualization, analysis, and interpretation.

CHAPTER EIGHT

From Niches to Distributions

Species' potential geographic distributional areas G_P often differ from their occupied distributional areas G_O. In this chapter, we discuss the conceptual bases for this discrepancy, and summarize methodological approaches to addressing the consequent problems. First, we discuss the meaning of the potential distribution G_P, and describe reasons why a niche model may not estimate it correctly. Next, we explore reasons why a species may not be at equilibrium with its potential distribution G_P (or G_A, see the following), but rather inhabits only some subset of areas suitable for it (see chapter 3; Araújo and Pearson 2005). In terms of the BAM diagram (see figure 3.1), nonequilibrium situations may arise for three reasons. The first is that the set **M** is frequently less than universal, meaning that the species will not be present in all suitable areas of the landscape. Second, **M** may change through time; here, dispersal limitation reduces G_P to G_O. Third, **B** may often differ from **M**; that is, in some situations, the Eltonian Noise Hypothesis does not hold, and negative interactions reduce G_A to G_P, or to G_O if **M** is not limiting. (Bear in mind that this reasoning is limited to cases of negative effects of biotic interactions, and that the situation will differ if interactions are positive.) In some cases, both **M** and **B** can contribute to nonequilibrium distributions. Finally, we outline procedures for further processing of a niche model, which expresses G_P or G_A, to yield an estimate of G_O. In this chapter, as in most of this book, we focus on the case of conditions when the Eltonian Noise Hypothesis is true, but attempt to note necessary modifications when it is not.

POTENTIAL DISTRIBUTIONAL AREAS

Estimates of species' potential geographic distributions G_P are critical to myriad niche modeling applications (chapters 10 to 15), and can also be used to estimate the species' occupied distribution, although under certain assumptions (see the following; Anderson and Martínez-Meyer 2004, Peterson 2006c). G_O is observable in nature, and is frequently of interest in theoretical and empirical studies, particularly applications in conservation biology where it is crucial to know the actual distribution of the species. In contrast, G_P is a theoretical construct (see chapter 3), and as such its full extent will probably remain unknowable in practice, although in theory it would be possible to perform broad suites of experiments to outline it. Now, remembering that that the environments of $\eta(G_O) = E_O$ constitute the occupied niche space, it is clear that those environments may occur outside of G_O. As we saw in chapter 3, a map of $\eta^{-1}(E_O)$ may well be larger than G_O extending into the potential distributional area. Niche modeling exercises often strive to estimate G_P (see chapter 3), but numerous factors may cause the predicted (modeled) potential distribution to be different from, and typically smaller than, the species' true G_P. In strategizing for model development, investigators should consider each of these factors and aim to eliminate, or at least minimize, as many of them as possible.

First, the calibration data G_{data} on which the model is based are drawn from G_O, which may not encompass the full breadth of environmental conditions that the species can inhabit (and may encompass some conditions where it cannot persist, of course, if sink populations are included). Unfortunately, the species' full biotically reduced niche $E_P = \eta(G_O \cup G_I)$, the projection of which onto geography would constitute G_P, is unknowable. G_O (or a representative sample from it) will underrepresent E_P and G_P if:

1. The data G_{data} used to estimate G_O fail to include a representative sampling of the environments that the species uses.
2. The algorithm $\mu(G_{data}, E)$ used to calculate G_O overfits to G_{data}.
3. The universe G is limited, such that M is limited with respect to $\eta(G_P)$.
4. The Eltonian Noise Hypothesis is not true, and some environmental conditions suitable for the species exist only in regions where biotic interactions do not permit the species to inhabit its full abiotically suitable distributional area G_A.

With correlative models (the main subject of this book), researchers must generally accept this reality (i.e., that niche models may not estimate G_P completely), and deal with the fact that the resulting models may be conservative—in other words, that $\hat{E}_P$ will generally be a subset of the true E_P (contra Soberón and Peterson 2005), and as such will identify as suitable an area smaller than the species' full potential distribution G_P. The only solution is to measure species' eco-physiological limits of tolerance experimentally; such approaches nevertheless identify E_A and not E_P, and the experiments necessary to consider all possible negative biotic interactions related to B remain impractical (see chapter 3). We discuss these four conditions in greater detail in the next paragraphs.

The occurrence data on which niche models are based may be drawn from an unrepresentative portion of G_O (Peterson 2005b). In this case, these occurrence data capture only a subset of the occupied niche space. This problem will occur when some conditions that the species inhabits do not exist in the subset of G_O from which the occurrence data were sampled. Furthermore, in cases where the Eltonian Noise Hypothesis is not true, the region from which the occurrence data were sampled may contain a community such that biotic interactions limit the species to a subset of the environmental conditions in which it occurs elsewhere in different community contexts (Baselga and Araújo 2009). To minimize such issues, occurrence data for model calibration should be drawn from throughout the species' occupied distributional area whenever possible, and may require testing for sampling bias in environmental dimensions (see chapter 5).

Similarly, even if occurrence records are sampled in an unbiased manner from across G_O, they may not necessarily capture even its full E_O, much less its full E_P. Generally, this sort of problem occurs with datasets containing few records, where sampling error becomes more important. Whenever possible, researchers should strive to obtain reasonably large numbers of occurrence records (see chapter 5). Numbers of records necessary to calibrate models will vary among modeling techniques, and likely also among species (Wisz et al. 2008). This point should not be taken as indicating that studies based on small numbers of records cannot hold great utility (Raxworthy et al. 2003, Anderson and Martínez-Meyer 2004, Pearson et al. 2007), but the resulting predictions must be interpreted as likely indicating conservative underestimates of the ecological breadth of the species (underestimating E_P and thus G_P).

Several methodological issues can lead to underestimates (or in some cases, overestimates) of species' niches and potential distributions. Certainly, overfitting to noise or to bias in the occurrence data will tend to shift estimates of the species' potential distribution toward an overly restricted estimate (see chapters 7 and 9). Similarly, if too many environmental variables are included in the model, again, overfitting will be common (Beaumont et al. 2005), and geographic predictions overly restrictive. (Note that use of too *few* variables will usually broaden estimates of the species' potential distributional area for lack of information with which to constrain estimates of distributional areas.)

Finally, choice of a study region (G) for model calibration that is too broad and inclusive can be problematic for all modeling techniques except presence-only techniques. In particular, G may be so overly broad that it samples G_I instead of only G_O and areas of genuine absence (i.e., those not in G_P, assuming that the Eltonian Noise Hypothesis is true); such models will frequently underestimate the breadth of a species' niche and potential distribution (Anderson and Raza 2010). As discussed in chapter 7, this confusion occurs because of false negative signals deriving from pixels containing *suitable* conditions, but from which the species is absent owing to dispersal limitations (and biotic interactions, if the Eltonian Niche Hypothesis is not true). Such a situation can lead algorithms to fit spurious responses. Care should be taken to avoid these problems, if the goal of the exercise is to estimate $\mathbf{G}_P$ (or $\mathbf{G}_A$).

NONEQUILIBRIUM DISTRIBUTIONS

Two major classes of factors can cause species to inhabit less than their full abiotically suitable distributional areas $\mathbf{G}_A$ (see chapter 3): dispersal limitations and biotic interactions (the latter of which can in some senses include human modifications of the landscape; Araújo and Pearson 2005)—we will refer to these situations as "nonequilibrium distributions." Detecting nonequilibrium distributions is critical to rigorous development and use of ecological niche models, and certainly, when such nonequilibrium is not considered, it can produce misleading results (Ganeshaiah et al. 2003, Peterson 2005b).

In the terms of the BAM diagram, the question is whether models estimate just $\mathbf{A}$, or whether they also take into account the effects of $\mathbf{B}$ and $\mathbf{M}$. Quite simply, niche models will estimate $\mathbf{A}$ when areas that *are* inhabited provide a good (i.e., unbiased) sample of environments across the species' $\mathbf{G}_A$, and the data $\mathbf{G}_{data}$ are unbiased with respect to E-space. However, if $\mathbf{M}$ or $\mathbf{B}$ happens to bias the sampling of E-space by the species, or if the sampling itself is biased in environmental dimensions, the calibration data (i.e., known occurrences) cannot summarize environments across the species' full scenopoetic existing fundamental niche, and the model will estimate something less extensive than $\mathbf{A}$, in the direction of $\mathbf{A} \cap \mathbf{B} \cap \mathbf{M}$ (Jiménez-Valverde et al. 2008). Alternatively, unbiased samples from $\mathbf{G}_O$, in comparison with absence (or background or pseudoabsence) data (if taken from the correct portions of $\mathbf{G}$; see the preceding discussion and chapter 7) can produce suitable models of $\mathbf{A}$ if response curves are not truncated in $\mathbf{G}_O$. We note, however, that restrictions in geography owing to $\mathbf{M}$ or $\mathbf{B}$ do not necessarily translate into restrictions in E-space.

DISPERSAL LIMITATION

Dispersal limitations can limit species to a $\mathbf{G}_O$ that is only a subset of its full $\mathbf{G}_P$, leaving out either disjunct unoccupied areas (which may remain unoccupied over the long term) or areas that are eventually accessible, but to which the species has not as yet spread, as in the case of species invasions and responses to large-scale environmental changes (Anderson et al. 2002a, Peterson 2003a,

Svenning and Skov 2004). Failure to occupy such areas may stem from accidents or vagaries of evolutionary and biogeographic history, or simply from lack of time to respond with population expansions. These factors may introduce stochastic elements in the distributions of species; biogeographers commonly refer to such phenomena as contingent historical factors (Patterson 1999).

It is important to note that dispersal limitation can be regarded over multiple timescales ranging from the present only to deeper, over evolutionary history. That is, present dispersal abilities speak to the ability of a species to colonize areas separated from inhabited areas. The key nature of such present-day dispersal limitations in constraining species' distributions is demonstrated clearly by the case of invasive species overcoming barriers with human assistance (NAS 2002). Dispersal limitation is also manifested over evolutionary timescales, with **M** representing the areas that have been available to the species for colonization over some longer period. Apparently, such long-term dispersal constraints are frequently coincident spatially among species, responding to macrogeographic features (e.g., continents, oceans, major mountain ranges), which is the basis for the field of historical biogeography (Crisci et al. 2003). The spatial coincidence in dispersal limitation (and to some degree environmental characteristics) across large numbers of taxa is the basis for establishment of biogeographic regions.

BIOTIC INTERACTIONS

Biotic interactions also, under some circumstances, may cause a species to inhabit less than its full G_A (Anderson et al. 2002b), although we expect that these interactive effects will generally take place on finer spatial scales, and not necessarily at biogeographic extents, as discussed in chapter 3 (Soberón 2007). Counterexamples are known, particularly in the case of suture zones between closely related species, but we expect that biotic interactions will generally be manifested on finer spatial scales (i.e., that the Eltonian Noise Hypothesis holds, at least at coarse resolutions). However, this question is empirical, and most likely will require in-depth field studies to resolve. Competition generally has been the focus of such considerations (MacArthur 1972), but other classes of interactions (e.g., predation, parasitism, and even mutualism) may shape ecological and geographic distributions as well (Araújo and Luoto 2007). Under competitive exclusion, one species reduces the density of another to zero, effectively reducing the geographic distribution of the other. Quite commonly, closely related species show parapatric (i.e., adjacent but nonoverlapping) distributions highly suggestive of niche conservatism and competitive exclusion (e.g., Hall 1981). Clearly, even distantly related species can compete and reduce distributions as well, but such cases are harder to identify. It is important to remember that species can compete for many resources that vary in space; therefore, outcomes of competitive interactions between species may vary across geography, which can complicate understanding of the distributional consequences of competition; this area is an arena for challenging but fruitful future research.

Other classes of biotic interactions can similarly reduce species' distributions. For example, a predator may be present in parts of the G_A of the species in question, preventing it from maintaining populations in those regions (Crooks and Soulé 1999, Kutz et al. 2005). The same is true for parasites and pathogens (Sutherst 2001), such as the intriguing constraints that tsetse flies (*Glossina* spp.) placed on early human distributions (Rogers and Randolph 1988). Interestingly, the opposite pattern holds true for mutualism and any other kind of facilitation (e.g., commensalism), which represents a limitation of the heuristic Venn representation of the BAM diagram, which allows only for reductive interactions. A critical mutualist may be missing in part of the G_A of the focal species—this absence causes absence of the species as well, even though the general abiotic environmental conditions are favorable (Heikkinen et al. 2007). Lack of consideration of effects of mutualistic interactions in niches and distributions is one the most blatant gaps in modern niche theory (see chapter 3; Araújo and Guisan 2006).

Finally, a special case is that of biotic interactions with humans, particularly competition and predation, which affects many, perhaps even most, of the species on Earth. Major human modifications of environments include deforestation, selective logging, disturbance of other nonforested habitats, hunting, fishing, and gathering of other species, not to mention modification of global climates (i.e., scenopoetic variables). Although these actions generally reduce species' distributions, some species benefit in the face of human presence (e.g., cockroaches, house mice, black rats, bedbugs, and head lice; Sánchez-Cordero and Martínez-Meyer 2000).

DETECTING AND PROCESSING
NONEQUILIBRIUM DISTRIBUTIONS

Ecological niche models that initially represent species' abiotically suitable distributional areas can be processed *post hoc* to take into account factors that cause nonequilibrium distributions, leading to closer approximations of the species' occupied distribution G_O. Such steps may also allow more realistic evaluations of model performance under certain circumstances (see chapter 9). We refer to these steps as "postprocessing" because they are taken after the initial niche model is calibrated. Several of the procedures for postprocessing require binary predictions (presence/absence predictions), so a suitable thresholding rule is often needed to turn continuous and ordinal predictions into binary ones (see chapter 7).

TESTING FOR ARTIFACTUAL ABSENCES

Before ascribing lack of occurrences in a given region within G_A to dispersal limitations or biotic interactions, researchers must first demonstrate that the species is genuinely highly likely to be absent there. That is, lack of records may derive simply from lack of adequate sampling in the region (Heyer et al. 1999, Soberón et al. 2007; see chapter 5). Such a situation represents "artifactual absence," where the species is not really absent, but rather the lack of records is an artifact of inadequate or nonexistent sampling. Protocols exist to test for this possibility, at least under certain circumstances (figure 8.1): (step 1) model output consists of or can be transformed to a binary prediction; (step 2) a particular area of G_A lacking occurrence records can be identified (typically a disjunct area); and (step 3) a suite of occurrence records for other species that are sampled effectively in similar ways (a "target group") is available as a surrogate for sampling effort (Anderson 2003).

These tests calculate the binomial probability of obtaining the observed pattern of occurrence records were the area in question actually inhabited. A low resulting probability indicates that sampling effort *has* been sufficient to detect the species were it present, thus demonstrating that the species is likely genuinely absent, at least with the confidence accorded by the probability value obtained. Otherwise (i.e., with a higher probability resulting from the test), the species' absence could be an artifact resulting from inadequate sampling (see figure 8.1). In such cases, and particularly in the case of rare species, additional sampling is necessary before conclusions regarding true absence can be considered robust. More realistic tests should be developed to permit assessment of artifactual versus true absences when predictions are continuous or ordinal (in the former case, likely by weighting the probability of a record falling into a given grid cell by the strength of the prediction there).

Data quantifying sampling effort directly (e.g., data documenting survey location and intensity for the species in question) can be used for such tests if they exist (Anderson 2003), but such information is rarely available. An alternative approach is to use records of species belonging to a more inclusive target group that can provide an index of sampling effort (Voss and Emmons 1996). The target group comprises all species that can be recorded using the same techniques that produce records of the focal species: these species do not necessarily form a monophyletic group (Anderson 2003)—e.g., shrews, salaman-

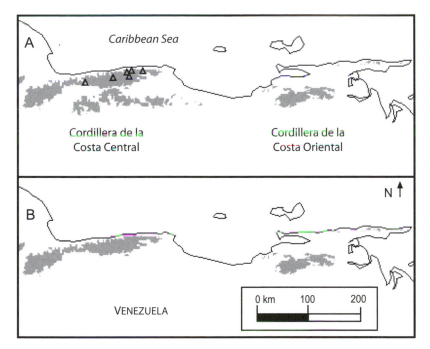

FIGURE 8.1. Example of tests of artifactual absences for the oryzomine rodent *Oryzomys albigularis* in northern Venezuela. (A) Binary representation (gray) of the modeled abiotically suitable area G_A; triangles indicate localities where the species was collected by the Smithsonian Venezuelan Project (Handley 1976). (B) Processed distribution, after removing areas where the species' absence was supported by statistical tests (note disjunct area in the east, in which no records exist, but where sampling has not been sufficient to support a conclusion of absence of the species). Modified from Anderson (2003).

ders, and small lizards all are captured in pitfall traps. The data for the target group must be derived from the same data sources that produced (or not) the occurrence records for the focal species, so that absences for the focal species are in spite of the survey effort expended. If data for a target group are unavailable, these tests can be accomplished based on proportional area of regions holding records versus those lacking records of the focal species—this approach, however, requires assumptions of equal sampling effort per unit area in both regions, which is unlikely in most systems (Anderson 2003).

Considering Dispersal Limitation
After establishing that absence from a region of G_P (or G_A) is statistically supported as genuine, researchers can remove those areas of predicted presence

(but demonstrated actual absence) by means of explicit assumptions. Generally, this process involves identifying regions of G_p that are disjunct and unoccupied. Disjunct regions of suitable but unoccupied habitat (i.e., those separated from known areas of occupied distribution by intervening unsuitable areas) are then removed from the prediction (Anderson and Martínez-Meyer 2004).

Alternatively, some researchers have used maps summarizing biogeographic regions (developed previously and independently based on patterns of distribution of many other species) to delimit areas lacking records that can be removed from the prediction (Peterson et al. 2002b). In such cases, the possible artifactual nature of the absences should still be assessed before any reductions of the species' G_p. Whenever possible, however, a species-specific approach to delimiting areas for removal is preferable, rather than broad and inflexible assumptions based on faunal or floral summaries. All of these approaches assume that the species cannot disperse across intervening unsuitable areas, which is likely reasonable, at least for organisms that are poor dispersers.

It is also possible to model dispersal processes explictly, rather than the simpler (but less realistic) assumption that the species has not crossed and cannot cross areas of unsuitable habitat. Development of such approaches is quite important in the context of anticipating species' distributions under future climatic scenarios. Such an approach takes into account the spatial configuration of suitable patches within an unsuitable matrix (e.g., Collingham et al. 1996, Pearson and Dawson 2005, Svenning et al. 2008, Engler and Guisan 2009). These approaches permit consideration of spatial patterns of relative suitability across the study region, rather than just binary maps. In this context, niche model results can provide information regarding suitability both for population persistence *and* for dispersal. The inverse of suitability can be used as a "cost surface" (Lee and Stucky 1998) for modeling dispersal, and can be input into simulations of the likelihood of dispersal derived from landscape ecology tools (Bunn et al. 2000, Ray et al. 2002); some types of dispersal barriers may best be estimated via other means (e.g., the effects of rivers as barriers may best be measured by river flow volume).

For any of these approaches, a very general challenge is to distinguish between lack of dispersal (ever) to the unoccupied region and a past population that went extinct. For example, the Island Scrub-Jay (*Aphelocoma insularis*) is endemic to Santa Cruz Island off southern California; this island was connected to Santa Rosa Island only 20,000 years ago, but whether the jays were ever on Santa Rosa Island is up in the air completely. Only fossil, subfossil, or anthropological (e.g., cave paintings, written records, or reliable oral traditions) data can establish the latter situation with certainty (e.g., Timm et al. 1997), and such information rarely exists. As such, lack of positive records confirming occurrence in a particular region in past time periods may represent another kind of artifactual absence, unless abiotic conditions have changed. If appropriate data were to be available, the same tests mentioned earlier (Anderson 2003) could be employed to determine whether sampling for the past period was sufficient to demonstrate absence in the region in question in that time period (i.e., that lack of past records in the region is not an artifact of inadequate sampling for records in past time periods). Unfortunately, however, researchers will seldom be able to obtain sufficient amounts of data for these tests. Overall, these methods provide a framework for reconstructing and understanding spatial patterns of dispersal limitation.

Considering Biotic Interactions

It is also possible to assess constraints on species' distributions caused by biotic interactions (again, in tandem with tests demonstrating the species' absence statistically). As mentioned previously, competition represents a primary biotic interaction limiting species' distributions (Udvardy 1969, MacArthur 1972, Heikkinen et al. 2007), although the roles of other negative (and positive also) interactors have not been fully appreciated. Whereas a competitor may reduce density of a focal species without reducing its populations to zero (see chapter 3), we here examine the extreme situation in which the competing species excludes the focal species completely from some areas of **A**.

The idea of competitive exclusion and competitive release has seen extensive discussion and exploration, as perusal of a good Ecology textbook will show (e.g., Begon et al. 2006), and the idea has been applied to real species' distributions at least in some simple situations (Anderson et al. 2002b). It is important to bear in mind that these spatial tests cannot demonstrate competition conclusively: experimental field tests, particularly involving removal or exclusion of putative competitor species, are necessary for definitive conclusions regarding competition (Koplin and Hoffmann 1968, Brown 1971). However, because such experiments are not frequently possible, geographic tests may constitute the best opportunity to address the phenomenon (Costa et al. 2008). Extending these approaches to other classes of biotic interactions (i.e., predation, parasitism, and mutualism) will be an important step forward in understanding the dimensions and role of biotic interactions in limiting species' distributions in **G** and **E** (Araújo and Luoto 2007, Heikkinen et al. 2007).

Tests for competition are available for two-species interactions, and are applicable only under certain conditions (Anderson et al. 2002b). First, binary predictions are required, and distributions should be parapatric, such that geographic contact occurs without the possibility of broad coexistence. What is more, the species' respective G_A's must overlap, indicating areas of potential sympatry (i.e., within G_A for both species) in three situations: (1) a real contact zone between known distributional areas for the species, (2) regions where only one of the two focal species is present, and (3) regions where only the other species occurs. The first condition ensures that outcomes of interactions between the two species can be observed, whereas the latter two allow tests of competitive release, and ensure that the niche models for each species are informed regarding the full suite of environments that it can inhabit without the restrictions imposed by competition with the other species.

The approach in these tests consists of comparing the geographic and ecological "behavior" of species in area (1) as compared with areas (2) and (3); here, we present the logic of tests for the effects of competition only (see Anderson et al. 2002b). Competitive exclusion by the first species is indicated when it is unexpectedly (i.e., more than chance expectations) more common than the other within (1), and yet the excluded (i.e., the other) species inhabits those environments in (3). In cases where these tests indicate competitive exclusion (and note that the distributional areas in these discussions are in actuality the modeled estimates of each area), G_A of the inferior competitor may be reduced by removing areas of its G_A that are also within the G_A of the superior competitor and lie along real contact zones. No areas along real contact zones should be removed from G_A of the superior competitor, and areas (2) and (3) should not be changed because of competition, although they represent areas where dispersal limitation (M) likely reduces G_P to G_O (see the preceding discussion). In reality, though, the intensity and outcomes of competitive interactions may vary across geography, calling for more complicated analyses, and similar approaches may allow insights when not all conditions described earlier are met (Anderson and Martínez-Meyer 2004, Costa et al. 2008).

CONSIDERING HUMAN MODIFICATIONS OF THE ENVIRONMENT

Clearly, humans have impacted distributions of countless species, and these impacts must be considered in estimating G_O. Two paradigms exist for such studies. The first is to include environmental data reflecting human actions directly as predictor variables in the modeling process; the other considers human effects on species' distributions via *post hoc* processing steps.

In the first case, human-affected variables are included in the modeling process itself (e.g., Thuiller et al. 2004a). Various remotely sensed datasets, such as land cover, vegetation indices, and maps of human presence across landscapes, lend themselves nicely to this approach—quite simply, occurrence data from the same time period as the human-affected environmental data are used to generate models of $A \cap B$ ($= G_P$) directly. However, it should be borne in mind that anthropogenic effects and abiotic factors may frequently act at different scales, necessitating careful consideration of potential biases. Also, human-influenced variables may frequently covary with climate, leading to statistical problems and underestimation of the importance of the human-influenced variables (Thuiller et al. 2004a). Perhaps the most significant drawback of this approach, however, is that occurrence records must correspond temporally to environmental data, which will frequently limit occurrence datasets to prohibitively small subsets (see chapter 5), which themselves are more likely to be prone to collection biases.

The second approach is to build the niche model based only on environmental data that do not include effects of human presence but then remove areas where human presence dominates (Sánchez-Cordero et al. 2005, Peterson et al. 2006b). Here, both historical and recent records are used to calibrate models, so the models will often be more robust to the vagaries of sampling. Subsequently, the human "footprint" is removed from the predicted potential distribution G_P to estimate G_O (of course, consideration of the role of M is also necessary). Simple approaches just remove those areas using a GIS via what has been termed a "cookie cutter" approach, although identifying areas for removal may require additional biological information, perhaps from recent occurrence records, natural history information, or expert opinion (Anderson and Martínez-Meyer 2004, Sánchez-Cordero et al. 2005). For example, for a species known to inhabit only primary forests, all deforested or substantially disturbed areas can be removed from the prediction (Ortiz-Martínez et al. 2008). More complex approaches develop models based only on longer-term variables (i.e., that do not reflect human presence, such as "original vegetation"), but then transfer models onto human-influenced versions of the landscapes (Peterson et al. 2006b).

SUMMARY

An important and consistent result of niche modeling exercises has been that abiotically suitable areas generally exceed occupied distributional areas. A first step is to consider sampling to date to determine whether it has been sufficient to establish the species' absence from a predicted region lacking records of the species. This inequality results from the effects of dispersal limitations and/or biotic interactions in constraining the distribution of the species to a subset of the full distributional potential. Key steps for taking these considerations into account are reviewed, including means of reducing estimates of the abiotically suitable area to estimates of the occupied distributional area.

CHAPTER NINE

Evaluating Model Performance and Significance

Evaluating the predictive performance and statistical significance of a model constitutes a critical phase of niche modeling, and researchers should demonstrate that their models are of sufficient quality to meet the needs of the project at hand before using or interpreting them in any way (Peterson 2005a). In chapter 4, in the process of clarifying modeling concepts and developing basic mathematical notations, we also provided an overview of key principles of model evaluation. Here, we develop the topic in considerably greater depth, and discuss a framework for selecting appropriate evaluation strategies for a particular study.

We begin by reviewing key concepts, but now in the light of input data characteristics discussed in chapters 5 and 6, which lead us to explore limitations inherent in most occurrence datasets available for model evaluation, and to comment on ways in which they can influence evaluation adversely. The picture contrasts rather sharply with the optimistic panorama that we painted in chapter 4, but shows clearly the need to discuss methods for selecting evaluation data carefully. We begin by presenting commonly used quantitative measures of model performance and significance. Because the aims of modeling projects vary (see chapter 3), no single "best" approach to evaluation exists (just as we saw in chapter 7 that no single "best" modeling algorithm is likely to exist); however, we can outline approaches that are more or less suited to rigorous model evaluation. Hence, we discuss various evaluation approaches in light of when they are likely to be appropriate. Finally, we set out a vision of the research agenda as regards model evaluation, highlighting areas in need of theoretical and/or methodological advances.

PRESENCES, ABSENCES, AND ERRORS

We recall a few critical principles that were outlined in chapter 4. There, we took for granted several rather optimistic assumptions that may be incorrect or untenable in the arena of modeling species' ecological niches $\mathbf{E}_A$ or $\mathbf{E}_P$ (particularly in contrast to the challenges of evaluating the related but distinct species distribution models). Foremost is the elementary assumption that evaluation data are sufficient to allow for unequivocal, transparent empirical observation of $\mathbf{Y} = 1, 0$. This situation, however, is rarely the case: occurrence data can be presence-only, presence/background, presence/pseudoabsence, or presence/absence, and even in the latter case the meaning of absence data is manifold (see chapter 5). Although most studies evaluate models based on the same kinds of data used in model calibration, this situation is not necessarily always the case. For example, Elith et al. (2006) developed models based on presence-only or presence/background data, but evaluated them with presence/absence data. Furthermore, even when more kinds of data are available for evaluations, some model evaluation strategies use only presence records.

In fact, special considerations arise because the notion of an "absence" is questionable in niche modeling (see chapter 5). That is, omission errors ($\mathbf{Y} = 1$, $\hat{\mathbf{Y}} = 0$) are usually genuine in indicating model failure, except when identification or georeferencing errors, sink populations, or other misleading factors cause problems. On the other hand, problems surrounding the concept of commission error ($\mathbf{Y} = 0$, $\hat{\mathbf{Y}} = 1$) are pervasive in niche modeling—in fact, much of the "error" ascribed to commission may not be erroneous at all in such applications. A distinction, at least conceptually, and operationally to the extent possible, between real and apparent commission error components is paramount in model evaluation (Anderson et al. 2003).

Apparent commission error does not reflect real error in model calibration, but rather may derive from incomplete evaluation data, inappropriate selection of the evaluation region, or both. Two major factors contribute to apparent commission error: incomplete biological sampling across the landscapes being used to evaluate models (which is universal), and nonequilibrium distributions (e.g., owing to dispersal limitations and possibly to biotic interactions), creating absences in areas in which the species could maintain populations (see chapter 8).

In figure 3.1, it is clear that $\mathbf{G}_P$ may exist outside of $\mathbf{M}$; put another way, $\mathbf{G}_I$ is rarely or never empty, meaning that some areas within $\mathbf{G}_P$ will be uninhabited by the species. Moreover, even within $\mathbf{M}$, few or no taxonomic groups have been sampled thoroughly across their entire geographic distributions (Soberón et al. 2007), so even within $\mathbf{G}_O$ one should expect to see some (often many) undocumented map cells. Demonstrating absence for a species is particularly difficult for analyses at the relatively coarse resolutions typical in niche modeling studies (see chapter 5; Anderson 2003). Hence, lack of a record of a species in a grid cell that the model predicts as suitable does not necessarily indicate that the prediction was in error (i.e., commission error or a false positive)—rather, the species may truly be there, but the grid cell has seen no sampling or sampling has been inadequate to detect the species there (see chapter 5). This problem exists for presence/background and presence/pseudoabsence evaluation datasets, but not for genuinely presence-only evaluations that do not use any of the negative data. Furthermore, even with high-quality presence/absence datasets, most "absence" records cannot be taken as such with certainty, because probability of detection is typically <1 (Boulinier et al. 1998, Mac-Kenzie et al. 2002); further discussion of these issues is provided in chapter 5.

As mentioned earlier, the situation is most extreme for presence/background (or presence/pseudoabsence) datasets. With such datasets, a sample of the study region is taken to characterize the study region as a whole and serve as a comparison with locations where the species is known to be present. For most taxa and regions, sampling is nonexistent or at best insufficient in the vast majority of grid cells (Prendergast et al. 1993, Bojórquez-Tapia et al. 1995, Peterson et al. 2002a, Soberón et al. 2007). Hence, even though they may be appropriate for model calibration, background or pseudoabsence pixels typically are not valid for use in evaluation (though note, for example, some uses in presence/background ROC, later; McNyset 2005). To provide realistic evaluations, resulting estimates of commission error must be downweighted relative to omission error to a degree appropriate to the analysis at hand (unfortunately, determination of an appropriate weighting scheme remains an arbitrary process; see chapter 4 and the section "Future Directions" later in this chapter). Background or pseudoabsence pixels, however, should never be used in evaluation of binary models (see the following).

The second major factor that may contribute to apparent commission error is the nonequilibrium nature of many species' distributions. As detailed in chapter 8, few species inhabit the full spatial extent of their environmental potential G_P (or G_A, if the Eltonian Noise Hypothesis is true); as a result, in many cases, the spatial complexities of nature and geography contribute importantly to apparent commission error. Nonequilibrium distributions derive from dispersal limitation and/or biotic interactions (including human impacts) that limit species' distributions, and may contribute more to apparent commission error than incomplete sampling (depending, of course, on the spatial extent of the analysis). For presence/background and presence/pseudoabsence evaluation datasets, the relative importance of these two factors will be species- and landscape-specific, and will depend critically on the size of the study region used for model evaluation relative to the extent of the potential distributional area of the species within that region (Anderson et al. 2003, Phillips et al. 2006, Barve et al. 2011).

These factors lead us to a series of guidelines for selecting appropriate regions for evaluating models. Evaluations will be more valid for taxa that are closer to equilibrium with contemporary environmental conditions, which is often the case for generalist and good disperser species, and for species occurring in regions where dispersal barriers or limiting biotic interactions are minimal. Many times, it is difficult to separate when species are not in equilibrium versus when lack of sampling creates gaps, especially for rare species (see chapter 8). Because effects of biotic interactions may be more pronounced at finer resolutions, and they may not even be manifested noticeably at coarser resolutions (see chapter 2), consideration of dispersal limitations becomes principal (see in-depth treatment later). Put another way, if this reasoning is true, $\mathbf{M}$ in the BAM diagram generally should define the region of analysis (Barve et al. 2011).

The problem, of course, is that no good, objective approach is available for delineating $\mathbf{M}$ for any particular species (see detailed treatment in Barve et al. 2011). Occurrence data for recently arrived invaders may be particularly prone to problems of apparent commission error because $\mathbf{M}$ is so difficult to estimate. Model evaluations may be most realistic when limited to areas within $\mathbf{M}$; however, although this tactic remains rare, theoretical treatments of these issues exist (see chapter 8), as do empirical examples (Anderson et al. 2002a, Anderson 2003). It is no accident that the same principles for selecting appropriate study regions for calibration (see chapter 7; Anderson and Raza 2010) coincide with those for choosing regions for realistic model evaluation.

CALIBRATION AND EVALUATION DATASETS

A second implicit assumption in chapter 4 is the availability of two pools of occurrence data: data for calibrating models, and data for evaluating model predictions. These two datasets are assumed to be statistically independent. If data points are to be used for estimating E_{val} and E_{ver} (see chapter 4) correctly, we expect that both datasets will be independent samples from the *same* distribution for $(\mathbf{Y}, \mathbf{X})$. Ideally, this distribution is equal to the distribution of "typical" values $(\mathbf{Y}, \mathbf{X})$ that one hopes to reconstruct via modeling. When datasets fail to be representative in this sense, we speak of sampling bias produced by vagaries of the biological sampling that produced the occurrence records at hand (although the dataset also might not be representative of $\mathbf{G}_A$ or $\mathbf{G}_P$, because of limited dispersal or biotic interactions in the region of model calibration). In both such cases, the distribution being sampled may not coincide with the typical values. Sampling bias in calibration occurrence datasets may or may not be present in the evaluation datasets as well. Throughout this book, unless otherwise noted, we use "bias" to refer to the effects of biased sampling, rather than to estimation bias possibly inherent in the model-fitting methods themselves (i.e., the idea that $\hat{f}$ may consistently over- or underestimate f, even if data are unbiased in terms of sampling).

Given discussions in chapters 5 and 6, we should expect that the data at hand are likely to be biased, but the effects of this bias may be insidious. Biased data that violate stated or implicit assumptions may throw off modeling methods and produce unreasonable results. Whereas such biases reflect problems inherent in the calibration data, some modeling methods may be more or less susceptible to these problems. For instance, a model may fit a subtle pattern in the calibration dataset that is not found in the evaluation dataset; such problems are not just shortcomings of the data, but also reflect a lack of robustness of the modeling method to minor patterns in the calibration data. For some reason (e.g., too many degrees of freedom in a GAM, too-high-order polynomials in a GLM, or too many nodes in hidden layers in a neural network), the algorithm fits relationships that are more complex than the real ones. With a slightly different calibration dataset, relationships would not be fit in the same way. Conversely, if it is the evaluation dataset that is biased, unrealistic assessments of model quality may result. Finally, if the same biases are present in *both* calibration and evaluation datasets (e.g., when these datasets are created by splitting a single sample), the evaluation cannot detect the bias or any overfitting to it, and hence may lack generality when the model is challenged with genuinely independent data.

OVERFITTING, PERFORMANCE, SIGNIFICANCE, AND EVALUATION SPACE

A fundamental concept described in chapter 7 is that of overfitting. Overfit models show close fit to calibration data, but are less able to predict fully independent data (see chapter 7). Overparameterization (i.e., excessive model complexity) and/or insufficient sample size are generally the culprits for overfitting. Models may also show a small E_{ver} relative to E_{val}, owing to overfitting to biases in the calibration data, as described earlier.

An important goal of model evaluation is to detect such overfitting or sensitivity to bias, which in any case indicates a poor or unreliable model. In this section, we clarify strategies for detecting these two kinds of problems in model calibration, especially when data have both bias and noise. Of particular interest is understanding situations in which automated modeling methods are able to fit models with little or no monitoring required.

Model evaluation includes two distinct endeavors: quantification of *performance*, and tests of *significance*. Measures of performance per se generally do not test statistical hypotheses, but rather characterize how well or poorly the model achieves a particular goal. They include quantifications of omission error or commission error and indices that combine omission and commission (see chapter 4). Some measures of performance indicate a model's ability to rank presences and absences correctly, while others assess a model's goodness of fit to observed presences and absences. It should be noted that different sorts of studies with different needs may require distinct performance measures.

In contrast, tests of significance determine whether observed predictions of evaluation data differ from null expectations with a particular level of probabilistic confidence. Tests of significance are often based on some measure of performance. Commonly, significance tests assess whether model predictions of records in the evaluation dataset are "better than random" regarding the prediction and evaluation data. In other cases, tests can determine whether two models differ from each other as regards some measure of performance. In the end, it is the user's final goal in the study that determines which of the two evaluation concepts is relevant to a particular challenge—for an example from disease diagnosis, a test for hepatitis infection may perform statistically much better than a random model, but a performance evaluation might indicate that it is yields correct predictions in only 54% of tests, which is probably not an acceptable level of performance. Because of inadequacies in the data typically available, we must also learn to distinguish between situations in which measures of performance and significance can be interpreted literally, and others in which they cannot be taken at face value.

Finally, an issue worthy of discussion is whether model evaluation is best carried out in E-space or G-space. Typically, and almost without exception (Martínez-Meyer and Peterson 2006), researchers conduct evaluations of models in geographic space (i.e., in terms of prediction of geographic patterns of occurrence). This G-space evaluation occurs by applying the model to make a prediction for each pixel in the study region. The strength of the prediction is assessed for each cell holding an occurrence record (whether regarding presence or absence) in the evaluation dataset. However, it is also possible to evaluate models in environmental space—that defined by the environmental variables used in modeling. Although such approaches remain rare (see example in Martínez-Meyer and Peterson 2006), they may hold considerable advantages when the goal is to estimate ecological niches. Therefore, we look toward testing in environmental space as an optimal solution in niche modeling applications, but much work remains to clarify its advantages and peculiarities (see the section "Future Directions" at the end of this chapter).

SELECTION OF EVALUATION DATA

Meaningful evaluation of ecological niche models and associated geographic predictions depends on careful and appropriate selection of suitable occurrence datasets with which to calculate measures of performance and significance. Several related issues are of importance: the degree of independence between calibration and evaluation data in space and time, how best to divide occurrence records into calibration and evaluation datasets, and whether records are divided with respect to spatial position.

Perhaps most commonly, the evaluation data derive from the same study region and time period as the calibration data (Peterson 2001). Alternatively, evaluation data may be drawn from a different time period (e.g., Martínez-Meyer et al. 2004a, Araújo et al. 2005a, Peterson et al. 2005a), even though the geographic area remains the same for both calibration and evaluation. In some studies, however, the evaluation data come from a different region than the calibration data, although usually from the same time period (Peterson et al. 2009b). Studies of invasive species represent an example of this situation, with a species' model being calibrated on its native range, and then evaluated according to its ability to predict records of the species on the invaded range (Peterson 2003a; see chapter 13). Whereas nature divides records spatially into two pools in the case of invasive species, investigators can develop these spatially structured divisions themselves (see discussion of spatially structured evaluations, later). Finally, calibration and evaluation data may be drawn from both different areas and different time periods, but such cases are rare (Nogués-Bravo et al. 2008b, Peterson et al. 2009b). Whenever models are evaluated in regions or time periods distinct from those in which they were calibrated, an additional requirement will be consideration of whether the model is being projected onto conditions (E-space) not found in the calibration dataset, which may constitute a substantial problem (see discussion of model transferability in chapter 7).

Ideally, evaluation data will be fully independent from calibration data, allowing for true model validation (Araújo et al. 2005a). Unfortunately, fully independent datasets seldom exist, and confusion can arise between genuinely statistically independent data and nonindependent data that were collected independently. It is worth clarifying that "independence" can be understood in two ways: that (1) data points or that (2) entire datasets may be statistically independent such that estimates of E_{val} and E_{ver} are correct. However, we hasten to note that the idea of obtaining calibration and evaluation datasets from distinct sources or from separate surveys does not make them fully independent, as has been assumed, for example, in Elith et al. (2006). A more restrictive notion of independence means that points should be neither spatially nor temporally autocorrelated (Dormann et al. 2007); this point is important, for it leads to the question of biological independence—any occurrence of a population at a particular place is determined in some way by occurrences of some other population at some previous point in time. Researchers often attempt to obtain separate datasets for model calibration and evaluation (e.g., via new field collections or samples from data sources distinct from those that produced the calibration data), but care must be taken to assure that independence of samples is sufficient that model evaluations are reliable.

Data from such "different" sources may not in truth be independent. Such situations may occur because similar biases are present in the biological sampling underlying both the calibration data and the seemingly independent evaluation data. For example, because all researchers face similar constraints on access to study sites, specimen records in different museums often reflect similar geographic biases in sampling (e.g., more sampling near the same cities, roads, and rivers; Funk and Richardson 2002, Reddy and Dávalos 2003; see chapter 4). Evaluation and calibration data may also not be fully independent owing to the nonequilibrium distributions that can create common associations (see chapter 8). Similar biases, at least in geographic space, complicate evaluations (see the following).

Because fully independent datasets seldom exist, and because even relatively independent datasets can be quite costly to generate, most evaluation efforts are based of necessity on partitions of single samples into calibration and evaluation datasets. Many strategies for achieving this subdivision are available, but investigators have typically simply divided occurrence records randomly without respect to geography (Fielding and Bell 1997, Araújo et al. 2005a). Here, a subset of records is chosen randomly for the calibration dataset, and remaining records are set aside as the evaluation dataset (the "split-sample" approach of Fielding and Bell 1997). Because this particular approach has major drawbacks that will be discussed later, we present various alternative strategies for data splitting, including in particular use of spatially structured data subdivisions, which may offer important advantages.

One approach that extends the idea of a single random data split is termed "K-fold cross-validation" (sometimes referred to as cross-partitioning). Here, we echo the exposition and basic notation of Hastie et al. (2001), making specific adaptations to the case at hand when pertinent. This cross-validation procedure does not match our usage of the term "validation" (i.e., evaluation based on fully independent data), but its intent is the same: to quantify in some way the ability of a model to anticipate the behavior of the system in unsampled regions. Here, the investigator divides the n occurrence records into calibration and evaluation pools randomly. However, several (K) roughly equal-sized pools are created, rather than just two. Next, K models $\hat{f}_{-1}, \hat{f}_{-2}, \dots, \hat{f}_{-K}$ are built successively by setting aside (hence the "minus" subscript) one of the K pools to be used as evaluation data, and building a model using the remaining $K-1$ pools, which together constitute the calibration dataset for that step. Each pool is used as an evaluation dataset only once, and each occurrence record appears in an evaluation dataset exactly once. Let us denote by $k(i)$ the index of the evaluation pool corresponding to the i-th occurrence record: for example, if observation 25 ends up randomly in the third pool, then $k(25) = 3$. For any loss function as introduced in chapter 4, an estimate of prediction error is then given by

$$E_{\text{val}}^K = \frac{1}{n}\sum_{i=1}^{n} L[\mathbf{Y}_i, \hat{f}_{-k(i)}(\mathbf{X}_i)]. \tag{9.1}$$

When only few records are available, a special case of K-fold cross-validation can be useful. When $K = n$ (i.e., each pool consists only of a single record), K-fold cross-validation is equivalent to the standard "leave one out" or "$n-1$ jackknife" procedure (Hastie et al. 2001). An approach has been developed that allows researchers to assess statistical significance and performance, in spite of small sample sizes (e.g., <25) of occurrence records available (Pearson et al. 2007). However, with these approaches, it should be borne in mind that, in contrast to the E_{val} considered in chapter 4, the terms now being averaged in E_{val}^K are not independent, because every data point participates both in evaluation and calibration. This double role introduces estimation bias as defined earlier, in the sense that E_{val}^K can systematically overestimate or underestimate the true value of validation error. When $K = n$, E_{val}^K is approximately unbiased for the true prediction error, but its variance may be underestimated, because the n calibration datasets share many common entries (Shao and Wu 1989, Efron and Tibshirani 1993; see implementation in Anderson and Raza 2010). The computation involved under this approach can be onerous, as it implies n times more model calculations.

If K is too small (i.e., the pools are too large), on the other hand, E_{val}^K achieves lower variance, but it may be biased. It is hard to specify what value of K should be selected as an appropriate middle ground. Typical values of K are in the 5 to 10 range (Hastie et al. 2001). Without a doubt, in the end, a price is paid when occurrence data are not abundant: the price is that one must settle for more approximate answers and reduced confidence in them.

Another version of the jackknife can be used to develop calibration and evaluation datasets when larger numbers of occurrence records are available. A more general form for the jackknife is $n - d$, where d is the number of records withheld (deleted) from the calibration set and used for evaluation. This procedure amounts to sampling without replacement and is also sometimes termed subsampling (Efron 1987). Typically, many iterations (500–1000) of this sort of jackknife are done. Like K-fold cross-validation, the $n - d$ jackknife can produce composite predictions (see the following) as well as estimates of performance, significance, and uncertainty (after correction of variance estimates for lack of independence among calibration sets; Warren et al. 2008). However, when n is large, these approaches do not guarantee that every occurrence record enters into an evaluation dataset (in contrast to K-fold cross-validation, where each record does so exactly once). However, if large numbers of iterations are developed and d is even moderately large, all occurrence records will likely occur in one or more evaluation sets.

K-fold cross-validation holds two main advantages over the single split-sample strategy. First, because each occurrence record appears in one evaluation dataset, the exercise ensures consideration of the ability of the models to predict all environments known for the species; on the other hand, in the single split-sample approach, key atypical records may not enter in the evaluation dataset at all. Second, the resulting set of K models provides a useful indication of the sensitivity of the modeling process to random differences in the occurrence data available. The K models can be combined to yield estimates of the species' niche and distribution (e.g., mean value for each pixel) that are more robust to vagaries of division of occurrence records into calibration and evaluation sets than would be a single split; furthermore, the suite of K models can be used to estimate the error term E_{val} (Hastie et al. 2001, Araújo and Luoto 2007). Alternatively, a model calibrated using all available occurrence records can be used for interpretation, but evaluation statistics drawn from the K-fold exercise can provide measures of model performance, significance, and uncertainty.

Finally, a standard bootstrap procedure can be employed to create many calibration and evaluation datasets from a sample of n occurrence records. As in cross-validation and jackknifing, bootstrapping produces multiple calibration datasets. More precisely, if $(\mathbf{Y}_1, \mathbf{X}_1), \ldots, (\mathbf{Y}_n, \mathbf{X}_n)$ is the original dataset, one draws size n random sets $(\mathbf{Y}_{11}^*, \mathbf{X}_{11}^*), \ldots, (\mathbf{Y}_{n1}^*, \mathbf{X}_{n1}^*)$ through $(\mathbf{Y}_{1B}^*, \mathbf{X}_{1B}^*), \ldots,$ $(\mathbf{Y}_{nB}^*, \mathbf{X}_{nB}^*)$ with replacement from original data points, where B is a large number (in the hundreds or thousands, depending on available computing power) of repetitions. Each of these B replicate datasets is called a "bootstrap sample." Conceptually at least, one may fit a model to each replicate to obtain predicted values $\hat{f}^{*1}(\mathbf{X}_i)$ through $\hat{f}^{*B}(\mathbf{X}_i)$ for $i = 1, \ldots, n$. The estimate of prediction error is an average of averages of all losses computed with these bootstrap samples, or

$$E_{\text{val}}^{\text{boot}} = \frac{1}{n}\sum_{i=1}^{n}\frac{1}{B}\sum_{b=1}^{B} L[\mathbf{Y}_i, \hat{f}^{*b}(\mathbf{X}_i)]. \tag{9.2}$$

Again, this estimate may be affected by estimation bias, because several of the replicate samples may actually contain $(\mathbf{Y}_i, \mathbf{X}_i)$. One suggestion for reducing the tendency for bootstrapped models to overfit and carry estimation bias is to average only over samples that do *not* contain $(\mathbf{Y}_i, \mathbf{X}_i)$, or

$$E_{\text{val}}^{\text{boot2}} = \frac{1}{n}\sum_{i=1}^{n}\frac{1}{|C(i)|}\sum_{b\in C(i)} L[\mathbf{Y}_i, \hat{f}^{*b}(\mathbf{X}_i)], \tag{9.3}$$

where $C(i)$ is the set of bootstrap samples that do not contain the ith observation point. Other approaches exist to correct bias (see Hastie et al. 2001 and references cited therein).

Bootstrapping may be advantageous relative to the techniques discussed earlier, which result in calibration datasets with fewer records than the original dataset, which in turn may affect parameterization of algorithms for which implementation varies according to the number of records (Phillips et al. 2006). In contrast, bootstrapping yields calibration datasets with the same n records as the original, and thus avoids such problems. However, in bootstrapping, an individual record may be included in a given calibration set more than once, which may complicate application of some modeling applications, particularly when sample sizes are small. Because bootstrapping samples with replacement, it does have one important and intuitive drawback over the techniques discussed earlier: in each iteration, it does not produce an evaluation dataset composed exclusively of records withheld from the calibration dataset, so it does not lend well to assessment of performance and significance. Bootstrapping has received little use in modeling ecological niches (but see Guisan and Zimmermann 2000).

All of the preceding split-sample approaches, because both calibration and evaluation records are selected randomly from the same pool of data, suffer from at least two problems. First, they constitute nonindependent samples from the overall occurrence data, so evaluations of models based on such data will often yield overoptimistic assessments of model predictive ability. Indeed, measures of model performance based on such split samples of data may correlate little with measures based on genuinely independent evaluation data (Araújo et al. 2005a). Second, any biases (see chapter 5 and discussions earlier) present in the overall occurrence dataset are preserved in both the calibration and evaluation datasets, so the evaluation dataset cannot detect overfitting to those biases, because the evaluation records hold the same biases, and model performance is overestimated.

As a partial solution to the second problem, it is possible to filter occurrence records spatially (Iguchi et al. 2004). For example, an occurrence dataset suspected of having substantial environmental bias, perhaps owing to biased sampling in G-space, may be reduced by removing records falling close to each other in geography. Such a procedure may reduce (although likely not eliminate) environmental bias present in the dataset (Hidalgo-Mihart et al. 2004). However, these procedures have the drawback of reducing sample sizes available for modeling, and possibly even diluting the environmental signal of the niche, if filtering is too strong.

As an alternative to all the random split-sample approaches described earlier, spatially structured partitioning of occurrence data offers important advantages, in particular regarding the potential to identify overfitting to sampling bias. Spatial partitions separate occurrence records into calibration and evaluation datasets by geographic areas rather than at random (Araújo and Guisan 2006). Geographic regions for subsampling can be chosen in various ways: according to political units such as states or counties (McNyset 2005), via arbitrary geometric shapes such as squares or bands (Peterson et al. 2007c, Williams et al. 2008), or by calculation of p sets of areas so as to maximize the distance traveled between them (a p-maximum problem in contrast to the commonly framed p-medium problem; Araújo et al. 2001).

The idea behind spatial subsampling is to provide a more stringent test of model predictions than a random split: where evaluation and calibration records are closely intermingled (as in the latter case), spatial autocorrelation reduces their independence, and the same sampling biases will be present in both datasets. Because truly independent evaluation data seldom exist, spatially structured partitioning represents one of few means available to identify overfitting to bias, rather than just to noise in calibration datasets. Still, if biases permeate *all* of the occurrence data collected in a particular fashion—for example, many biologists accumulate the bulk of their records along roads and roads may consistently follow valleys (Reddy and Dávalos 2003)—such pervasive biases will not be eliminated even by spatially stratified subsampling. Among the data-partition strategies presented earlier referring to random splits, K-fold cross validation lends itself better to implementation with spatially structured splits than do random strategies.

Another potential drawback exists with spatially structured partitions, requiring care in their design and interpretation. By partitioning occurrence records spatially, the investigator may inadvertently insert artificial sampling bias into the calibration data, at least in G-space, and possibly in E-space as well (a potential solution to this problem is via calculation of p-maximum solutions, maximizing distances among data points, in both geographic and environmental space; Araújo et al. 2001). A modeling technique robust to such biases would perform well in evaluations of ability to predict records from regions from which no calibration records were available (Peterson et al. 2007c), which is precisely the reason why spatially structured partitions are desirable. Either explicitly or implicitly, lack of sampling bias represents an assumption of all techniques for modeling ecological niches: extreme geographic subdivision for calibration versus evaluation may violate this assumption more than what is reasonable for the purpose of allowing the investigator to identify overfitting to both bias and noise. Such artificial sampling bias introduced by spatially structured partitions is reinforced with successively larger numbers of occurrence records.

An optimal balance between relative size of the geographic shapes on one hand and numbers of occurrence records on the other must exist. Moderately sized geographic regions for partitioning may be desirable over large ones (e.g., a checkerboard approach rather than division into geographic halves; Peterson et al. 2007c). Similarly, evaluations based on spatially structured partitions will likely prove most realistic when numbers of occurrence records are not excessively large. For datasets with larger numbers of occurrence records, smaller subregions should be employed to avoid introducing excessive artifical sampling bias; for smaller datasets, larger subregions will likely be more appropriate to ensure that sufficient artificial sampling bias has been introduced. More generally, the objective should be to introduce a level of sampling bias that mimics that which is suspected in the overall occurrence dataset. Such decisions will invariably be specific to the particular situation at hand, and this area of research clearly remains open, with much empirical testing and exploration still needed.

EVALUATION OF PERFORMANCE

Many quantitative measures exist for evaluating model performance, which can be divided into two categories: those designed for use with binary predictions versus those relevant to nonbinary (e.g., continuous or ordinal) predictions, often referred to as "threshold-dependent" and "threshold-independent" measures, respectively (Fielding and Bell 1997). Because threshold-independent

TABLE 9.1. Summary of the confusion matrix,
as used in model evaluations.

		Actual status	
		Present	Absent
Predicted status	Present	a	b
	Absent	c	d

measures are frequently extensions of metrics used for evaluation of binary predictions, we begin by discussing the latter.

In terms of our general principles, it is useful to think in terms of models, $\hat{f}$, and their corresponding loss functions. Evaluation of binary predictions is based either on modeling methods that produce a single, binary prediction, or from a binary prediction that derives from application of a threshold u to a continuous or ordinal model (see chapter 4). In contrast, threshold-independent approaches in essence evaluate a whole nested set of binary models, $\hat{f}_u$, one corresponding to each value of the threshold u.

Most performance measures for binary models derive from elements of a matrix typically termed the "confusion matrix" (Fielding and Bell 1997; sometimes simply called an error matrix). The confusion matrix relates rows summarizing predicted presence and absence to columns indicating the true status (table 9.1). The four cells of the matrix thus indicate distinct combinations of prediction versus reality; the literature regarding confusion matrices uses by convention the letters a, b, c, and d to refer to particular elements of the matrix (note that these designations overlap with variable designations used elsewhere in this book for other purposes).

Elements a and d of the confusion matrix denote pixels corresponding to correct predictions: a represents known distributional areas correctly predicted as present, and d reflects regions where the species has not been found and which are classified correctly by the model as absent. Elements c and b, in contrast, indicate prediction errors (figure 9.1). Element c denotes omission error: cells of known distribution predicted absent by the model. Element b indicates commission "error": areas from which the species is not known but that are predicted present (but recall that commission "error" is complex, including both true misclassification and also apparent commission error; see discussion of apparent commission error earlier in this chapter).

Although much of the literature uses the a, b, c, and d notation, to integrate with discussions earlier, we present an alternative notation for the confusion

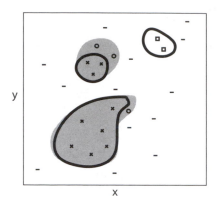

Occupied distributional area, $\mathbf{G}_O$

Areas predicted by an ecological niche model

×   True positive

−   True negative

○   False negative

□   False positive

FIGURE 9.1. Illustration of the confusion matrix in G-space, showing the two sorts of error that are possible in such models. Shaded areas indicate areas occupied by the species. Open shapes enclose areas predicted by a model as suitable. The occurrence data are as follows: dash = absences both of the species and of predicted suitable conditions ($= d$); X = presences of the species correctly predicted ($= a$); open circles = presence of the species incorrectly predicted as absent ($= c$); open squares = absence of the species in spite of a positive prediction ($= b$).

matrix and measures that are derived from it (see table 9.1). These measures all amount to indicating a specific loss function and corresponding rates that are the expected values of loss for each combination in the confusion matrix. That is, a function $L(\mathbf{Y}, \hat{\mathbf{Y}})$ exists implicitly for each of the four values that $(\mathbf{Y}, \hat{\mathbf{Y}})$ can jointly take (table 9.2).

The omission error rate constitutes a common measure of model performance, calculated from the confusion matrix as $c/(a + c)$. It provides the proportion of positive test occurrence records (i.e., localities of known presence for the species) falling outside the area predicted for the species. Because it is a proportion, it ranges from 0 to 1. The omission error rate also can be thought of as a measure of the rate at which the model incorrectly predicts absence, and hence its alternate name, the false negative rate (Fielding and Bell 1997). The omission error rate equals 1 minus a quantity called "sensitivity," which is the

TABLE 9.2. Values of the loss function that are implicitly specified when using different performance-evaluation measures based on a confusion matrix.

	L(0,0)	L(1,1)	L(0,1)	L(1,0)
Omission error rate, or false negative rate	0	0	0	1
Commission error, or false positive rate	0	0	1	0
Overall misclassification rate, or zero-one loss	0	0	1	1
A general, possibly nonsymmetrical loss	0	0	q	$1-q$

absence of omission error; "sensitivity" is drawn from the medical literature, where confusion matrices are used to indicate the efficacy of diagnostic tests, and will be used later in discussions of threshold-independent approaches. Sensitivity shows how responsive the model is to the condition being tested for: that is, how well it correctly indicates a positive result when the positive condition is true. For example, a perfectly sensitive pregnancy test *always* returns a positive result when the patient indeed is pregnant. Clearly, we desire sensitive models, which would then always predict suitability when the environment is suitable.

Virtually all niche modeling applications call for models with zero or low omission error rates. Rare exceptions involve cases where the investigator might desire a binary model identifying a subset of the areas suitable for the species, typically the "best" conditions for the species (Peterson 2006c). In addition, low but nonzero omission rates may be acceptable and even desirable in cases where sink populations, misidentifications, or georeferencing errors are likely present in the occurrence data (Peterson et al. 2008a). However, for most applications, low omission error is a necessary (but not sufficient) condition for a good model: a model that predicts the entire study region as suitable has zero omission error, but is not a useful prediction. Therefore, in addition to assessing performance using omission rates, evaluations must also test whether the omission rate is better than expectations under random predictions (Anderson et al. 2002a, Pearson et al. 2007; see the section "Assessing Model Significance" later in this chapter).

The commission error rate is another model performance measure based on the confusion matrix, as $b/(b + d)$. It provides the rate at which negative test occurrence records (localities of known or assumed absence for the species) fall in pixels of predicted presence for the species. Like the omission error rate, it ranges from 0 to 1. Another phrasing of the commission error rate is the rate at which the model incorrectly predicts presence, and hence its alternative name, the false positive rate (Fielding and Bell 1997). The commission error rate equals 1 minus a quantity called "specificity" (the sister term to sensitivity) that represents the absence of commission error. It shows how specific the model is to the condition being tested for, or how well it correctly indicates a negative result when the condition really is absent. For example, a perfectly specific pregnancy test *always* produces a negative result when the patient is not pregnant.

As discussed earlier, interpretation of commission error rates is more complex than that for omission error rates, for two reasons. First, few if any datasets provide realistic information documenting species' absences from localities, leading to apparent commission error. For presence-only datasets, such as those derived from information associated with specimens at natural history museums and herbaria, absence data simply are not included in the information that is preserved. Here, the only way to assess commission is via sampling the background (see the preceding discussion), or via assessment of sampling of similar species to establish whether sampling has likely been sufficient to detect the species (see discussions of testing absence in chapter 8). More simply, use of commission error rates for presence-only datasets assumes absence for background pixels, a false assumption for all pixels where the species is truly present, regardless of whether or not we have records of its presence there. For most species, unsampled areas represent a substantial and even overwhelming proportion of the study region. Even with presence/absence datasets, because detection probabilities are generally <1 (MacKenzie et al. 2002), the species may truly be present at many sites where the evaluation dataset indicates absence (see figure 5.1), which will inflate commission error estimates.

Nonequilibrium distributions are a second factor complicating interpretation of commission error rates, because many, and potentially all, species do not inhabit the entire geographic footprint of their ecological tolerances (see chapter 8). Many niche-modeling applications aim to estimate not the occupied distribution G_O, but rather the *potential* distribution G_P or even G_A of the species in question. In such cases, if the species fails to occupy suitable areas owing to dispersal limitations or biotic interactions (i.e., the distribution is not at equilibrium), use of commission error rates will penalize a model for *correctly* predicting suitable conditions there. Therefore, commission error rates should be employed only to evaluate predictions of species' true distributions (occupied distributional areas, G_O), in which case minimal commission error rates are indeed desirable. Commission error rates should not be applied in circumstances requiring estimates of G_A or G_P, a point that has not been appreciated sufficiently in the broader literature (e.g., McNyset 2005, Elith et al. 2006).

Using the cells of the confusion matrix, several measures can be derived that combine omission and commission errors into single overall measures (Fielding and Bell 1997). For the same reasons mentioned earlier regarding commission error rates, however, because they include the contribution of commission error, these measures are valid only for presence/absence evaluation datasets, and only when evaluating $\mathbf{G}_O$. The overall misclassification rate constitutes one such measure, calculated as $(b + c)/(a + b + c + d)$. Other, more complex measures, such as Kappa and normalized mutual information (NMI), also aim to characterize combinations of omission and commission errors (Fielding and Bell 1997). However, because of complications introduced by apparent commission error, they are misleading in the same situations as just discussed. Kappa, in addition, is strongly influenced by the prevalence of the species (i.e., the proportion of the study area covered by the species' distributional area; Manel et al. 2001, Liu et al. 2005).

More generally, in performance-evaluation measures that combine omission and commission, the two kinds of error need not (and probably *should not*) be weighted equally. The last entry in table 9.2 corresponds to an asymmetric loss that combines the omission and commission errors, but not necessarily with equal weight. A user may conclude that the loss function of interest is asymmetric; if a reason exists for adopting specific values, then direct substitutions for L in all of the preceding expressions for evaluating loss follow immediately. The difficult question, of course, is what the relative weights (i.e., the value q in the table) of the two types of misclassification errors should be. This decision is called "loss function elicitation" in the decision-theory literature, and is neither easy nor straightforward. This difficulty probably explains why this loss has not been considered in the literature explicitly: it is seldom, if ever, clear what the cost of a false positive is relative to that of a false negative, particularly when real absence data are lacking (which they generally are). For example, in conservation applications, false positives are associated with money lost in protecting a site unnecessarily, but the cost of losing a species by not protecting a site is not something easy to describe on a monetary scale. Of course, not all costs are described on monetary scales, but the example serves to illustrate the complex issues involved. Failing to address this lingering problem of balancing the two error components does not make it disappear, and no amount of mathematical or statistical reasoning can compensate for this failure—as such, use of performance measures assuming equal weights for omission and commission error rates is rarely appropriate in niche-modeling exercises.

ASSESSING MODEL SIGNIFICANCE

In addition to showing sufficient performance, a "good" model should be significantly better than random predictions. For example, recall the model that predicts the entire study region and shows zero omission (i.e., obtains a high performance for that measure), but clearly is not particularly useful. Testing for model significance thus comprises a critical element of evaluating binary models; these tests are one-tailed in nature, since predictions that are worse than random are neither desirable nor of interest in niche modeling applications.

THRESHOLD-DEPENDENT TESTS

For presence-only evaluation datasets, the goal is to test whether the model predicts positive evaluation occurrence records (localities of known presence for the species) better than a random prediction of the same proportional extent within the study region. This question corresponds to a binomial situation, because the model predicts only one of two possible outcomes for each evaluation record: suitable versus unsuitable. The probability of each individual positive record being predicted correctly (i.e., present) by a random model with the same proportional extent of the study region predicted suitable equals the proportion of the study region predicted suitable (e.g., probability of 0.5 for a model predicting suitability for half of the study region, probability of 0.25 for a model predicting suitability for a quarter of the study region, etc.). The binomial probability for the overall test indicates the chance that the total number of positive evaluation records predicted correctly by the model could have been achieved by a random model (Anderson et al. 2002a).

As an alternative to the binomial approach, chi-square tests have been used to approximate the binomial probability. Like all chi-square tests, counts are used (not rates or proportions). Similarly, when using a presence–absence *evaluation* dataset, a chi-square test can be used to assess prediction of both positive and negative test occurrence records; the contingency table for this test has two rows (predicted presence or predicted absence) and two columns (observed presence or observed absence), or vice versa. However, a frequent limitation of chi-square approaches is the requirement that expected values of all cells in the table be larger than 4 (Sokal and Rohlf 1995); as such, these approaches are best applied in situations with medium-to-large sample sizes.

Assessing significance becomes more complicated with approaches that include multiple iterations of model calibration. With the standard $n - 1$ jackknife strategy suggested for small sample sizes, significance can be assessed when using a presence-only evaluation dataset via a binomial test in which the probability of success varies among iterations (Pearson et al. 2007). To our knowledge, similar $n - 1$ jackknife procedures do not yet exist for presence/absence evaluations, nor for other methods of creating repeated divisions of occurrence records into calibration and evaluation datasets (e.g., K-fold cross partitioning, $n - d$ jackknife, bootstrapping). Unless already present in the literature, significance tests need to be taken from the general statistics literature and modified as necessary for these other methods, likely as modifications of the binomial test statistic for a single split sample.

As with any statistical test, the power to detect statistical significance (i.e., to reject correctly a false null hypothesis) varies according to the data at hand. First, small sample sizes of occurrence records reduce statistical power. Second, statistical power depends on how extensive the prediction is within the study region. For example, tests of omission with presence-only evaluation datasets have reduced statistical power as models predict presence across increasing proportions of the study region, because the expected probability of predicting individual test occurrence records equals the proportion of the study region predicted present. One consequence is that many more test occurrence records are necessary to demonstrate significant improvement over random for models predicting broad presence in the study region; in fact, these factors create the *appearance* of better predictions for species with small ranges (Hernandez et al. 2006). This effect also explains why significance on its own does not constitute a sufficient condition for a good model: models may be significant without having acceptable performance. For example, a model that shows a very restrictive prediction for a species, but correctly predicts half of the positive evaluation records, may be significantly better than random, yet its omission rate of 0.5 would not be considered as acceptable performance in most applications.

THRESHOLD-INDEPENDENT TESTS

These approaches, also usually based on measures derived from the confusion matrix, can be used to evaluate nonbinary predictions across various thresholds. Here, the user does not have to choose a threshold by which to convert the prediction to a binary output, so these approaches are often termed threshold-independent methods—however, a cost accompanies this advantage, as will be discussed in detail later. Because model output types vary (see chapter 7), the appropriate evaluation strategy and/or interpretation of evaluations can vary as well. Nonbinary predictions may be either continuous or ordinal, with higher values indicating higher suitability for the species in either case. Some of these outputs aim to provide predictions in which the value is proportional to environmental suitability or probability of presence (Ferrier et al. 2002), while others offer only a relative ranking of suitability. Hence, it is important to distinguish between evaluation tactics that measure how well model output matches suitability or probability of presence among map pixels (goodness-of-fit evaluations) versus evaluation approaches that do not (i.e., rank-based evaluations that measure only the performance of the model in ranking presence higher than

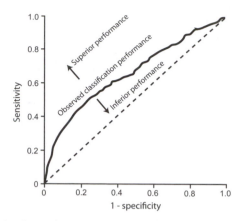

FIGURE 9.2. Example of a receiver operating characteristic (ROC) curve, plotting sensitivity against 1 – specificity. The expected performance of a random classifier (for presence/absence evaluation data) is shown as a broken line, whereas the observed performance in this hypothetical example is shown as a solid line.

absence—or background or pseudoabsence—pixels correctly). We begin with considering the latter, which are the most commonly used approaches.

All rank-based evaluation approaches known to us involve receiver operating characteristic (ROC) plots. ROC plots relate two measures derived from the confusion matrix, sensitivity and specificity: sensitivity on the y axis versus 1 – specificity on the x axis (figure 9.2). As outlined earlier, specificity equals 1 – (commission error rate), and sensitivity equals 1 – (omission error rate). Hence, specificity indicates absence of commission error, and sensitivity represents absence of omission error. Therefore, ROC plots derive from a plot of lack of omission error on the y axis versus commission error on the x axis.

In contrast to the threshold-dependent evaluations earlier, which calculate error rates at particular thresholds of the prediction, ROC curves derive from multiple measures of omission and commission error rates across the range of strengths of the prediction. The two resulting rates are plotted for each threshold across the ramp of predictive values, and the points are connected to form a curve. For a continuous prediction, the ROC curve typically contains one point for each test instance, while for discrete (ordinal) predictions, ROC curves typically contain one point for each different predicted value. ROC analyses have a long history in distributional modeling with presence/absence evaluation datasets (Boyce et al. 2002).

Recent work has addressed use of ROC curves in cases where absence data are unavailable, allowing their application to presence/background evaluation data at least under some circumstances (Wiley et al. 2003, Phillips et al. 2006, Peterson et al. 2008a). Here, large samples of background or pseudoabsence pixels are used for calculating commission error (both real and apparent) for the x axis of the ROC curve. With large samples of background pixels, estimates of commission error converge on the proportion of the overall study area predicted present by the model at that threshold (Anderson et al. 2003, Phillips et al. 2006, Peterson et al. 2008a). Interpretation of such presence/background ROCs differs from traditional usage with presence/absence evaluation data (see the following).

A model that discriminates well between presences and absences would have low omission (high sensitivity) and low commission (high specificity; see figure 9.2). Such models would be plotted in the upper portion and the left portion of the ROC plot, respectively. The endpoints of the full curve are by necessity (0, 0) and (1, 1). As such, a model with excellent discrimination between presences and absences produces a curve that increases quickly along the y axis from (0, 0) to very near to (0, 1), and then continues to (1, 1). The area under the curve (AUC) of an ROC plot represents an overall measure of the model's performance across all thresholds and strengths of the prediction. Specifically, AUC ranges 0–1, and summarizes the model's ability to rank presence localities higher than absence localites (or higher than a sample of random background pixels, in the case of presence/background testing). Although high AUC values are desirable, it should be noted that, for reasons associated with apparent commission error, the maximum achievable AUC is below unity in presence/background evaluations (Wiley et al. 2003, Phillips et al. 2006).

In deriving global measures of model performance across multiple thresholds, significance measures are obtained as well. Indeed, AUC carries with it an interpretation under a null hypothesis because it can be used as a test statistic. The value AUC = 0.5 corresponds to the expected performance of a random classifier, although this interpretation rests on assumptions that must be considered carefully. First, an implicit assumption is a situation of zero-one loss (symmetric loss), which is not often suitable to the details of evaluation datasets. That is, an AUC > 0.5 indicates significance only for true presence/absence evaluation data, a critical point that is frequently misunderstood. Second, the *desirability* of quantification of model predictive behavior over all possible thresholds is assumed. Given the data and questions at hand, these assumptions may frequently not be reasonable.

As mentioned earlier, ROC tests a different null hypothesis when presence/background evaluation datasets are used. In this arena, the problem is to distinguish presence from random, rather than to separate presence from absence (Phillips et al. 2006), so interpretations also must differ. The difference springs from the point of asymmetric loss, with the consequence that apparent commission error overestimates commission error rates, moving the AUC curve artificially to the right (figure 9.3). This estimation bias produces an artificially low AUC value (see the preceding discussion), particularly for models predicting high suitability across large portions of the study region. In contrast, models indicating high suitability only for small proportions of the study region lead to high (and typically inflated) AUC values, an unfortunate artifact caused by background pixels being interpreted as negative results, which coincide with the low predictions throughout the study region (Lobo et al. 2007).

Some statistical interpretations are clear if the assumptions mentioned earlier are met, and additional rules of thumb for interpreting AUC values under some circumstances can be outlined. In ROC, AUC $= 0.5$ is the expected performance by a random classifier, and AUC values <0.5 correspond to predictions that are worse than random and are not of interest. For presence/absence evaluation datasets, AUC values >0.5 are generally classed into (1) poor predictions (0.5 to 0.7); (2) reasonable predictions (0.7 to 0.9); and (3) very good predictions (>0.9; see Swets 1988, for interpretation of model quality based on the value of AUCs from evaluations). However, we emphasize that these guidelines are subjective, and contingent on what one is trying to predict: if the goal is to predict G_A or G_P, measures that penalize models for predicting suitable areas that lack presence of the species because of factors related **B** and/or **M** are not appropriate. Furthermore, these guidelines are not relevant for ROCs based on presence/background evaluation data, where maximum achievable AUC values depend on the proportional presence of the species and its potential distribution across the study region; hence, such AUC values are species- and region-specific (Wiley et al. 2003, Phillips et al. 2006). For these reasons, convenient rules of thumb do not exist for presence/background AUCs. For AUCs calculated with presence/background evaluation data, comparisons must be controlled carefully, and significance is assessed via bootstrapping or other randomization approaches (Raes and ter Steege 2007, Peterson et al. 2008a).

Hence, with presence/background evaluation datasets, AUCs cannot be used to erect valid universal comparisons of model performance, for several reasons. First, AUCs are not comparable among species, because different species' potential distributions will cover different proportional areas of the study region (Phillips et al. 2006). Similarly, comparisons of AUCs between regions will not be valid, because potential distributional areas of species inevitably cover different proportions of different regions (Lobo et al. 2007). Even comparisons between models built using the same occurrence data in the same region but based on different model-building algorithms and approaches may not be valid (see the following), depending on the aim of the evaluation. Differences of

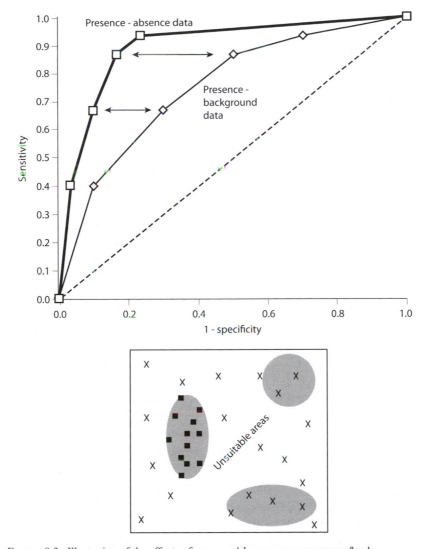

FIGURE 9.3. Illustration of the effects of presence/absence versus presence/background data as bases for model calibration, as well as of different hypotheses regarding mobility constraints (**M**) in model evaluation via receiver operating characteristic tests using the two kinds of data. (Note that models are not necessarily calibrated and evaluated using the same kinds of data.) Calibration with presence/absence data from throughout **G** (see bottom panel) will lead to exclusion of invadable areas **G**$_I$ from the model's predictions, whereas presence/background data will not, because niche models commonly identify areas that are suitable according to abiotic and biotic factors, but that are not inhabited due to dispersal limitations (i.e., they are not within **M**). Similarly, with presence/background *evaluation* datasets, pixels correctly predicted actually increase the apparent commission error (1 − specificity). These two factors act similarly to shift the ROC curve artificially to the right, reducing the estimate of model discriminatory power. In the worst case, if definition of the area for testing is ad hoc, and not linked to a hypothesis of **M**, almost any prediction can be made to appear significant simply by increasing the area of analysis (Barve et al. 2011).

opinion or variation in de facto usage exist in the literature regarding the appro-
priateness of AUCs in such comparisons either because different researchers
may desire evaluation of different properties of the prediction (see second assumption earlier), or simply because of misunderstanding of the methodology
and its assumptions. More generally, AUC is inappropriate in many comparisons of global model performance because of artifactual variation in AUC values
related to incomplete predictions across much of the spectrum of the ROC plot
by some modeling algorithms (Peterson et al. 2008a). It is worth noting that the
problems mentioned in this paragraph are also present (but perhaps less severe) in evaluations with presence/absence test datasets (see the discussion of
instances of apparent commission error caused by incomplete sampling, nondetection of species, and nonequilibrium distributions, earlier).

Despite the many situations in which AUCs are *not* valid with presence/
background evaluation datasets, a few reasonable uses of this approach do exist.
For example, AUCs are appropriate in comparisons of performance between
and among different calibration occurrence datasets (e.g., spatially filtered versus unfiltered occurrence datasets), different environmental datasets (e.g., climatic data versus remotely sensed data), and different parameterizations of a
given modeling algorithm (e.g., Phillips and Dudík 2008). Furthermore, AUC
can be useful and appropriate for more diverse comparisons, for example
among model-building algorithms and approaches (e.g., BIOCLIM versus
generalized linear models) when a global comparison *across the full spectrum
of prediction is desired*—see the following (Peterson et al. 2008a).

Some of the concerns about AUCs can be addressed via modifications of
traditional ROC approaches. Specifically, one can consider the area under only
a selected portion of the curve, thus emphasizing subsectors of ROC space
(Dodd and Pepe 2003, Peterson et al. 2008a). Considering omission error (i.e.,
the y axis), many investigators will want to ignore performance at thresholds at
which models show high omission rates, rather than to create a "global" performance measure (Lobo et al. 2007, Peterson et al. 2008a), so analysis is limited
to portions of the ROC curve that provide predictions with acceptable levels of
omission error (figure 9.4). Similarly, regarding proportional area predicted
present (i.e., the x axis), two or more models may differ in whether their predictions cover the entire spectrum or just a subset of it; if evaluation of performance is desired over the portion of the spectrum where both models actually
make predictions, the ROC plot must be restricted to the area in which the
models actually make predictions (figure 9.4). Another option is to weight the
two axes of the plot individually, thus changing the dimensions of the plot, but
providing a means of weighting omission and commission errors differentially
(figure 9.4). This sort of modification of spaces over which ROC plots are de-

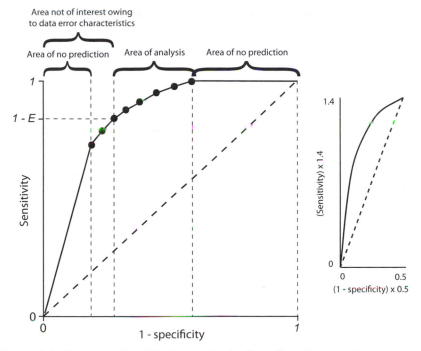

FIGURE 9.4. Illustration of modifications to the "traditional" receiver operating characteristic plot testing approaches to make applications to ecological niche modeling challenges more appropriate. Sectors of the ROC plot are eliminated because the model makes no explicit discrimination in predictions, and/or because of specific tolerance regarding error in the occurrence data. The right-hand portion of the figure illustrates how a traditional ROC plot can be modified to weight omission and commission errors differentially, based on consideration of an expected amount of error inherent in input occurrence data, E, as a means of identifying portions of the ROC plot that are of interest and those that are not. Adapted from Peterson et al. (2008a).

veloped and interpreted can influence results dramatically (Peterson et al. 2008a).

As mentioned earlier, ROC plots assess the ability of models to rank presences versus absences correctly; other evaluation options include quantitative performance measures of suitability via goodness-of-fit approaches. A simple correlation approach has been used to evaluate the fit of model output to the likelihood of observing the species, calculating pixel-by-pixel Pearson product-moment correlations between model predictions and evaluation records in a presence/absence evaluation dataset (absence = 0, presence = 1; Ferrier et al. 2002). It is a parametric test (in contrast to the nonparametric AUC), evalu-

ating both performance and significance (Lobo et al. 2007). For some pres-
ence/absence calibration techniques, this correlation assesses the fit to probability of occurrence, but for presence-only, presence/background, and presence/
pseudoabsence calibration occurrence data, unless $\mathbf{G}_O = \mathbf{G}_P = \mathbf{G}_A$, it would
assess fit only to relative environmental suitability or to probability of suitable
environmental conditions (Ferrier et al. 2002; see chapter 7). Correlations and
AUC have produced similar rankings in recent comparisons of modeling techniques evaluated using presence/absence data (Elith et al. 2006, Lobo et al.
2007, Phillips and Dudík 2008). Unfortunately, given all of the complications
regarding apparent commission error discussed earlier and in chapter 8, this
method is inappropriate for application to any other type of evaluation data
(presence-only, presence/background, and presence/pseudoabsence).

When application of the correlation approach is appropriate, in interpreting
correlation coefficients, it is important to note that even excellent models will
not achieve values of 1, which may hinder such comparisons. First, although
model suitability values are often continuous, evaluation data are binary, which
can limit the fit of the relationship. Also, pixels holding evaluation data will
seldom exist in equal frequencies across the ramp of predictions: if proportion-
ately more evaluation data are concentrated in the middle of the suitability
spectrum, for example, it will decrease the correlation coefficient artifactually.
Hence, even for presence/absence evaluation datasets, correlation should not be
used in comparisons among species and regions, in spite of its recent application to such comparisons (Elith et al. 2006).

FUTURE DIRECTIONS

We suspect that the reader will already have realized that model evaluation is
a part of this emerging field that is in particular need of further development.
Major misunderstandings exist in the field, and we have tried to explain many
of the pitfalls of various techniques. In addition, major rearrangements and
shifts in these methodologies have been proposed recently, and we envision
that more will be forthcoming. As such, we lay out in this final section a series
of avenues that we consider important for exploration and development.

MODEL EVALUATION IN ENVIRONMENTAL SPACE
The discussions developed earlier lead us to explore the possibility that ecological niche models should better be evaluated in E-space rather than across
geography, as is customary. In theory, at least, E-space-based evaluations have
the advantage of speaking directly to what is being modeled (i.e., the ecologi-

cal niche), and in reducing effects of mobility and dispersal ability (i.e., the **M** in the BAM diagram). Whereas testing in G-space is useful for some applications, such as when the focus is on modeling spatial distributions rather than ecological niches, given that the most important aspects of use and application of niche models are driven by their representation in environmental space, these models ultimately should be evaluated in environmental space. This point has been appreciated in the literature only rarely (see, for example, Peterson et al. 2007d).

One justification for this assertion is the point made in chapter 2, that the number of cells in the geographic grid that is our study area ($|\mathbf{G}|$) may exceed the number of environmental combinations ($|\mathbf{E}|$) represented across the same region, particularly when few or relatively simple environmental variables are used as the basis for model development. As a consequence, $|\mathbf{G}| \geq |\mathbf{E}|$, and the degree of inequality between the two will depend on the complexity of both the landscape in question and of its environmental characterization. When $|\mathbf{G}| > |\mathbf{E}|$, grid pixels will exist across **G** that have exactly the same vector of environmental values $\vec{e}_g$: in such cases, pixels would be counted as independent when they are not, and statistical power would thus be inflated artificially in light of the increased sample sizes, if the goal of the modeling exercise is to characterize the environmental dimensions of ecological niches.

We can develop threshold-dependent and threshold-independent evaluations of performance and significance in **E** analogous to those described earlier for **G**. In this sense, a threshold-dependent, **E**-based analog to the cumulative binomial probability approach would be as follows. The binomial is based on k, the number of successes, out of n, the number of trials, and p, the probability of a success in a given trial. Whereas p was estimated as $|\mu(\mathbf{G}_+, \mathbf{E})|/|\mathbf{G}|$, or the proportion of grid cells in the overall study area predicted present, we must seek an analog in environmental space. We use $|\eta[\mu(\mathbf{G}_+, \mathbf{E})]|/|\eta(\mathbf{G})|$, which is the number of elements (environmental combinations) in the niche space corresponding to the geographic prediction of the model within **M** divided by the number of elements in the niche space corresponding to the entirety of **M**.

Translating symbology into more accessible prose, we are attempting to estimate the probability that a random point *in E-space* will fall into the set of environments predicted as suitable for the species in question. We assume that the E-space has been reduced in dimensionality and standardized to avoid artificial inflation of distances across different sectors of the space (see chapter 6). The researcher must then make explicit assumptions regarding the dimensions of **M**, the geographic area accessible to the species (see chapters 3 and 8; see also the following discussion). The E-space associated with this assumed **M** is $\eta(\mathbf{M})$; the number of environmental combinations in this niche space, or its cardinality, is $|\eta(\mathbf{M})|$. This number is used along with the number of environmental combinations in the niche space associated with the geographic prediction of suitability by the model ($|\eta[\mu(\mathbf{G}_+, \mathbf{E})]|$) to estimate the probability of a success p. (Of course, a threshold would have to be chosen and applied to continuous or ordinal predictions, as discussed earlier.) Combining this estimated p with the number k of the n evaluation points falling inside the predicted area leads to a cumulative binomial probability for any binary prediction.

For threshold-independent testing, a ROC curve can be developed that plots proportional success in predicting independent evaluation records (y axis) against the proportion of $\mathbf{E}$ predicted present ($|\eta[\mu(\mathbf{G}_+, \mathbf{E})]|/|\eta(\mathbf{M})|$). As earlier, in many cases, the research question at hand may require that this ROC curve be reduced to only the range of proportions of $\mathbf{E}$ predicted by the models in question (x axis), and to the range of predictive success considered to be useful in that particular application (y axis; extending Peterson et al. 2008a).

Although these procedures are equivalent to G-based testing when $|\mathbf{G}| = |\mathbf{E}|$, when $|\mathbf{G}| > |\mathbf{E}|$ testing in E-space has conceptual advantages over the usual approach, in that environmental variation is emphasized, and geographic sites presenting identical (or nearly identical) environments are not overcounted. Indeed, we suspect that an improvement to the preceding suggestions that are based on raw counts of environmental combinations would be to base calculations of p on frequency-based abstractions of E-space. That is, E-space could be divided into equal-frequency bins, and probabilities calculated based on proportional occurrence in these bins. Such an approach would speak to the odd differences in density of pixel representation that are frequently observed across E-spaces.

EXPLICIT PRESENTATION OF M

Although treated conceptually earlier, another key step forward in improving niche model evaluations will be that of testing models only within the spatial realm and associated environments defined by $\mathbf{M}$ (and of course the area actually sampled); a first major step in development of these thinking frameworks has now been published (Barve et al. 2011). Put simply, the only sites and environments relevant to niche model evaluation are those that the species has "sampled"—these sites are either within the species' niche or not, but $\mathbf{M}$ is the critical area containing these relevant sites. This restriction has rarely been appreciated in the niche-modeling literature, in spite of its critical importance— indeed, almost any model prediction can be "made" significantly better than random if the area of analysis is simply expanded, which tends to add unsuitable areas for the species (consider this situation as parallel to that depicted in figure 9.3). Stating the issue more explicitly, presence or absence of the species in areas outside of **M** is patently irrelevant to the validity of the model, yet how this critical area has been estimated (if it has been considered at all) is rarely if ever stated explicitly in niche-modeling applications. This idea of **M** as a guiding area in niche analyses becomes clear when designing indices based on the idea of marginality (Doledec et al. 2000, Hirzel et al. 2002). Similar principles apply to **B** when the Eltonian Noise Hypothesis is not true; alternatively, models can be processed to consider the effects of **M** and **B** (see chapter 8) before conducting evaluations (see Anderson and Raza 2010 and Barve et al. 2011).

ADDITIONAL CONSIDERATIONS

Myriad other areas relevant to model evaluation require careful consideration and analysis, both theoretically and methodologically. Some of these areas relate to issues of data, whereas others involve development of new quantitative evaluation methods. With regard to data, improvements are needed for both occurrence data and environmental data.

First, researchers should address criteria for reducing bias in occurrence datasets used for evaluating models. Here, spatial autocorrelation (i.e., biases in G-space) is paramount. Records that lie close together are not independent of calibration data or of other records in the evaluation dataset. "Close together," of course, is an ambiguous term dependent on resolution, and implicitly also on the heterogeneity of the environment and dispersal ability of the species. Furthermore, biased data (i.e., that result from biased biological sampling in G-space) may also be biased in E-space. One possible avenue for ameliorating these problems is via spatial filtering of occurrence records: for example, if many records of a species are clustered spatially due to clustered sampling, occurrence records could be filtered by removing records constituting clusters of lying close to one another within a particular distance (Araújo and Guisan 2006). Determination of suitable distances for such filtering could be based on the spatial autocorrelation structure of the landscape and the particular environmental dimensions to be used (Veloz 2009), but further exploration and experimentation is required to take into account biological realities such as individual movements.

Second, further work is needed regarding selection of strategies for spatially structuring evaluation data. Such partitions are independent of calibration datasets, subject to certain assumptions discussed earlier, and as such may allow researchers to detect overfitting to sampling bias and to tune models adequately. Such approaches also represent one of few avenues to development of detailed tests of model transferability (Araújo and Rahbek 2006, Peterson et al. 2007c), since evaluation datasets from other time periods are rare. However, care must be taken to structure partitions such that they mimic amounts of sampling bias likely present in the overall occurrence dataset, so experimentation toward guidelines for such choices is needed.

Despite the growing body of literature regarding quantitative evaluations of model performance, much work remains in this area. Major areas needing advances include characterization and consideration of uncertainty in model predictions and new tests for significance (see preceding). Regarding uncertainty in predictions, some papers have discussed the steps in the modeling process at which uncertainty enters the picture (Heikkinen et al. 2006, Araújo and New 2007). Similarly, methods for assessing model significance apparently remain lacking for many of the more-complicated strategies for partitioning occurrence records into calibration and evaluation datasets for niche modeling, such as K-fold partitioning, $n - d$ jackknifing, and bootstrapping. Ideally, for example, methods comparable to that developed by Pearson et al. (2007) would be available for each of these data partitioning strategies.

Other "next steps" may change model evaluation more drastically. First, a clear understanding of best practices for differential weighting of omission and commission error in model evaluation would circumvent many problems of apparent commission error, while still allowing some evaluation of commission. Similarly, processing geographic predictions to take into account nonequilibrium distributions before evaluation would permit many evaluations that are not appropriate currently (see preceding discussion of evaluation within **M**).

Ultimately, we suspect that model evaluation may evolve in directions not particularly related to where the field lies currently. Certainly, evaluating models in ecological dimensions is a possibility that offers distinct advantages. Beyond that, however, we see considerable promise in evaluation procedures that use randomizations to characterize intrinsic variation in model development, as has been described for niche comparisons (Warren et al. 2008). Finally, some promise lies in use of null models, as has recently been developed to assert that bird species' distributions are not frequently determined by climatic factors (Beale et al. 2008, although we would argue that this particular implementation is inappropriate; see Araújo et al. 2009, Peterson et al. 2009a, Jiménez-Valverde et al. 2010). Comparisons against null models, once properly implemented, however, may offer additional insight for evaluation of niche models.

SUMMARY

Perhaps the most complex set of issues regarding ecological niche modeling revolves around the process of evaluating the predictions. This field is also—

we believe—among those in which error and misconception are most prevalent in the literature to date. Perhaps the most critical point is that the weights accorded to errors of omission versus errors of commission should not generally be equally balanced in model evaluation: presence data are generally trustworthy (with some exceptions), whereas absence (or background or pseudoabsence) data in many or perhaps even most cases do not signify true absences. As a result, model evaluation procedures must be considered carefully to avoid traps that may be seriously misleading. Although evaluation strategies vary, occurrence data typically must be split into calibration versus evaluation subsets—this subsetting can be conducted randomly, or can be stratified spatially, temporally, or by other means. Model evaluation consists of quantifying performance (e.g., measuring omission and commission error rates), versus testing model significance (i.e., often in comparison with a random model). Two other important distinctions are between threshold-dependent approaches (e.g., omission) versus threshold-independent approaches (e.g., AUC/ROC) and between ranking approaches (e.g., AUC/ROC) versus those that assess goodness-of-fit (e.g., correlation). In all evaluation exercises, consideration of principles of the BAM diagram is key, in order to evaluate models in a study region where biological realities match assumptions, just as in selecting the study region for model calibration.

PART III
APPLICATIONS

CHAPTER TEN

Introduction to Applications

So far in this book, we have set out a theoretical framework for modeling ecological niches and estimating abiotically suitable, potential, and occupied distributional areas (chapters 2 to 4). Then, in a more practical mode, we have described issues related to the practice of modeling, including the particulars of occurrence and environmental datasets (chapters 5 and 6), aspects of how to estimate different niches using diverse correlational modeling methods (chapter 7), the process of modifying raw model predictions to estimate geographic distributions (chapter 8), and methods by which to evaluate model performance and significance quantitatively (chapter 9). In this final section of the book, we describe a range of applications of these approaches, which include challenges in biogeography, conservation biology, ecology, evolutionary biology, and public health.

Before discussing particular applications in detail, however, it is important to refer back to the theoretical framework presented earlier in the book—indeed, it is crucial to understand the theory behind the models, if one is to apply them appropriately. In chapters 3 and 7, we saw that correlative models lacking true absence data are unlikely to capture either the occupied niche ($\mathbf{E}_O$) or the scenopoetic existing fundamental niche ($\mathbf{E}_A$) perfectly; rather, these models estimate that portion of the niche represented by known occurrence records, which is likely to fall somewhere between the two extremes. Similarly, we saw that, when projected onto geographic space, correlative models identify parts of the abiotically suitable ($\mathbf{G}_A$) areas or even, in some cases, only the occupied area $\mathbf{G}_O$; in most cases, these models appear to estimate more than $\mathbf{G}_O$, likely including much of $\mathbf{G}_I$. The observation that ecological niche models are not likely to predict the full extent of either $\mathbf{G}_O$ or $\mathbf{G}_A$ has been cited as a critical limitation of correlative modeling approaches (Woodward and Beerling 1997, Lawton 2000, Hampe 2004). In this section of the book, however, we illustrate how these methods can be applied to interesting challenges to yield highly useful results, provided that the researcher understands exactly what is being estimated based on which data.

Let us consider some potential uses of the types of predictions in $\mathbf{G}$ illustrated in figure 10.1. Of course, this scenario is idealized, whereby we assume that models fit neatly to known occurrence records, that occurrence records are sampled only from $\mathbf{G}_O$, and that none of the other complications described in chapter 7 exists. Nonetheless, this illustration characterizes three types of prediction that may be obtained. First, environmentally informed spatial interpolations "fill the gaps" around known occurrence localities, thus providing an improved estimate of the occupied area $\mathbf{G}_O$ that is likely more informative than a minimum convex polygon based on coordinates, or than other spatial approaches that do not take underlying environmental variation into account (see predicted area labeled 1 in figure 10.1; Getz and Wilmers 2004, Mace et al. 2008). Furthermore, if predicted areas that are isolated from known occurrences by dispersal barriers are removed (see chapter 8; Peterson et al. 2002b), then model predictions provide a further improvement to estimates of $\mathbf{G}_O$. This prediction may be useful, for example, in conservation planning (see chapter 12). Notice that we are not expecting the model to predict the full extent of $\mathbf{G}_O$, but the approach certainly yields more information than is available from raw occurrence records alone (Rojas-Soto et al. 2003).

Second, we move to spatial transferability predictions, which identify parts of the occupied area $\mathbf{G}_O$, or even $\mathbf{G}_P$ or $\mathbf{G}_A$, for which no occurrence records have been collected (i.e., this part of $\mathbf{G}_O$ or $\mathbf{G}_P$ is unknown). Although the model does not predict areas of $\mathbf{G}_O$ that have environmental conditions not represented among known occurrence records, this type of prediction (labeled 2 in figure 10.1) can be used to guide field surveys toward areas with high probabilities of holding new occurrence records. Accelerating discovery of unknown populations in this way has already proven particularly useful in landscapes where species' distributions are poorly known (see chapter 11). Similarly, this approach holds considerable potential for identifying unknown vector or reservoir populations of zoonotic diseases (see chapter 14).

Third, an extension of this reasoning (see figure 10.1) is that of estimating the portion of the abiotically suitable area ($\mathbf{G}_I \subset \mathbf{G}_A$) that is environmentally similar to sites where the species is *known* to occur, but which is not necessarily inhabited (see predicted area 3 in figure 10.1). This area is the invadable distributional area $\mathbf{G}_I$. This type of prediction can be used to identify sites where a species may become invasive if it overcomes dispersal barriers and possible biotic barriers (see chapter 13). Such predictions can also be useful for identifying sites suitable for reintroduction of endangered species (see chapter 12). Of course, niche models cannot be expected to identify the full extent of $\mathbf{G}_A$ because areas probably exist under conditions that are abiotically suitable

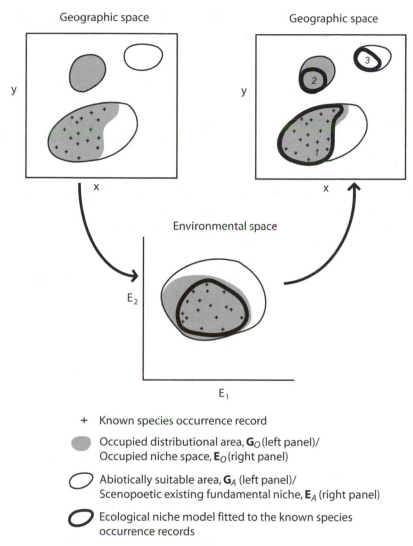

+   Known species occurrence record

⬤   Occupied distributional area, $\mathbf{G}_O$ (left panel)/
     Occupied niche space, $\mathbf{E}_O$ (right panel)

⬭   Abiotically suitable area, $\mathbf{G}_A$ (left panel)/
     Scenopoetic existing fundamental niche, $\mathbf{E}_A$ (right panel)

⬭   Ecological niche model fitted to the known species
     occurrence records

FIGURE 10.1. Hypothetical example of the process of modeling ecological niches from observed presences. Observed presences $\mathbf{G}_+$ are shown as +'s; the occupied distributional area $\mathbf{G}_O$ and the occupied niche space $\mathbf{E}_O$ are shown as gray shading; and the abiotically suitable area $\mathbf{G}_A$ and scenopoetic existing fundamental niche $\mathbf{E}_A$ are shown as nonbold open outlines (as in figure 3.4). The bold shape encloses the areas and environments reconstructed by the hypothetical model as suitable, given the available observed presences. Labels 1 through 3 refer to different types of predictions yielded by the model, as explained in the text. Reproduced from Pearson (2007).

but not represented by known occurrence records; still, again, these models yield information much more useful than simple occurrence localities alone.

As a consequence, in this final section of the book, we describe critically a variety of applications of correlative niche models. We do not attempt to provide an exhaustive literature review for each application, but rather strive to explain and evaluate theoretical principles on which the applications are based, and to present selected examples to illustrate the approach. As such, we begin each chapter by describing key questions that the models address (e.g., Where are unknown populations likely to be present? Which areas are most susceptible to invasion by nonnative species?), and then outline the theoretical basis and key assumptions of the application. In each application, we relate the desired model output to the types of predictions just discussed. We also describe practical considerations for implementing each application, including the most appropriate modeling methods and input data, with illustrative examples and discussion of future directions and challenges.

In sum, in this brief introduction, we have attempted to give examples of types of model predictions that can yield useful information. It is important to reemphasize that, as with most modeling approaches, great opportunity exists to *mis*apply these methods, with inappropriate assumptions or poor implementation. In particular, the risk exists that model users will be swayed by the ease of use of some methods with default settings, without critically evaluating how reasonable they are for the application and data at hand.

Many approaches described earlier utilize advanced computational technology [e.g., evolutionary-computing tools such as artificial neural networks (ANNs) and genetic algorithms (GAs)], along with large databases of digital environmental layers. It is tempting to think that advanced methods and large datasets will yield robust predictions, but such will be the case only if the conceptual underpinnings of the model application are sound. Quite simply, no substitute exists for a good understanding of the natural history and distributional ecology of the species in question, the theoretical strengths and weaknesses of different methods, and the realities of the data available for the problem at hand, which may or may not match the assumptions of the application and/or the algorithm.

CHAPTER ELEVEN

Discovering Biodiversity

Ecological niche models may be exciting principally because they provide a predictive basis for novel inferences about biodiversity and its distribution in space, time, and environment. One way in which this predictive understanding can be put to good use is that of anticipating distributions of new elements of biodiversity (populations and species) that are not as-yet known or documented. Conceptually, the idea is quite straightforward, and a few initial applications have been developed; however, this application of niche models begs further exploration. If the initial promise continues to translate into further success, this application may rank among the most interesting uses to which these tools can be applied, enabling further discovery and documentation of biodiversity. In this chapter, we describe the conceptual basis of using ecological niche modeling for discovering new elements of biodiversity, and review applications of this approach undertaken to date. We then outline both limitations and frontiers for this field.

Generally, current knowledge of the diversity and distribution of biological species on Earth is remarkably poor, with the great majority of species yet to be described and catalogued scientifically. This problem has two key elements, which may be termed the "Linnaean" and "Wallacean" shortfalls (Whittaker et al. 2005). The "Linnaean Shortfall" refers to lack of knowledge of how many and what kind of species exist—the term is a reference to Carl Linnaeus (1707–1778), who laid the foundations of modern taxonomy in the eighteenth century. The "Wallacean Shortfall" refers to inadequate and incomplete knowledge of geographic distributions of species, referring to Alfred Russel Wallace (1823–1913), who, as well as contributing to the early development of evolutionary theory, studied geographic distributions of species long before it became popular. Ecological niche modeling offers a powerful tool with which to address both of these shortfalls.

DISCOVERING POPULATIONS

The idea of using ecological niche models to guide searches for and discovery of unknown populations of species is perhaps the simplest of the "discovering biodiversity" applications. Given that few or no species have been sampled exhaustively, one of the original motivations behind the development of niche modeling tools was exactly this: filling in gaps among known occurrence sites by means of interpolation that is informed by environmental conditions. Hence, discovering as-yet undocumented populations that are disjunct from known populations (i.e., more complete documentation of G_O) is another important functionality of ecological niche models. This approach takes advantage of "type two" model predictions illustrated in figure 10.1: the model identifies areas environmentally similar to sites where the species has already been found, but from which no occurrence records are available. New surveys targeting these areas should have increased chances of discovering unknown populations, in comparison with unguided, randomly placed surveys.

The idea of niche suitability providing an indication of presence of unsampled populations links to questions of inventory completeness that have been the focus of a suite of studies (e.g., Moreno and Halffter 2000). As biodiversity knowledge accumulates for a particular site, existing knowledge can be used to anticipate how many additional species remain to be detected there, and consequently how complete the inventory is at any point (Colwell and Coddington 1994, Soberón et al. 2007). Niche models can provide an independent source of information on the question of how many additional species remain to be detected at a site, by superimposing individual niche models for each of multiple species. Ideally, such analyses should be conducted after processing distributional predictions to consider the limitations of **M** and **B** and, as a consequence, estimate G_O more closely (see chapter 8). Several analyses have tested the predictive nature of niche projections for anticipating community composition with some degree of success (e.g., Feria and Peterson 2002, Graham and Hijmans 2006).

Conversely, the existence of barriers and interruptions in landscape suitability, combined with absence of a particular species from isolated suitable regions, can be a means of discovering dispersal barriers. When a species *could* be present (i.e., conditions are suitable) but is not (and this absence is demonstrated via tests of sampling adequacy, as discussed in chapter 8; Anderson 2003), we may interpret the situation as one of dispersal limitation (i.e., site is within G_P, but not G_O; in other words, it is in G_I) or biotic limitation. A few explorations of these ideas have now been developed (Anderson et al. 2002a,

Kambhampati and Peterson 2007), but these ideas will be treated in greater detail in chapter 15, which focuses on how niche models can inform studies of evolution and biogeography.

DISCOVERING SPECIES LIMITS

An important issue in systematics and taxonomy is the fact that many currently recognized "species" actually represent complexes of morphologically similar species (e.g., Burg and Croxall 2004). The usual signals of such unrecognized species limits are differentiation in phenotypic or molecular characters, but ecological characters can potentially offer additional, complementary information (Wiens 2007). Coarse-resolution ecological characteristics related to scenopoetic variables offer the opportunity to discover populations that are effectively allopatric, isolated from remaining populations of the species by unsuitable environments (Wiens and Graham 2005). Also, niche modeling exercises can allow detection of populations that are differentiated in ecological features from other populations (Peterson and Holt 2003, Warren et al. 2008).

If other character sets indicate possible species-level differentiation for a set of populations, allopatry and ecological divergence (although not required for recognition as a distinct species) may provide additional evidence in that direction. More generally, to date, ecological dimensions have generally been considered in decisions regarding species status only in very general, nonquantitative terms, and the approaches explored herein offer a means of improving and enriching these discussions. These ideas have been reviewed and explored by Raxworthy et al. (2007), Rissler and Apodaca (2007), Martínez-Gordillo et al. (2010), and Cadena and Cuervo (2009). In addition, they provide some basis for studying the geographic and ecological processes involved in speciation (e.g., Peterson et al. 1999, Graham et al. 2004b; see chapter 15).

However, the expectations of niche model comparisons in conspecific versus multispecific situations are not clear under different species concepts. That is, when no ecological differences are manifested, forms in allopatry may represent distinct species that have undergone vicariant speciation, or may simply represent undifferentiated forms of a single species (Peterson et al. 1999). When ecological differences are manifested, however, once again the meaning is unclear—differentiation in allopatry has represented a long-term problem for decisions under some concepts of species (Zink and McKitrick 1995).

DISCOVERING UNKNOWN SPECIES

To the extent that ecological niches are conserved across evolutionary time periods (see chapter 15), niche models can be developed for known species and used to "prospect" for unknown species. That is, if a species has an unknown sister species (or other close relative), and if ecological niche evolution has been minimal, the ecological characteristics of the known species may be able to inform us about the geographic distribution of the unknown species. This approach makes use of the "type three" predictions illustrated in figure 10.1: areas are identified that are unoccupied by the species being modeled, but where closely related species occupying similar portions of environmental space may be found. In this way, ecological niche models—at least under some circumstances—offer the opportunity to discover presently unknown elements of biodiversity.

The basics of this idea—the assumption of niche conservatism—were established in an early paper (Peterson et al. 1999) that showed that sister species of birds, mammals, and butterflies distributed on opposite sides of the Isthmus of Tehuantepec in southern Mexico tend to have similar ecological characteristics. This result has seen both support (Kozak and Wiens 2006) and nonsupport (Peterson and Holt 2003, Graham et al. 2004b, Knouft et al. 2006) in subsequent analyses (again, see chapter 15). Warren et al. (2008) pointed out, however, that these conclusions depend on the null hypothesis being tested—some studies have tested the null hypothesis that the two species have ecological niches that are more similar than would be expected at random (Peterson et al. 1999), whereas others have tested the null hypothesis that the two species have identical niches (Graham et al. 2004b, Knouft et al. 2006). More detailed discussion of these points is provided in chapter 15, which focuses on niche evolution; here, we assume that some unspecified degree of niche conservatism exists that provides some degree of predictibility across related taxa and makes the use of niche models for discovering biodiversity possible.

CONNECTION TO THEORY

Biodiversity discovery applications of ecological niche modeling are feasible under certain circumstances. This use involves using occurrence data sampled (with all of the usual biases and gaps) from the occupied distributional area G_O to create a niche model that characterizes the occupied area in detail. The assumption is made that $\eta(G_+) \approx \eta(G_O)$ or $\eta(G_P)$ (the latter in the cases of population and species discovery)—that is, that the ecological niche reconstructed from the points derived from the occupied area is at least roughly equivalent to the niche that applies to the occupied and invadable distributional areas (Guisan and Thuiller 2005). If this assumption holds, then the niche model can be projected onto the landscape to identify a broader potential distributional area G_p for the species or lineage.

A further assumption, at least in applications oriented at discovering unknown species, is that dispersal constraints constitute a significant restriction on distributions of species in the system under study. That is to say, if the dispersal abilities of the species in question are good relative to the extent of the region under study, then the sort of isolation required for allopatric speciation to have occurred is unlikely to exist, and any new populations to be discovered are likely not to be differentiated markedly from known populations. In quantitative terms, in the BAM diagram discussed in chapter 3, if $M \subset A \cap B$, then the presence of differentiated populations would be unlikely. However, to assess connectivity among populations on evolutionary timescales requires reconstructions of paleoclimates and the assumption of niche conservatism over longer timescales. For example, even if sets of populations occur over continuously distributed suitable habitat today, they are frequently strongly differentiated genetically, owing to independent evolution (and perhaps speciation) in prior times when they were distributed allopatrically (Avise 2000) or to local adaptation along environmental gradients.

Finally, also for applications focused on discovery of unknown species, niche conservatism is necessary if predictions of likely distributional areas are to prove realistic. If the known and unknown sister species have been evolving independently for a time span t, then the total evolutionary change accumulated over $2t$ years (i.e., the total time of independent evolution between the two lineages) must not be so great that no predictivity among their respective distributional areas is obtained. This assumption, of course, remains untested in most cases, as one of the two species in question has yet to be discovered. In cases where niche evolution has been marked, niche modeling is unlikely to assist in discovery of currently unknown species—see Peterson and Navarro-Sigüenza 2009 for an exploratory test of these eventualities, which identified distributional areas for simulated "unknown" species with some success.

PRACTICAL CONSIDERATIONS

The methods, data, and approaches most useful in applications of ecological niche models oriented at discovering biodiversity are not particularly restrictive. One requirement is that the geographic extent of analysis (although likely not that of calibration; see chapter 7) must be broad so as to include potential distributional areas of the related species. Rather than selecting one modeling algorithm or another, the focus should be on implementing best practices with the algorithm(s) used. In general, the focus, of course, is on discovering the potential distribution of the species G_p. As a consequence, any possible overfitting to noise and/or sampling bias in model calibration must be avoided; by a similar token, overprediction (i.e., predicting too broad an area) can fail to identify specific areas, and as such can distract search efforts. Evaluating and filtering models based on subsets of occurrence data (preferably with evaluation data spatially stratified and thus more independent of calibration data) will reduce the likelihood of overfitting during model development (see chapters 7 and 9).

The key step methodologically is that of seeking undocumented or uninhabited areas that are suitable for the species, but in particular suitable areas that are disjunct from known distributional areas of the species. The simplest such areas will be clearly isolated from known populations—for instance, by a substantial distance. In other cases, uninhabited areas may be close to or even contiguous with known distributional areas—in these cases, populations to be discovered will likely not be distinct from known populations, unless some fine-scale barrier (e.g., a river) or suture zone between two taxa exists in the midst of continuous suitable habitat (Costa et al. 2008).

In terms of still-more practical considerations, and particular emphases in model development and processing, several points should be weighed. In particular, occurrence data from known species/populations should be drawn from as much of the occupied distribution of the species (i.e., G_O) as is feasible to represent its niche dimensions as completely as possible. Environmental datasets should be as broadly generalizable as possible—that is, they should be in no way tied to particular landscapes or time periods; rather, direct variables such as climatic dimensions will most probably be best able to anticipate the niches and distributions of unknown species and populations, because the correlations between them and the species' response are less likely to change in other regions (see chapter 6). Similarly, as discussed in chapter 7, for algorithms that use background, pseudoabsence, or true absence samples, the study region should be selected so as to avoid inadvertently inserting effects of history (or possibly sampling bias) into the modeling process, leading to underestimates of the species' potential distributional area (Anderson and Raza 2010). Finally, thresholding decisions and evaluation statistics should emphasize minimization of omission error and should not overly penalize commission error (see chapter 9), as the latter is the focus of most inferences drawn from niche models (and indeed, it is not necessarily "error" in this context).

REVIEW OF APPLICATIONS

We consider biodiversity discovery to be one of the areas in which niche model potential has been least well explored. That is, although prototypes have been developed for several applications, we believe that much more remains to be achieved and learned in this realm.

DISCOVERING POPULATIONS

Applications of ecological niche modeling to discovery of unknown populations have been explored in a number of cases. For instance, a recent analysis (Guisan et al. 2006) applied general additive models (GAMs) to 92 occurrences of a rare plant species in Switzerland (*Eryngium alpinum*). A map from the publication (figure 11.1) illustrates the sampling and predictions that resulted from this effort. The modeling and field efforts in this study revealed several previously unknown populations, and improved search efficiency somewhere between twofold and fourfold.

Such analyses have now been used on several other occasions, basically confirming the predictive power of ecological niche models in discovery of new populations of known species. Jarvis et al. (2005) used the distance-based niche modeling algorithm FloraMap (Jones and Gladkov 1999) to anticipate the distribution of a rare wild pepper (*Capsicum flexuosum*). They discovered an additional six populations of the species. Bourg et al. (2005) used classification and regression tree (CART) niche model estimates to discover eight additional populations of a rare forest plant (*Xerophyllum asphodeloides*) in the southeastern United States. Finally, Siqueira et al. (2009) based a very simple, distance-based prediction on a single known occurrence point of a rare plant in the southern Brazilian cerrado (*Byrsonima subterranea*), and discovered six additional populations of the species via searches in high-probability areas.

DISCOVERING DIFFERENTIATED FORMS

Discovering differentiated forms that are "lost" in synonymy of known species can be facilitated by two classes of information from ecological niche models—confirmation of allopatry (Wiens and Graham 2005), and documentation of ecological differentiation of the candidate form (Raxworthy et al. 2007). Each of these elements of evidence can be derived from ecological niche models, and can then be considered along with phenotypic and molecular evidence to arrive at educated decisions regarding species status. Although, strictly speaking, the new forms were already "known" to science (just not recognized as specifically distinct), this approach nonetheless serves to detect their isolation

Observations
★ new
● confirmed
◆ not visited
✖ not confirmed
■ introduced or cultivated

Predicted areas
■ probability > 0.80

Subareas
▨ unsuitable
▧ uncertain
□ suitable

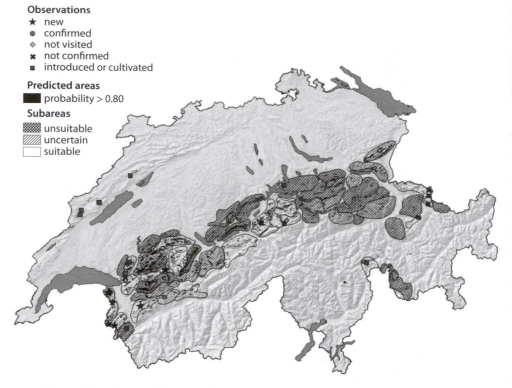

FIGURE 11.1. A map of the potential distributional area $\mathbf{G}_p$ of the rare and endangered plant *Eryngium alpinum* in Switzerland (Guisan et al. 2006). The darker gray shading indicates area predicted by the models, and stars indicate new populations discovered as a consequence of the model predictions.

and/or differentiation, and as such to assist in appreciating the terminal units of biological diversity when combined with phylogeographic analyses.

Raxworthy et al. (2007) developed this reasoning further in explorations of the day gecko genus *Phelsuma*. For the "species" *P. madagascariensis*, they showed that ecological niche models based on all populations of the group combined was overly broad, extending into areas well outside the species' distribution. Individual niche models for each of three differentiated subspecies, however, produced predictions that were considerably tighter, and the authors included this information in their argument for species status for each of the three forms. Similar analyses within the *P. dubia* group contributed to the formal elevation to species status of a previously unrecognized species. However, these analyses would benefit from tests of niche identity that consider explicitly the similarity of niches and the landscapes that are accessible (**M**) to each form (Warren et al. 2008), and the expectations under different species concepts are unclear (Rissler and Apodaca 2007). This analysis is fascinating in that many present-day species-level taxa around the world are known to be overly "lumped" (Peterson 2006b), and as such would benefit greatly from analyses of this sort.

DISCOVERING SPECIES

Perhaps most fascinating among "discovering biodiversity" applications of ecological niche modeling is an analysis of chameleons in Madagascar (figure 11.2; Raxworthy et al. 2003). The authors developed niche models for 11 species in three chameleon genera, and for each species identified disjunct areas of overprediction in which the species was not known to occur. In particular, three

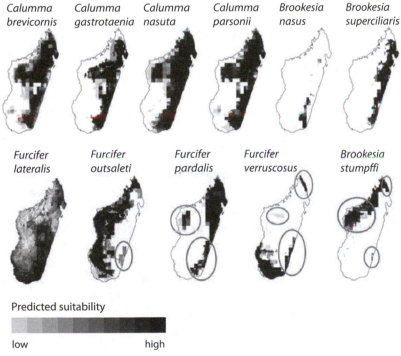

Predicted suitability

low   high

FIGURE 11.2. Example of the use of ecological niche models to anticipate the distributions of species not yet known to science. Eleven species of chameleons were modeled based on known occurrences, with the goal of estimating the abiotically suitable area G_A. Areas of suitability that were not known to hold populations of the species in question are indicated with ellipses—these areas were the targets for surveys. Adapted from Raxworthy et al. (2003).

disjunct, uninhabited areas were identified as consistently suitable. Subsequent field surveys in those areas yielded seven new chameleon species, whereas surveys in many other areas did not yield more than a single new chameleon species. Hence, this initial exploration used the idea of ecological niche conservatism to identify likely distributional areas for related, but as-yet unknown, species. We know of no other such explorations, but greatly look forward to seeing them as they are developed.

DISCUSSION

The preceding examples should illustrate the promise of these applications of niche models. Nonetheless, the set of examples is relatively sparse, with much more exploration remaining. Some considerations for such future studies follow.

CAVEATS AND LIMITATIONS

As mentioned earlier, an important assumption in applications oriented toward discovery of species or populations is that niche evolution in the lineages in question is minimal, either through time if the goal is to discover species or across space if the goal is discovery of populations (Peterson and Holt 2003). If these basic assumptions are not met, then niche models will fail or be less effective in identifying distributional areas for the populations or species of interest.

A more subtle caveat, however, is that existing knowledge can constrain which parts of unknown diversity can be discovered. Given the cautions mentioned in the previous paragraph, new discoveries will focus on taxa or populations similar to those already known, so if greater ecological niche diversity is present among unknown forms, only the part that is similar to known forms will be discovered preferentially using these methods. Of more concern, although unavoidable for practical reasons, is that if new populations of a rare species with a broad niche are discovered based on their environmental similarity to known occurrence points, then our understanding of the species' niche will be underestimated, because the environments of the first localities constrained discovery of populations to those in similar environments. Indeed, several studies in this realm have used an iterative process of model development, exploration and addition to knowledge, and then further model development, etc. (Guisan et al. 2006, Siqueira et al. 2009)—this approach may be particularly vulnerable to such constraints regarding which forms or populations are likely to be discovered.

Future Directions and Challenges

The suite of niche modeling applications treated in this chapter is clearly still in only the early stages of exploration and development. Methodologies are still preliminary, and comparative evaluations are only beginning to be published (Peterson and Navarro-Sigüenza 2009). Certainly, though, the opportunity for such evaluations exists, as one could set aside known species, and test the ability of different approaches to "predict" the existence of the "unknown" species. Such exploratory analyses would serve as an important benchmark for further explorations of the utility of niche modeling tools to assist in biodiversity discovery. Also, confirmatory surveys in such studies need to include sites *not* predicted by the niche model so as to explore further its discriminatory ability.

More generally, however, it is more or less clear that niche modeling tools have a great deal to offer to the process of biodiversity discovery. Individual species may be understood in greater detail, or close relatives discovered; iterating analyses over many species, the same techniques can be used to pinpoint areas of highest priority for broad-spectrum biodiversity sampling. Once the concepts of ecological niches are clarified, and the capacities of the various tools for characterizing them are understood better, the present storehouse of biodiversity information can prove enormously useful in developing and completing a more comprehensive picture of landscape- and regional-level biodiversity patterns.

CHAPTER TWELVE

Conservation Planning and Climate Change Effects

The field of conservation biology seeks to provide scientific guidance for halting or slowing the current extinction wave and degradation of the planet's biological diversity. To achieve this goal, conservation biologists attempt to answer fundamental questions, such as what to conserve, where best to conserve it, and how best to conserve it (Primack 2006). Can niche models help to address these questions? We believe that the answer is yes, particularly by helping researchers answer the "what" and "where" questions. However, using niche models to address conservation questions requires a solid understanding of the underlying concepts and methods.

Inappropriate interpretation of the underlying theory and methods can lead to mistakes and potentially misleading interpretations of niche model outputs. Therefore, in this chapter, we introduce briefly the conceptual aspects of the "what" and "where" questions in conservation biology, and discuss how niche models can help address these questions. Topics addressed include inferences about extinction risk, identification of regions for species reintroductions, conservation reserve network planning, and considerations of how climate change may affect species' distributions. Each of these conservation applications is discussed with respect to the conceptual framework laid out in chapters 2 and 3, and practical recommendations regarding calibration and evaluation of niche models are also offered.

GENERALITIES

If given the opportunity, conservation practitioners would certainly target and manage all genetically distinct populations of all species on Earth. In practice, however, biodiversity conservation must coexist with competing human interests (Primack 2006). Effective conservation action thus entails a difficult, but unavoidable, process of prioritization of limited opportunities and resources.

In addition, scientific and technical difficulties further complicate the situation: only a small proportion of global biodiversity is known (the Linnaean Shortfall) and distributions and abundance patterns of elements of biodiversity are also poorly understood (the Wallacean Shortfall; Whittaker et al. 2005). Arguably, even less understood are the networks of interactions existing among organisms and between organisms and physical and environmental systems—gaps in understanding that might be termed the Eltonian and Grinnellian shortfalls.

The idea of measuring biodiversity value as the relative contribution of a particular biodiversity object to an overall set of objects was coined as the "complementarity principle" (Vane-Wright et al. 1991). Although several definitions of complementarity exist, the most general defines it as "a property of sets of objects that exists when at least some of the objects in one set differ from the objects in another set" (Williams 2001). Even though complementarity is most often associated with prioritization of areas for conservation (Margules and Pressey 2000), the original formulation was broader, providing a rationale for ranking species as well as areas regarding relative biodiversity value (Vane-Wright 1996). These ideas have had greatest impact on prioritization of areas for conservation, probably because of theoretical (Erwin 1991) and technical (Faith 1993) difficulties with implementation of the principle for organisms.

The idea of urgency of conservation action was discussed by Norman Myers in his famous allusion to the concept of triage, in which wounded soldiers are assigned priority based on who can be saved, who can probably survive without attention, and who will die regardless of how much attention is received (Myers 1979). As such, extinction risk is a common currency of conservation priority. One example of prioritization at the site level is mapping biodiversity hotspots, a concept championed by Myers et al. (2000) and promoted in particular by Conservation International (Mittermeier et al. 1998).

CONNECTION TO THEORY

We have argued that conservation problems can be classified into broad questions—the "what" question entails mainly measurement of conservation priorities at the level of species or populations, whereas the "where" question entails both measurement of value (complementarity) and priority at the level of sites. The diversity of these challenges makes for distinct considerations when applying niche modeling to conservation. In light of this diversity, we review each sort of application separately.

Estimating Extinction Risk for Species

The process of extinction is one of population reduction, which is manifested as range loss and eventually extinction of the species. Some mechanisms leading to such losses involve reduction of quantity and quality of suitable areas for a species, a process that can be modeled fairly easily. For example, Brooks et al. (1997) used species-area relationships derived from island biogeography theory (MacArthur and Wilson 1967) to predict losses of forest species following deforestation of insular systems in the Philippines and Indonesia. Thomas et al. (2004a) extended this idea to make predictions of extinction risk for individual species in response to modeled future changes in the spatial footprint of climate envelopes of species. In both cases, generalizations are made about relationships between area and extinction risk, but the general principle—that range loss follows population reduction and increases extinction risk—is hardly controversial.

Mechanisms leading to extinction of small populations may be difficult to model with correlative approaches like niche models, because they frequently involve stochastic rather than deterministic processes (Lande et al. 2003). Furthermore, they entail processes that are more demographic or genetic (e.g., demographic stochasticity, unbalanced sex ratios, inbreeding depression, genetic drift) than geographic. Nevertheless, integration of deterministic processes leading to population decline (e.g., habitat reduction, estimated via niche modeling and current environmental data, e.g., from remote sensing) is possible—for example, Keith et al. (2008) used niche models to project changes through time in climatic suitability for 234 plant species, and coupled the model outputs with spatially explicit stochastic population models. With this approach, they were able to explore interactions of mechanisms causing population decline with stochastic factors driving population fluctuations. Although the results were not tested with real data, the study demonstrated the possibility of coupling niche and population models in a common framework. A simpler study that employed an elegant test of model predictions focused on *Aloe dichotoma*, a tree species of southern South Africa, in which dead individuals are clearly visible and identifiable at long distances—as a consequence, when mortality rates were compared with niche model predictions for changing suitability over coming decades, a significant result was obtained (Foden et al. 2007).

Niche models, when applied to estimating extinction risk, must then estimate changes in species' occupied geographic distributions (that is, G_O before and after some change event, which we will term ΔG_O; table 12.1), rather than their potential distributions (G_P). In the simplest sense, these estimates can be based on before-and-after estimates of species' distributions. However, realis-

TABLE 12.1. Dimensions of the challenges in conservation biology that are potentially illuminated by ecological niche modeling approaches.

Topic	G-space	Comments
Extinction risk	$\Delta \mathbf{G}_O$	Requires explicit modeling of changes in $\mathbf{G}_O$, e.g., by coupling niche-based predictions with spatially explicit stochastic population models.
Introductions	$\mathbf{G}_A$ or $\mathbf{G}_P$	Niche models are not asked only to fit the data well. Instead, they should characterize the abiotically suitable area.
Conservation planning	$\mathbf{G}_O$	Suitability indices are estimated and overlaid on observed species occurrences.
Climate change	$\Delta \mathbf{G}_A$ or $\mathbf{G}_P/\Delta \mathbf{G}_O$	Generally, changes in potential distributions are required, but sometimes it is changes in observed distributions that matter. Solutions depend on the specific goals.

tic extinction models will require spatially explicit, mechanistic simulations of dispersal probabilities, biotic interactions, and key population-level parameters unique to each species (but that are often generalized among ecologically similar species or sister taxa). To address this problem, niche models can be coupled with spatially explicit dynamic population models by providing resource-function estimates, which can then be used as surrogates for carrying capacity in population models (Anderson et al. 2009). Resource function estimates are usually taken from estimates of probabilities of occurrence or suitability indices, but with a closer tie to population processes. Initial explorations of the use of niche models to predict extinction probability or local abundances are yielding promising results (Araújo et al. 2002, VanDerWal et al. 2009).

Finally, simple statistical models allow estimation of total abundance given occupied area and an index of aggregation of individuals in space (He and Gaston 2000). Using our notation along with equation 5 in He and Gaston (2000), an estimate of total number of individuals $\hat{N}$ in a region of area $\mathbf{G}$, if the distributional area of the species is $\mathbf{G}_O$, would be

$$\hat{N} = \frac{|\mathbf{G}|}{a}\left[\left(1 - \frac{|\mathbf{G}_O|}{|\mathbf{G}|}\right)^{-1/k} - 1\right], \tag{12.1}$$

where a is the resolution of the cells in the grid, and k is the negative binomial parameter describing degree of "clumpiness" in the spatial distribution of individuals (small k means highly aggregated distributions; note other uses of these same symbols in previous parts of this book).

CONSERVATION PLANNING

Reserve networks are designed ultimately to ensure persistence of species or other valued ecosystem attributes (e.g., habitats or vegetation types). To this end, reserve networks should be designed to represent enough individuals of the valued species to ensure viable populations. How many is enough is a discussion beyond the scope of this book (see Lande 1988, Rodrigues and Gaston 2001). When the geographic distribution of a species is poorly documented, it is tempting to use niche models to predict possible suitable and presumed occupied locations that are not documented owing to insufficient sampling. Of course, this process of estimating G_O is not simple and straightforward (see chapters 7 and 8), and all of the usual precautions regarding estimating G_O and distinguishing it from G_P must be considered. However, this sort of estimation prior to conservation planning can potentially inform prioritization efforts greatly (Peterson et al. 2000).

Strictly speaking, unless niche models are coupled with dispersal modeling (e.g., Collingham et al. 1996) and incorporate interactions with other species as potential modifiers (e.g., Anderson et al. 2002b, Araújo and Luoto 2007, Heikkinen et al. 2007), model outcomes should be interpreted as G_A or G_P (see chapter 8). When making predictions with ecological niche models, we have no guarantee that the species occurs at a particular site or will ever occur there. Therefore, decisions as to whether to use such estimated distributions for reserve selection are contingent on the types of error that conservation planners are willing to accept. In most cases, planners are cautious about incurring commission error (i.e., protecting an area for a species that is not, in fact, present), but may be willing to accept omission error (except for the case of very rare species), in which some occupied sites are left out of the prioritization exercise.

An alternative, and potentially more helpful, use of niche modeling in conservation planning is that of providing suitability scores (see chapter 7) as a surrogate for persistence probabilities (e.g., Araújo and Williams 2000). If niche theory is correct and applicable (see chapter 2) and the data (see chapters 5 and 6) and methods used to estimate niches (see chapter 7) are adequate, then areas predicted as highly suitable for a species can be identified. Overlaying sites of known occurrence, candidate reserves can be selected with emphasis on populations known to be present and predicted most likely to persist. Of course, suitability is being used as a surrogate for persistence, so ideally, more direct, process-based estimates of population viability would be derived from models explicitly simulating extinction processes (perhaps taking into account effects of land use change, climate change, and dispersal, in addition to patch size and configuration), although it is often not feasible to construct such models across large areas or for many species.

Species Reintroductions

A further application of niche models in conservation activities is that of prioritizing for reintroduction efforts, by identifying the most suitable areas. Although many GIS-based analyses of reintroduction sites have been developed (Howells and Edwards-Jones 1997, Bright and Smithson 2001, Carroll et al. 2003), ecological niche suitability considerations have been taken into account explicitly in only a few studies to date (Pearce and Lindenmayer 1998, Martínez-Meyer et al. 2006). Here, clearly, the focus is not on G_O, but rather on G_I, as the species is not present at the sites prior to reintroduction.

Factors unrelated to A and B (see chapter 3) are hypothesized to have led to the species' absence in such areas. In many cases, these factors will be human presence, anthropogenic habitat fragmentation, or hunting pressure that may have extirpated the species' populations from the area in the past (although strictly speaking, competition with humans falls under B), so the niche models would provide information regarding which sites hold optimal conditions for reestablishment of populations, after the additional negative factor has been removed or mitigated. In other cases, knowledge of original distributions may be very incomplete, or conservation-based introductions may even place the species at sites where it did not originally occur, i.e., where M is important (Cade and Temple 1995). At any rate, planning for reintroductions clearly must focus on G_I. Niche modeling should constitute a critical first step in the planning process for a reintroduction: once environmentally suitable but unoccupied areas (G_I) are identified, field work can focus on other factors, such as availability of food resources, threat by human activities, and attitude of local residents to introduction, to make a final decision on where to introduce a species.

Climate and Land Use Change

Evidence that climate governs the distributions of species comes from examination of two features of species' range limits (Root 1988, Gaston 2003): (1) distributional limits are often correlated with particular combinations of climatic variables, and (2) they often change through time in synchrony with changes in climate. Examples of the second point include latitudinal shifts in species' distributions in response to glacial cycles during the Pleistocene (Wells and Jorgensen 1964), or recent poleward expansions of plants (Walther et al. 2005), breeding birds (Thomas and Lennon 1999), butterflies (Thomas et al. 2004b, Parmesan 2006), and several other groups (Hickling et al. 2006). If climatic factors control species' distributional limits, then range shifts should be easily anticipated using methods focused on species-climate relationships. Such a situation requires careful and complete estimation of the full dimensions of the ecological niche of the species, because different sectors of the niche may be important to making an area "suitable" in different sectors of the distributional area.

Several generalities can be stated regarding the form that climate change effects will likely take for species. For example, with warming climates, particular climate regimes will—in general—be expected to shift upward in elevation and poleward in latitude (Parmesan 2006). Results of niche modeling studies suggest that an additional "general" property will be that flatlands systems will see more dramatic spatial effects of changing climates than montane systems (Peterson 2003b). However, and perhaps most critically, niche modeling studies (Iverson and Prasad 1998, Peterson et al. 2002b, Thomas et al. 2004a) and empirical analyses (Parmesan 2006) both indicate that species will often show individualistic responses to changing climates, so the generalities will hold only on average, and not in particular cases. Hence, climate change considerations will require species-by-species consideration of likely effects, which can be forecast via use of niche-modeling approaches.

PRACTICAL CONSIDERATIONS

Applications of niche modeling to conservation challenges are diverse, and as such the particulars required of model estimation for these applications are similarly varied. In particular, applications to conservation planning (i.e., protected area planning) require careful estimates of G_O, and understanding extinction risk requires detection of change in G_O over a given time period. In contrast, applications to planning species reintroductions and climate change forecasting require careful and complete estimates of G_P, and in the latter case careful consideration of dispersal potential of the species (M) to track or detect and colonize those areas. These diverse needs will be reflected in diverse methodologies, and of course all of the caveats and precautions discussed in previous chapters will apply.

One particular difficulty is that models projecting distributions of species under climate change scenarios are often asked to make predictions onto future conditions that are not available across the region on which the models were calibrated (i.e., extrapolation in E; see chapter 7). We will frequently face questions like, "if the species of interest ranges 10–18°C, and the maximum temperature represented on the calibration landscape is 18°C, then what will the species do under conditions of 20°C?" In theory, and probably in practice as well, correlative models are incapable of providing robust projections for non-analog climate conditions, and models fitted with different mathematical functions will extrapolate potential distributions of species in very different ways

(Thuiller 2004, Pearson et al. 2006). Making predictions in such situations requires assumptions regarding the shape of species' responses to unknown conditions, and various assumptions are possible (see chapter 7; Williams and Jackson 2007, Anderson and Raza 2010).

Although we know of no easy fix for this problem, models should be calibrated with as much of the species' E_O represented as possible. For example, if one is interested in modeling likely responses of plants in Great Britain to future climate change, models can be calibrated (with occurrence and environmental data) at the extent of all of Europe (see chapter 7; Pearson et al. 2002). These more complete representations of niche estimates may be more representative of the species' potential response.

In climate change studies, particularly those looking into the future (as opposed to the past), model validation is often impossible (Araújo et al. 2005a). Partial validation of predictions by "hindcasting" (i.e., calibrating models with contemporary data, and evaluating with past data before forward-projecting) is a possibility (Araújo et al. 2005a, Martínez-Meyer and Peterson 2006), but will be applicable to only a small number of species owing to the incomplete nature of the historical record. (Note that molecular systematic analyses may provide additional opportunities for tests.) Other partial model validations are possible by testing across space instead of over time (Fielding and Haworth 1995, Randin et al. 2006, Jakob et al. 2009), but the degree to which spatial dynamics mirror temporal dynamics in determining species' distributions has not been evaluated carefully. Although these validations are at best indirect, they nonetheless are at least relevant to the question of niche conservatism through time; the only other option is model verification (see chapter 9), but this strategy evaluates only internal consistency of models based on single time periods.

Assuming the lack of an objective basis for selecting the most appropriate algorithm for modeling species' range shifts under climate change, an alternative paradigm is to calibrate ensembles of models (via various algorithms) and explore the resulting range of uncertainties (reviewed by Araújo and New 2007). It has been shown, both theoretically (Bates and Granger 1969) and empirically (Clemen 1989, Araújo et al. 2005b), that combining ensembles of models can yield forecasts with better predictive ability than the individual forecasts (although see the discussion of limitations of ensemble approaches in chapter 7). With the starting assumptions that projections are equally correct (a big "if") and that they collectively delimit the range of uncertainties associated with the future range of the species (also not clear), a "majority-vote" criterion can be made by assigning higher probabilities to median projections. Alternatively, different projections from different models may be weighted differently based on Bayesian posterior probabilities or other measures of model quality

(Hastie et al. 2001) to avoid diluting powerful and insightful model signal with noise and error from less useful models.

REVIEW OF APPLICATIONS

As should be clear from the preceding paragraphs, applications of niche modeling to conservation are both numerous and diverse. As such, we explore each type of application separately. Several of these applications are quite specialized, and the examples we feature are selected from among dozens of examples that are being developed by researchers around the world.

Estimating Extinction Risk for Species

The simplest applications in this category consist of evaluations of the distributional characteristics of species of particular interest. For example, a recent study evaluated the distributional characteristics of the Sierra Madre Sparrow (*Xenospiza baileyi*), which has a sparse and perhaps declining distribution in central and western Mexico (Rojas-Soto et al. 2008). A particular question for this species was whether the three apparently disjunct known populations are genuinely disjunct (they appear to be), or whether additional populations might exist in intervening patches of suitable habitat (few or none appear to exist; figure 12.1). The study produced the first quantitative estimates of the distributional area of the species, and identified a few areas for additional field work in western Mexico. Other, similar studies have illuminated distributional areas of diverse species around the world (Ortiz-Pulido et al. 2002, Gaubert et al. 2006, Peterson and Martínez-Meyer 2007, Peterson and Papeş 2007, Vanak et al. 2008).

More complex analyses of extinction risk of species have incorporated land cover information into development of refined risk estimates. With a simple methodology, Sánchez-Cordero et al. (2005) estimated range loss owing to deforestation for mammals across Mexico—they simply estimated G_P based on niche modeling in climatic dimensions, and reduced that to an estimate of G_O by means of "cookie-cutting" based on explicit assumptions regarding species' associations with particular land-cover types (see chapter 8). An improvement on the initial estimates was to link classes of original vegetation maps (for model calibration) with the same classes of current land-cover maps (for model projections to estimate "current" distributions) in niche models to yield quantitative estimates of range loss resulting from land use transformation (Peterson

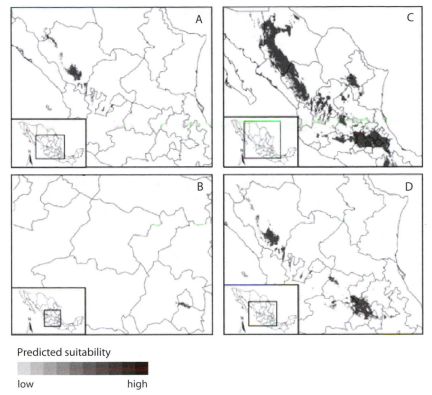

Predicted suitability

low   high

FIGURE 12.1. Ecological niche model estimates of distributional areas of the Sierra Madre Sparrow (*Xenospiza baileyi*), which has disjunct distributional areas in central and western Mexico, showing the effects of the numbers of environmental variables used in model calibration. Panels A and C were based only on occurrence data from northern populations, while B and D were based on occurrence data from southern populations. Panels A and B show model results based on an overly dimensional environmental space, whereas C and D were based on simpler environmental spaces. Panel D is that which is closest to the occupied distribution of the species. Adapted from Rojas-Soto et al. (2008).

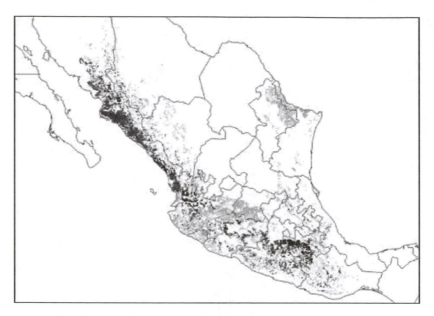

FIGURE 12.2. Illustration of detection of areas of likely population extirpation. This figure tallies population losses (greater losses shown in darker shades of gray) over Mexican species of jays (Corvidae), from Peterson et al. (2006b), based on niche model results tied to temporally explicit summaries of spatial extent of primary habitats across Mexico.

et al. 2006b, Soberón and Peterson 2008). These extrapolations offer the potential for development of species-specific and site-specific projections of likely range loss in considerable detail (figure 12.2).

CONSERVATION PLANNING

Perhaps the oldest applications of this ilk are the distributional mapping efforts developed as part of the U.S. National Gap Analysis Program (Scott et al. 1993 and 1996). These efforts intended to combine spatial interpolations (informed and refined by information on land-use patterns) with conservation reserve network information to identify gaps in protection. Unfortunately, this endeavor fell short of expectations owing to poor quality of the distributional information and modeling methods used, which were largely binary vegetation-surrogate approaches (Peterson and Kluza 2003, Peterson 2005a).

More recently, numerous studies have used niche modeling approaches to estimate distributions of suites of species, processed them into estimates of G_O, and used this distributional information in analyses of complementarity to prioritize areas for conservation. These efforts, as mentioned earlier, are vulnerable to many problems with error propagation—omission or commission error in individual species' predictions can be propagated through the procedure to yield prioritizations that do not give the desired results. At the same time, the distributional estimates—if developed rigorously—have the potential to avoid many problems with sampling effort and completeness (Rojas-Soto et al. 2003). The resolution of the analysis is also critical: the species may not be found in all areas of the pixel, given distributional constraints operating at finer resolutions. Nonetheless, at least as a heuristic tool, this approach has been applied broadly and has yielded some interesting results (e.g., Godown and Peterson 2000, Peterson et al. 2000, Chen and Peterson 2002, Loiselle et al. 2003, Ortega-Huerta and Peterson 2004, Moilanen 2005, Ramírez-Bastida et al. 2008, Toribio and Peterson 2008, Koleff et al. 2009).

Niche models have also been used to evaluate the effectiveness of current protected areas in protecting biodiversity under scenarios of future climatic conditions (Hannah et al. 2007). For the more detailed prioritization exercises mentioned earlier, in which extinction probability and suitability are estimated using niche model results, only initial steps have been taken. For example, using distributional data for breeding songbirds in Great Britain, Araújo et al. (2002) showed that probabilities of occurrence of species derived from a niche model were related to local probabilities of extinction over a 20-year period (1970s to 1990s). It was also shown that local extinctions in the 1990s could have been reduced if conservation areas had been selected to maximize species' probabilities of occurrence in the 1970s, suggesting that extinctions could be minimized if even simple treatment of estimates of persistence were incorporated into reserve selection procedures. In another example, Kremen et al. (2008) incorporated niche models (termed by them "species distribution models") in an analysis to identify priority areas for conservation in Madagascar. In that study, estimates of G_O were generated from G_P by "clipping" model predictions to a minimum convex hull fitted around the known occurrence localities for each species.

SPECIES REINTRODUCTIONS

A few studies have begun to incorporate niche modeling ideas into prioritization for reintroductions of species (Pearce and Lindenmayer 1998, Martínez-Meyer et al. 2006). In these studies, the researcher takes advantage of the analogy between reintroductions and species' invasions (Bright and Smithson 2001). That is, the site of interest for reintroduction is outside of G_O at present, but if it is within G_P, then it may be suitable for reintroductions, or even introductions to sites new for the species.

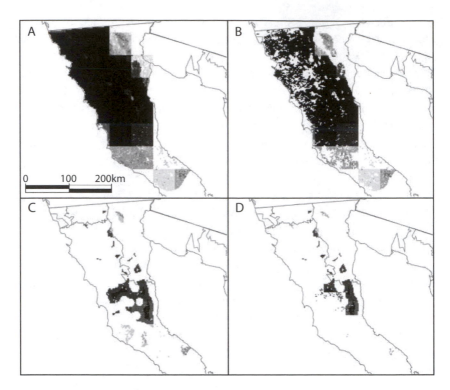

Predicted suitability

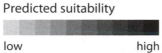

low   high

FIGURE 12.3. Example prioritization of areas for reintroductions of California Condor (*Gymnogyps californianus*) in Baja California, Mexico, from Martínez-Meyer et al. (2006). Panel A shows a crude and coarse niche model result, cut in B to reflect the distribution of primary vegetation, and in C by removing areas exposed to humans (i.e., roads and settlements). Finally, in D, areas from C are identified that are also robust to likely climate change effects.

For example, Martínez-Meyer et al. (2006) assessed sites across Mexico for suitability for potential reintroductions of California Condor (*Gymnogyps californianus*) and Mexican Wolf (*Canis lupus baileyi*; figure 12.3). The authors evaluated sites to ensure that they (1) fit the ecological niche of the species (i.e., in G_P); (2) were within the historical range of the species (i.e., in G_O); (3) were relatively remote from human presence; (4) were in relatively large extents of suitable habitat; and (5) were robust to changing climates (see the following). A parallel endeavor was presented by Pearce and Lindenmayer (1998), who identified potential reintroduction sites for the endangered Helmeted Honeyeater (*Lichenostomus melanops cassidix*) in Victoria, southeastern Australia.

CLIMATE AND LAND USE CHANGE

Studies investigating climate change impacts on species' distributions rank among the most prominent applications of niche models (e.g., Iverson and Prasad 1998, Peterson et al. 2002b, Thomas et al. 2004a, Thuiller et al. 2005a, Araújo et al. 2006, Huntley et al. 2008). The potential utility of this application of ecological niche models has been debated, with attention focusing on the importance of many factors including changing biotic interactions, dispersal limitation, and potential for rapid evolutionary adaptation (e.g., Davis et al. 1998, Pearson and Dawson 2003, Hampe 2004, Thuiller et al. 2004b, Beale et al. 2008).

Despite uncertainties, and in contrast to earlier treatments (Dobson et al. 1989, Peters and Myers 1991–1992), niche models offer the potential to forecast future potential distributional areas for species under explicit climate change scenarios, a considerable improvement over earlier, subjective treatments. Single species' possible distribution shifts have been evaluated to detect populations likely to remain stable over coming decades, as compared with others that are likely to be lost, and these projections have been assembled over faunas, floras, and biotas to build composite future projections (e.g., Peterson et al. 2001, Berry et al. 2002, Erasmus et al. 2002, Midgley et al. 2002, Peterson et al. 2002b, Midgley et al. 2003, Siqueira and Peterson 2003, Téllez-Valdés and Dávila-Aranda 2003, Peterson et al. 2004c, Thomas et al. 2004a, Anciães and Peterson 2006).

In another application, Araújo et al. (2004) used niche models to test the ability of spatial conservation planning methodologies to consider both current and future (changed climate) abiotically suitable distributional areas (G_A) of European plant species. They concluded that no single conservation area would ensure persistence of all species over a 50-year period, and that clustered conservation areas would lead to loss of most species under climate change. In contrast, conservation areas selected in high-quality habitat species-by-species would lose fewer species. Interestingly, they also showed that uncertainties accrued from dispersal were less important than uncertainties associated with choice of the place-prioritization technique. Niche models have also been used for identification of future protected areas and biodiversity corridors (Williams et al. 2005, Phillips et al. 2008).

CHAPTER THIRTEEN

Species' Invasions

Invasive species are a global phenomenon with massive consequences, both in biological and economic realms (Williamson 1996, NAS 2002). In human economic arenas, invasive species affect agricultural productivity, transportation systems, communication systems, disease transmission, recreational fishing, hunting, and birdwatching, and many other dimensions (Pimentel et al. 2004), with economic costs that mount into the billions of U.S. dollars annually. Indeed, a recent calculation was that the annual cost of invasive species in the United States alone reaches $120 billion annually (Pimentel et al. 2004). In natural systems, invasive species can be transformational, affecting not only ecosystem services (Zavaleta et al. 2001) but also endangering or extirpating native species (Chapin et al. 2000, Clavero and García-Berthou 2005). As such, invasive species—and the population-level and biogeographic processes that lead to invasions—are of considerable interest and importance.

In the field of ecological niche modeling, invasive species are of particular interest. This application makes use of the "type three" predictions illustrated in figure 10.1: areas are identified that are currently unoccupied by the species but are likely to be susceptible to invasion if limitations on dispersal are removed. Niche modeling has been applied most commonly at broad spatial extents (Soberón 2007)—as such, it is difficult to test with experimental manipulations, due to the considerable effort required to perform ecological experiments at biogeographic extents. The broadening degree of human movements and activities, however, provides what is effectively a series of experiments—what happens to a species' geographic distribution when its dispersal capabilities (**M** in the BAM diagram) are expanded? ENM and associated theory encapsulated in the BAM diagram yield clear, quantitative predictions that can be tested by means of such unplanned experiments, so species' invasions offer a fascinating arena in which to explore the processes underlying geographic ecology.

In effect, both the practical (applied) interest and the theoretical (conceptual) interest in species' invasions distill down to the same question: to what degree can the geographic course of species' invasions be anticipated based on scenopoetic variables and biotic interactions? As will become clear in this chapter, the answer is neither simple nor straightforward, but the effort has certainly been informative and educational.

CONNECTION TO THEORY

Several points from chapters 2 and 3 earlier in this book are central to the invasive species question. First, species do not generally occupy the entire spatial footprint of their potential distributions ($\mathbf{G}_P$), but rather are limited to some subset of that potential by historical barriers to dispersal (related to $\mathbf{M}$) and potentially by biotic limitations (related to $\mathbf{B}$). This subset is $\mathbf{G}_O$, and we expect that $\mathbf{G}_O \subseteq \mathbf{G}_P$. Indeed, when the spatial extent of the analysis ($\mathbf{G}$) is broad, the expectation is that $\mathbf{G}_O$ will be much smaller than $\mathbf{G}_P$, because species' distributions are highly constrained by dispersal limitations, especially at continental or global extents.

A key insight is that understanding species' invasions in a niche modeling context may require inclusion of the role of biotic interactions in shaping species' distributions. That is, $\mathbf{B}$ in the BAM diagram plays a critical role in determining whether a model based on a species' distribution and ecology in one region will be able to anticipate its distribution in a novel region (e.g., a different continent). If $\mathbf{B}$ is so broad as essentially not to limit the species' distribution (the Eltonian Noise Hypothesis), or if the environmental structure of $\mathbf{B}$ (i.e., the community context) is comparable between regions, then a niche model calibrated in one region should be predictive of its distribution and ecology in another. This assertion is of course subject to the conditions that the models are well-calibrated and not substantially biased, and that the genetically determined elements of the scenopoetic existing fundamental niche $\mathbf{E}_A$ of the species do not differ between the native-range population and the invasive populations (meaning that scenopoetic niches are conserved). The fact that many niche models have shown excellent predictivity regarding the geographic potential of species' invasions suggests that, at biogeographic extents and coarse spatial resolutions, $\mathbf{B}$ does not frequently present a strong constraint on species' distributions. On the other hand, the massive and sometimes pervasive effects that some invasive species have on elements of native biotas attest to the opposite.

PRACTICAL CONSIDERATIONS

The basic approach to understanding species' invasions using ecological niche modeling is simple. The idea is to use occurrence records of the species in one region to calibrate models, and to then project those niche models onto other regions, where the species may or may not be invasive at present. These steps may be applied to regions where the species has already invaded, to test for changes in the scenopoetic existing fundamental niche or in **B** (Broennimann et al. 2007), or they may be applied to regions where the species has not yet invaded, to assess risk and identify areas particularly susceptible to that particular species based on **A** (López-Darias and Lobo 2008).

Most invasive-species applications of niche modeling so far have been based on models calibrated on native-range occurrence data and environmental data (Peterson 2003a), although some efforts basing models on invaded-range areas are beginning to appear (Anderson et al. 2006, Roura-Pascual et al. 2006, Broennimann et al. 2007, Fitzpatrick et al. 2007, Kluza et al. 2007). Calibration on the native range serves to identify the overall dimensions of the species' ecological niche, and is particularly attractive because the species' distribution is more likely to be closer to equilibrium in the native range than in invaded areas (Peterson 2005b).

If the purpose of the analysis is to identify susceptible areas in a novel region, however, incorporation of additional information into the analysis may be useful and informative (e.g., information on life history traits and dispersal abilities; Dullinger et al. 2009). Many invasive species have multiple invasive ranges, in other regions or on multiple continents. To the extent that occurrence data from the native range are representative of the fundamental scenopoetic niche, that the Eltonian Noise Hypothesis holds, and that **B** is similar among areas, the niche model will be the same, whether or not occurrence records from invaded areas are included in model calibration. However, failing these ideal conditions, calibrating models on both invaded and native ranges can inform models more completely, since occurrence records from the invaded range may reveal parts of the fundamental scenopoetic niche not represented on the native range (Beaumont et al. 2009). In contrast, models calibrated using records *only* from the invaded range, or a portion of it, are unlikely to capture the full extent of the scenopoetic existing fundamental niche $\mathbf{E}_A$, and this incomplete characterization must be taken into account in interpreting model results (Anderson et al. 2006). Inclusion of true-absence data in model calibration has appeared to improve predictions in invaded ranges (Vaclavik and Meentemeyer 2009), although the many caveats expressed in chapter 5 regarding absence data must be considered carefully.

Perhaps an ideally comprehensive analysis would incorporate all of these dimensions in a single broad consideration (Jiménez-Valverde et al. 2011). A model calibrated on the native range can outline the general dimensions of the scenopoetic existing fundamental niche of the species. This model can be validated initially within the calibration region via spatially stratified subsetting, and then projected to the various invasive distributional areas; comparisons among modeled niches in those areas allows a clear test of the hypotheses that niche dimensions have not changed (Warren et al. 2008). Once the model is evaluated and the level of niche conservatism in the particular species is understood, final models can incorporate both native-range and invaded-range information, with validation via something akin to k-fold cross validation among these distributional areas (see chapter 9; either with unequal sample sizes in each pool, or by achieving equal sample sizes via rarefaction). Projection of this validated suite of models to areas of particular interest then makes possible detailed and validated inferences regarding the invasive potential of the species.

REVIEW OF APPLICATIONS

Predictions of the invasive potential of species and tests of the ability of ecological niche models to anticipate this potential are now numerous in the scientific literature (e.g., Panetta and Dodd 1987, Honig et al. 1992, Richardson and McMahon 1992, Scott and Panetta 1993, Beerling et al. 1995, Martin 1996, Higgins et al. 1999, Golubov et al. 2001, Hoffmann 2001, Soberón et al. 2001, Welk et al. 2002, Papeş and Peterson 2003, Peterson et al. 2003a and 2003b, Peterson and Robins 2003, Iguchi et al. 2004, Robertson et al. 2004, Fitzpatrick and Weltzin 2005, Hinojosa-Díaz et al. 2005, Thuiller et al. 2005b, Mohamed et al. 2006, Nyári et al. 2006, Benedict et al. 2007, Peterson et al. 2007d). As such, this functionality has gathered steam in application after application, based at least in part on its success so far.

A recent analysis of the global invasion by the tiger mosquito *Aedes albopictus* (Benedict et al. 2007) illustrates the modeling process (figure 13.1). A survey of the scientific literature accumulated occurrences of this species from across its putative native range in Southeast Asia and offshore islands, and an ecological niche model was built using GARP based on that information and climatic datasets. This prediction of a likely native occupied distributional area G_O extended across tropical and subtropical China, west into tropical India, and southward into Indonesia.

Projecting this ecological niche model globally identified a hypothesized G_I (now effectively part of G_O post-invasion), and the authors then overlaid available invaded-range occurrence data from several regions—the example of the United States is shown in figure 13.1. Clearly, the coincidence between the prediction (based only on the ecological niche model built from occurrence patterns in Southeast Asia) and the independent test dataset (i.e., U.S. counties where

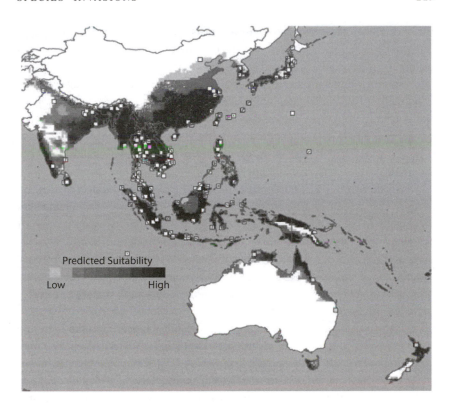

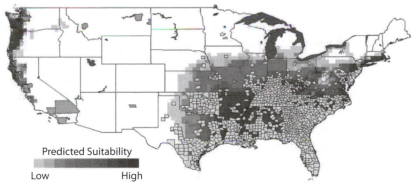

FIGURE 13.1. An analysis of the global invasion by the Asian tiger mosquito *Aedes albopictus* (redrawn from Benedict et al. 2007). Upper panel: White squares indicate known occurrences from the native range, and dark shading of land areas shows the areas predicted as suitable by a niche model calibrated using these sites. Lower panel: Projection of the native-range niche model to North America identified areas matching the species' niche ecologically in North America (i.e., belonging to the potential distributional area G_P; shown as dark shading); known occurrences at the level of counties known to hold populations is shown superimposed.

the species has been detected as invasive) is excellent and much better than random expectations. Particularly attractive about this analysis is that, by focusing on invaded areas outside $\mathbf{G}_P$ (i.e., sink populations), the authors "identified" counties in which the species was detected but did not persist—that is, essentially all of the counties where the species was detected outside of the limits identified in the niche model based on its native range occurrences did not support viable long-term populations. As such, in effect, this species obeys the same ecological "rules" in its invaded range in the United States (and elsewhere) as it does in its native range in Asia.

In another example, Thuiller et al. (2005b) used niche models to identify areas of the world that are potentially susceptible to invasion by 96 plant species native to South Africa. The accuracy of the predictions was assessed for three example species using presence-only records from their non-native distributions in Europe, Australia, and New Zealand. For each of the test species, predictions of suitable areas outside South Africa showed considerable agreement with observed records of invasions, again supporting the notion that niches in native and invaded ranges are similar.

A few ostensibly negative counterexamples have appeared in the literature in recent years. In particular, recent analyses of fire ants (*Solenopsis invicta*; Fitzpatrick et al. 2007) and spotted knapweed (*Centaurea maculosa*; Broennimann et al. 2007) have revealed what was interpreted by the respective authors as lack of predictivity between ecological niche models calibrated on native and invaded distributional areas. In each case, the authors attributed the lack of predictivity to evolutionary change in the species' ecological niche between the two sets of populations.

However, in both cases, methodological and artifactual explanations also exist. For example, the fire ant analysis (Fitzpatrick et al. 2007) was built based on 19 climate dimensions. A subsequent analysis (Peterson and Nakazawa 2008) using a simpler environmental space $\mathbf{E}$ (i.e., one that was less highly dimensional), however, found excellent predictivity between native and invaded ranges of the species (figure 13.2), suggesting an alternative explanation to the original conclusion of evolutionary differentiation. Rather, the picture appears to be one of overfit models that could not "transfer" effectively to the novel distributional area. A situation of this sort can be mitigated by stratifying validations spatially within the native range prior to any exercises involving transferability (see chapter 9).

Another reason why conclusions regarding niche shifts may be artifactual is that of the reduction of the fundamental niche to different existing manifestations in different regions of the world where the structure of climatic space is different. For example, in transferring niches from a native distribution that

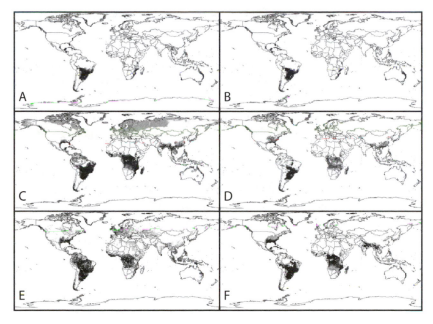

Predicted suitability

low   high

FIGURE 13.2. Analyses of the global potential distribution G_P of the red fire ant (*Solenopsis invicta*), showing the results obtained by analysis with different environmental datasets. Panels A and B show results from models calibrated in 19 climatic dimensions and show considerable underprediction and omission error (Fitzpatrick et al. 2007); however, panels C to F are calibrated in simpler environmental spaces and show broader predictions that accord much better with real-world invasion extents. Redrawn from Peterson and Nakazawa (2008).

covers all of North America to an invaded distribution that is the island of Oahu only, even though the fundamental niche remains unchanged, the portion of that niche that is manifested across North America versus on Oahu is quite different. Techniques such as inertia analysis (Doledec et al. 2000; as employed in Broennimann et al. 2007) will detect these differences as niche shifts, but the attribution of evolutionary change in causing the shifts is premature.

Niche shifts—be they the result of genuine evolutionary change in niche dimensions (i.e., change in E_A) or the result of differences in B between continents (i.e., producing changes in E_{lp})—are certainly possible. However, considerable methodological difficulties remain in characterizing niche shifts, particularly concerning the potential to "overfit" models (see, e.g., Peterson and Nakazawa

2008) or using misleading variables (Rödder et al. 2009), and further exploration of this topic is required. Perhaps the most convincing example is that of marine and freshwater invasions by diverse invertebrates, where physiological and genetic studies have revealed the effects of natural selection on ecological tolerances of invading populations (Lee 2002, Lee et al. 2007), although these situations have not yet been analyzed in terms of coarse-resolution environmental variables such as those generally used in niche modeling applications. As such, while rapid niche shifts are certainly possible (Boman et al. 2008), and indeed are quite fascinating when found, much work remains to establish protocols for documenting this phenomenon appropriately.

A further step in exploring the utility of ecological niche models in understanding species' invasions is to use the predictive power of ecological niche conservatism to anticipate likely changes in the distributional potential of invasive species under scenarios of changing environmental conditions. Roura-Pascual et al. (2005) analyzed likely effects of global climate change on the potential global distribution of the Argentine ant (*Linepithema humile*). Calibrating niche models on the species' native range in South America, the authors then projected that model onto global climate estimates derived from general-circulation model outputs for the year 2055 to produce a map of G_P for the ant at that time. The resulting maps identified areas of possible further invasion over coming decades by this species (figure 13.3).

These approaches can be developed further into what are, in essence, risk analyses associated with invasion of particular regions by species from other regions. For example, the Australian government implemented a broad program of evaluation of invasive potential of Australian invasive plants (Australian Weed Committee 2008), which provides visualizations of the potential distribution of each species. Similar risk assessments have been developed at global scales. For instance, Thuiller et al. (2005b) summed predicted distributions for all 96 South African species that they studied, and produced a global map of invasion risk by species of South African origin. Sites identified as being at high risk may be prioritized for monitoring, and quarantine measures could be put in place to help avoid the establishment of these species. This approach thus in general shows enormous promise for anticipating threats to human and natural systems from invasive species.

CAVEATS AND LIMITATIONS

A U.S. National Academy of Sciences report (NAS 2002) assessed the degree to which species' invasions are "predictable." The study examined questions

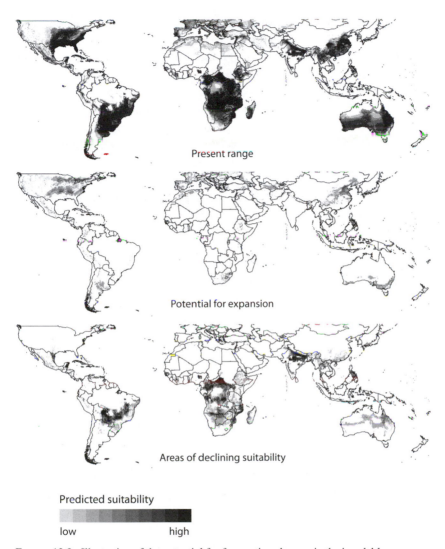

FIGURE 13.3. Illustration of the potential for forecasting changes in the invadable distributional area $\mathbf{G}_I$ through time, with changing climates, showing the example of the invasive Argentine ant *Linepithema humile* (Roura-Pascual et al. 2005), including areas currently suitable and areas of projected potential retraction and expansion.

such as which species are likely to become invasive, when will they arrive, how fast will they spread, and how damaging or dominant they are likely to become. The panel's conclusion was that species' invasions are mostly unpredictable, a conclusion with which we largely concur.

A subsequent review, however, aimed to clarify the situation (Peterson 2003a). It is true that the overall course of species' invasions is extremely difficult to anticipate, and that detailed questions such as those mentioned earlier are unlikely to be answerable. Nonetheless, one feature of species' invasions is frequently predictable—the possible extent of the invasion, or G_I—that is, given the scale distinctions between Grinnellian and Eltonian niches (Soberón 2007), and given the pervasive nature of ecological niche conservatism, at least on relatively short timescales, ecological niche models offer considerable predictivity as to where the species is likely to be distributed *if and when* it arrives and becomes established in the region.

The keys to appropriate application of niche modeling ideas in this realm are those of understanding clearly the assumptions inherent in such applications, and of evaluating model performance rigorously in the native range prior to interpretation of the results. That is, the assumptions of conservatism in niche dimensions and **B** being either very broad or based on comparable interactions among regions must be tested to whatever extent possible in a given situation, and models should be interpreted only after such evaluation has demonstrated the model's predictive ability.

FUTURE DIRECTIONS AND CHALLENGES

A clear priority in this realm of applications of niche modeling is that of identifying and understanding genuine exceptions to ecological niche equivalency between native and introduced ranges of species. The several studies that have asserted niche shifts between native and introduced populations (Broennimann et al. 2007, Fitzpatrick et al. 2007, Medley 2010), as argued earlier, may not be convincing demonstrations of niche shifts, as they do not reject the alternative possibility of methodological artifact (Peterson and Nakazawa 2008). Nonetheless, if examples of evolutionary niche shift can be found, and documented rigorously, careful examination of those cases would be enormously informative. Certainly, clarification of the circumstances under which genuine niche evolution occurs would be interesting, as a counterpoint or corroboration to the theoretical results that suggest that they might evolve most easily in species with certain population characteristics (Holt and Gaines 1992).

A further future challenge is that of transgenic organisms. Such organisms are now commonly developed and used in agricultural applications (Tiedje et al. 1989), and these situations require careful analysis and detailed risk assessment (Regal 1994). Although traditional niche modeling approaches may not be entirely applicable because no prior experience (i.e., occurrence data) with the "species" is available, there are nonetheless possibilities—the idea of modifying niche models of nontransgenic ancestors to mimic the modifications to the genomes may hold some promise. Certainly, though, addressing this challenge will be an important step forward as prototypes are developed.

CHAPTER FOURTEEN

The Geography of Disease Transmission

Zoonotic diseases (i.e., diseases that circulate in the animal world, occasionally affecting humans or other species of interest) are by definition a phenomenon of interactions among species. That is, the pathogen itself is a virus, bacterium, fungus, protozoan, or other small-sized species. Another (usually) larger-bodied species often serves as the reservoir species for that pathogen, holding a long-term pool of pathogen populations in a cycle of transmission and infection. Finally, other species (often arthropods or mollusks) may serve to move the pathogen from one individual of the reservoir species to another, or from the reservoir to humans—these species are termed "vectors."

An example disease transmission cycle, that of plague (*Yersinia pestis*), is shown in figure 14.1. As such, disease applications often differ from other applications of ecological niche modeling in that interactions among species can be very important and complex—that is, the Grinnellian and Eltonian perspectives intermix in this arena very frequently (Peterson 2008a). It is important to note that each of the interacting species likely has a different scenopoetic existing fundamental niche $\mathbf{E}_A$, and differences of scale may be enormous (pathogens may be capable of movements of millimeters, whereas some reservoirs are capable of movements on global scales; Kilpatrick et al. 2006, Peterson et al. 2007a). Thus, disease applications of niche modeling present some very real complications that challenge effective modeling and predicting.

These considerations lead to two different approaches to the challenge. First, it is certainly possible to integrate across the entire transmission system, treating it effectively as a "black box," and simply analyze the ecological and geographic distribution of disease occurrence (in essence modeling the "niche" of the disease occurrence in humans or other species of interest as if it were a species; e.g., Yeshiwondim et al. 2009). This approach subsumes all of the ecological requirements of the individual component species, as well as any ecological biases in their interactions—as such, key details may be lost in the process. However, in some cases, illustrated later, human case locations are the only information that is available, so analysis of the "niche" of the entire

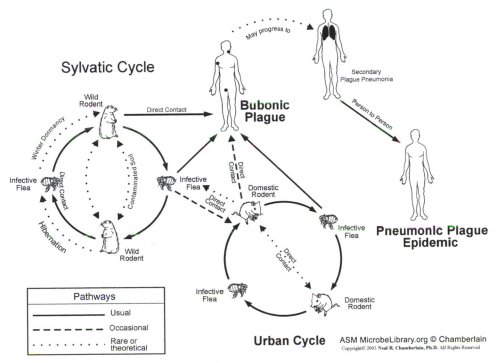

FIGURE 14.1. An example disease transmission cycle, showing how plague (*Yersinia pestis*) is transmitted in North America, and illustrating the involvement of various elements of biodiversity in the transmission. Image courtesy of Neal R. Chamberlain, PhD, A.T. Still University/Kirksville College of Osteopathic Medicine.

transmission system is the only option available (Peterson et al. 2004a, Ron 2005, Reed et al. 2008, Williams et al. 2008).

Perhaps more satisfying, however, is the idea of parsing the overall transmission cycle into the ecological niches of the individual component species. This approach offers the opportunity to distinguish different reasons for presence or absence of disease transmission in an area: transmission may be absent for lack of the pathogen, for lack of an appropriate vector, or for lack of an appropriate reservoir (Peterson 2007a), or because of rarity of any one of them. For example, in situations in which appropriate vectors and reservoirs are in place, introduction of the pathogen can lead to immediate transmission and spread, as was the case with the arrival of West Nile virus in North America in 1999 (Komar 2003). Certainly, these techniques can also be applied to the hypothetical situation of "bioterrorism" in the form of introduction of novel pathogens into a region with the intent to do harm to humans or other species of interest (Bhalla and Warheit 2004).

Applications of ecological niche modeling approaches to the challenge of understanding the geography and ecology of disease transmission are in an early stage. In general, most present applications fall in the category of "black box" analyses (e.g., Peterson et al. 2006a), as the necessary occurrence data are more readily available. A few efforts, however, have treated component species independently (Peterson et al. 2002c, Peterson et al. 2004b), which has potential to offer considerable novel insight into the ecology and geography of the transmission of the diseases in question.

Niche modeling has a lot to offer to the field of public health and epidemiology. Particularly relevant, the field of "spatial epidemiology" or "landscape epidemiology" has emerged in recent years, and a standard suite of tools and approaches has been achieved (Elliott et al. 2000). Typical spatial epidemiological applications include mapping geographic patterns of disease transmission risk, identification of risk factors (spatially or not), and assessment of populations at risk of infection. Tools used for these analyses include spatial regressions, smoothing procedures such as splining and kriging, and more conventional multivariate regressions, all developed chiefly in spatial dimensions (Kitron 1998). As readers of this book will appreciate, these tools do not capture the full complexity of the phenomenon of disease transmission because they are fitted in purely geographic dimensions, and as such distill complex ecological and distributional phenomena into broad spatial trends (Elliott and Wartenberg 2004), with expectedly unsatisfactory results.

Even when spatial epidemiological analyses are conducted more appropriately (i.e., including environmental drivers underlying geographic distributions of disease phenomena in the analyses), tools are generally not used in such a manner that they emphasize generality and full representation of geographic distributions of species and other biological phenomena (Peterson 2007a). A clear illustration is that of analyses of the geographic distribution of *Anopheles gambiae*, an important malaria vector in Africa. Rogers et al. (2002) presented analyses of the distribution of this species that failed to predict its full G_O, particularly in areas not sampled thoroughly; an analysis of the same dataset by Levine et al. (2004), however, painted a more complete picture—Rogers et al. (2002) underestimated the probability of presence of this vector species in areas of sparse sampling, whereas the niche model-based maps from Levine et al. (2004) are better able to predict across regions that have received different levels of sampling effort. Hence, if spatial epidemiology is to capture and communicate the geographic details of disease transmission, the full complexity of drivers of species' distributions must be considered, and tools must be used in such a manner that they capture the full relevant niches of all components of the system that, in turn, can be projected onto the landscape and produce more realistic hypotheses of relevant distributional areas, even in areas that are not sampled thoroughly.

CONNECTION TO THEORY

The possible applications of niche modeling to understanding disease transmission are diverse, covering much of the full gamut of applications of niche modeling in general. Under some circumstances, the focus is on characterizing G_O (i.e., what is the current distribution of transmission of this disease?), whereas under other circumstances, the focus would be on characterizing G_I (e.g., what is the potential for a particular pathogen to spread in this novel region?). Hence, applications of niche modeling to disease questions are very diverse and somewhat difficult to characterize specifically.

What is clear, however, is that disease transmission is by definition a situation in which biotic considerations are more dominant than in most other situations analyzed in this book (Peterson 2008a)—that is, $\mathbf{A} \cap \mathbf{B}$ may be small with respect to $\mathbf{A}$ in disease applications. Indeed, in some cases, $\mathbf{A}$ may prove so broad as to be irrelevant: consider influenza transmission among humans, which appears to be able to occur under extremely diverse environmental (in scenopoetic dimensions) circumstances (Brankston et al. 2007). As such, we could consider each element in a disease transmission system (species $1, 2, \ldots, n$) to have its own particular version of $\mathbf{A}$ (which we can denote $\mathbf{A}_i$). Disease transmission would then occur only within the area delineated by $\mathbf{A}_1 \cap \ldots \cap \mathbf{A}_n$. From the perspective of any single species in the transmission system for the disease, the combined intersections of the $\mathbf{A}$'s for all component species in the system are integrated into a single $\mathbf{B}$, making the simple BAM diagram framework much more complex (Peterson 2008a).

PRACTICAL CONSIDERATIONS

The diversity of questions and challenges in applications of ecological niche modeling to disease-related questions makes for the need for incorporation of diverse methodologies and practical considerations. No comparative studies of modeling approaches have as yet focused on disease systems, so this question remains relatively little explored.

Many disease applications of ecological niche modeling (as in all other applications) are, in the end, severely constrained by the availability and quality of occurrence data. For example, occurrence data that could be obtained for analyses of the geography of potential transmission of Marburg virus across Africa varied from relatively precise (e.g., accurate within 1000 m) to laughably coarse (i.e., somewhere in the country of Zimbabwe; Peterson et al. 2006a), yet the latter occurrence turned out to be critical in understanding the biotic interactions of this virus (Peterson et al. 2004b, Peterson et al. 2007b). In other situations, such as when the focus of modeling is on small-bodied, ephemeral arthropod vector species, *temporal* resolution ends up limiting the analyses that are possible. Such was the case in recent analyses of dengue virus vector distributions in Mexico (Peterson et al. 2005a), in which mosquito species' distributions were dramatically different from month to month over the course of a single year.

A recent commentary attempted to propose a solution to problems with coarse geographic referencing of disease occurrences, suggesting refinement of reporting procedures regarding locations of disease occurrence (Eisen and Eisen 2007). Current geographic referencing procedures for "reportable" diseases in the United States (note that comparable problems exist globally) involve noting simply the state and county (or equivalent tertiary political divisions) of origin (e.g., where the sick person lives), but county extents (1) are often quite coarse spatially, and (2) vary regionally in their resolution (e.g., counties in the western United States are much larger than those in the eastern United States) and environmental heterogeneity—this resolution has proven limiting in past attempts to model such diseases (Nakazawa et al. 2007). Nonetheless, the solution proposed (Eisen and Eisen 2007) was that of shifting from counties to either zip code areas or census tracts, which (although smaller) are similarly polygon-based (i.e., not point-based) and uneven in size of the spatial footprint. A better solution would be to move to a point-based georeferencing approach with uncertainty expressed as an error radius (Wieczorek et al. 2004), which permits flexible use of the occurrence data across diverse spatial resolutions (Peterson 2008b). The result would be a data resource that would be much more supportive of detailed analyses at many spatial extents and resolutions, limited of course by considerations of confidentiality as may be necessary.

REVIEW OF APPLICATIONS

CHARACTERIZING DISEASE ECOLOGY

For many diseases and associated causal pathogens, little is known of the basic environmental factors associated with their transmission. Such diseases can have very few known occurrences, limiting what can be learned from simple geographic intuition. For example, Marburg virus is known only from about seven occurrences since its original appearance in the 1960s in laboratory monkeys imported into Germany. Ecological niche modeling approaches helped to clarify the ecological circumstances under which Marburg has emerged (even based on such minimal sample sizes), to the point that the likely geographic point of origin of the southernmost known occurrence was revised by several hundred kilometers (Peterson et al. 2006a).

In a different context, ecological niche modeling has been used to characterize ecological niches of four related, putatively distinct species of triatomine bugs that are vectors of Chagas disease in Brazil (Costa et al. 2002). Ecological differences in niche were added to the list of phenotypic features of these bug populations that differentiate them, apparently at the level of species (Costa et al. 1997). These distinct populations differ in their capacity as vectors of Chagas disease, so these results have important implications for mitigation measures. Other ecological niche modeling studies have similarly characterized ecological niches of poorly understood diseases (Levine et al. 2007, Reed et al. 2008).

CHARACTERIZING DISEASE DISTRIBUTIONS AND RISK MAPPING

A parallel challenge for ecological niche modeling applications to disease systems is that of characterizing a full distributional picture for poorly understood disease systems. For example, the disease blastomycosis is caused by the endemic (which in the epidemiology literature means locally established for long time periods, as opposed to epidemic—present only for relatively short bouts of time) dimorphic fungus *Blastomyces dermatitidis*, and is known to occur broadly across eastern North America. In spite of its occurrence in relatively prosperous regions, the details of where it occurs—and where transmission to humans can occur—have remained obscure. Figure 14.2 shows the results of application of niche modeling to anticipating the geographic distribution of this disease across Wisconsin (Reed et al. 2008). Other such applications have focused on spatial distribution of transmission of Ebola and Marburg viruses (Peterson et al. 2004a, Peterson et al. 2006a), Chagas disease (Beard et al. 2003, López-Cárdenas et al. 2005), cutaneous leishmaniasis (Peterson and Shaw 2003, Peterson et al. 2004d), visceral leishmaniasis (Thomson et al. 1999), monkeypox (Levine et al. 2007), malaria (Moffett et al. 2007), highly pathogenic avian influenza (Williams et al. 2008), frog fungal diseases (Ron 2005), and even plant diseases (Kluza et al. 2007). A recent analysis of the ecological characteristics of the only two known sites of occurrence for a potential Chagas disease vector, *Triatoma sherlocki*, yielded predictions that led to the successful detection of at least five additional populations of the species (Almeida et al. 2009).

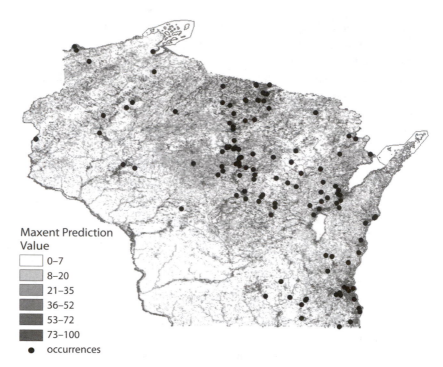

Figure 14.2. Example of prediction of the potential distributional area G_P (grays) for human exposure to the dimorphic fungus *Blastomyces dermatitidis* across Wisconsin, in the United States. Black dots are sites of known human infections. Redrawn from Reed et al. (2008).

Niche modeling applications to such questions require careful validation, so that the models indeed have predictive power when applied to novel landscapes (i.e., are transferable). Of particular importance, given the often-sparse occurrence data available for such applications, is spatial stratification into calibration and evaluation datasets, such that models must predict reliably into areas from which occurrence data are not used in calibration (see chapter 9). Then, niche models and the associated spatial predictions can be evaluated in terms of their predictive ability prior to their potentially being used in any particular public health remediation measure.

True risk-mapping, however, has a more realistic human flavor—that is, while niche models may be used to characterize the geographic distributions (e.g., G_A, G_P, G_I) of a particular pathogen, vector, or reservoir, risk maps must include the additional factors that modify the translation of the presence of those species into human exposure and genuine disease transmission (Lawson et al.

1999). To our knowledge, no disease has seen a complete risk mapping effort based on niche model results.

CHARACTERIZING POTENTIAL DISTRIBUTIONS

An additional suite of applications of niche modeling to disease biology is in understanding potential distributions G_p of pathogens or other individual component species in transmission systems. It is, of course, possible also to apply this line of reasoning to "black box" situations (i.e., the entire transmission system), but the link to environmental conditions becomes increasingly indirect, component species in the transmission system may or may not be present, and predictive ability may consequently be limited. That is to say, although environmental conditions may match closely, disease transmission systems nonetheless can depend critically on the presence or absence of suites of individual species, which likely have distinct **M**s and thereby augment the complexity of the BAM diagram considerably. As such, characterizations of potential distributions of disease transmission systems are ideally based on niche models of individual species, as discussed earlier.

These analyses, hence, are very similar to the numerous examples reviewed in chapter 13, which treats niche modeling applications to invasive species. Among the best such applications to disease-relevant species is the analysis by Benedict et al. (2007) of the invasion by Asian tiger mosquitoes (*Aedes aegypti*) into North America. This work is described in detail in chapter 13.

Recent concerns regarding the potential for intentional introduction of pathogens (Broussard 2001, Cunha 2002, Bhalla and Warheit 2004) further emphasize the importance of such analyses. In effect, many areas may be ecologically suitable for transmission of particular diseases, with competent vectors and hosts already in place, and lacking only the presence of the pathogen for transmission to initiate. For example, in 2003, monkeypox virus was introduced into North America via African small mammals imported for sale as exotic pets (Reed et al. 2004). The virus rapidly "jumped" into native mammals also in captivity for sale as pets, and has every appearance of being able to spread in wild North American mammals. Similar examples of the potential for pathogens to behave as invasive species are provided by the recent outbreaks of West Nile virus and SARS in North America. Niche models can in principle be used to assess such potential for spread of novel pathogens.

FORECASTING CLIMATE-MEDIATED SHIFTS

Another dimension, albeit very little explored, of the geographic ecology of disease transmission is that of how current disease transmission patterns are likely to change in the face of ongoing global climate change. This area may

Human population   *Anopheles gambiae*   *Anopheles arabiensis*

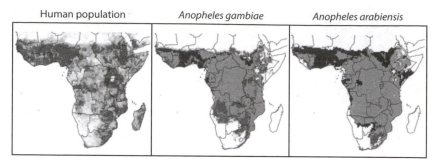

FIGURE 14.3. Summary of projections of likely changes in the occupied distributional areas of the two most important mosquito vectors of malaria across Africa, *Anopheles gambiae* and *A. arabiensis*. The first panel presents human population density, showing clear concentrations in West Africa, as well as parts of eastern and southern Africa. The second and third panels show projected shifts—black areas are areas of reduced future suitability, dark gray areas are areas of increased future suitability, and middle gray areas see little change. Redrawn from Peterson (2009).

prove particularly important given that transmission of many diseases depends on small, ephemeral arthropod vectors that are likely to prove particularly sensitive to changing climatic conditions. Once again, given the complex environment-transmission chain in disease systems, niche modeling approaches may be as useful in "black box" situations. As might be expected, considerable speculation has been offered in the literature as to how these shifts might be manifested (Kovats et al. 2001, Hunter 2003), although only a few actual analyses have been developed to date (Hay et al. 2002, Peterson and Shaw 2003, Van Lieshout et al. 2004, Nakazawa et al. 2007, Peterson 2009, González et al. 2010).

Peterson (2009) analyzed the spatial footprint of distributions of the two most important vectors of malaria across Africa, the species *Anopheles gambiae* and *A. arabiensis*, over scenarios of future climatic conditions. He documented the expected shifts of potential distribution (future $\mathbf{G}_P$) both poleward and upward in elevation (figure 14.3), and then related the areas of likely reduced distribution (future $\mathbf{G}_P$ in comparison with current $\mathbf{G}_O$) and areas of likely invasion to human population distributions (future $\mathbf{G}_I$ in comparison with current $\mathbf{G}_O$). The result was a picture in which 11% to 30% fewer people overall will likely be exposed to malaria in coming decades, but reductions and increases are focused in different regions: malaria exposure is likely to decrease in West Africa but increase in eastern and southern Africa. Such analyses can begin to provide specific quantitative estimates, whereas previous literature was restricted largely to speculation and generality.

DISCOVERING UNKNOWN INTERACTORS

As discussed earlier, disease transmission can be conceptualized as the integration of the various ecological niches and distributions of each of the species participating in the transmission cycle. Under this view, when one of those species is unknown, ecological niche modeling can be used in a more exploratory sense to narrow possibilities and hypothesize which species may be involved.

For example, considering Chagas disease transmission across Mexico, the *Triatoma protracta* species group of true bugs includes several important vectors (Usinger et al. 1966). However, the rodent hosts of several of these bug species remain unidentified or poorly documented. Peterson et al. (2002c) related modeled distributions of triatomine species in this complex to those of their putative hosts—*Neotoma* packrats. They used distributional coincidence among the $\mathbf{G}_P$ for each species estimated from niche models as a measure of niche similarity, and successfully "anticipated" all five already-documented bug-rodent associations. Similar analyses have focused on associations between filoviruses (Ebola and Marburg viruses) and diverse African mammal species (Peterson et al. 2004b, Peterson et al. 2007b).

CAVEATS AND LIMITATIONS

Disease applications of niche modeling are particularly challenging for a number of reasons. We have already mentioned the special features of these situations, in which biotic interactions are probably much more important in driving distributions in space and time than in many other applications of niche modeling. Those considerations demand several methodological adjustments to avoid useless or misleading results.

First, it is important to consider carefully the biases inherent in public health data. Detection biases can be strong, emphasizing rich versus poor regions or countries, or cities versus rural situations. Indeed, even political considerations enter the picture—for example, the People's Republic of China has often been reticent to make public its disease emergences (Breiman et al. 2003), which can bias ecological analyses based on available data. Second, public health data often have only a very coarse spatial resolution (Eisen and Eisen 2007), leading to numerous difficult challenges for attempts to produce predictive models at spatial resolutions sufficiently fine as to be useful (see discussion earlier regarding occurrence data). These complications have led to some unorthodox methodological approaches in niche modeling that nonetheless may be of some broader utility, including using random points as representatives of disease occurrences when occurrences are referenced to polygons or in situations of uncertain georeferencing (Peterson et al. 2006a, Nakazawa et al. 2007).

More relevant is that with such real-world challenges, being *correct* is extremely important. For example, an overfit risk map may give statistically significant evaluation statistics with a randomly split sample approach; yet, such a result, as it may not be fully predictive of the geographic potential of the species, may mean that mitigation efforts are not extending over the full potential distribution of the disease. Effectively, some people might end up going unprotected, as in the example of overfit malaria vector models described earlier (Rogers et al. 2002), or being put at risk unnecessarily, simply because a model was not calibrated or evaluated appropriately.

Still more challenging is the complexity with which diseases themselves interact. For example, Gyapong et al. (2002) produced a prediction (black box style) of the likely distribution of lymphatic filiariasis in West Africa; another paper examined the distribution of the related disease loa loa (Thomson et al. 2000). Getting the relative distributions of these two diseases correct is critical— good prophylaxis exists that protects against filiariasis, but administration of those drugs to a person infected with loa loa causes reactions that are frequently fatal (Thomson et al. 2000). As such, niche model applications must correctly identify areas where filiariasis *is* transmitted but loa loa is *not*—such situations, in which human lives depend on being "right," should give pause to the prospective modeler planning an analysis.

FUTURE DIRECTIONS AND CHALLENGES

Applications of niche modeling techniques and ideas in the public health and disease arena need to progress along at least three lines. First, real-world, practical applications that demonstrate concretely the potentially enormous utility of accurate distributional and ecological information will go a long way toward "selling" the idea of careful modeling based on species-level ecological considerations and biogeographic thinking. These "proofs of concept" speak for themselves—when well executed, disease experts can ponder the insights gained, and respond accordingly.

Second, it is important to compare and contrast the relative advantages and disadvantages of niche modeling approaches as compared with the purely spatial techniques currently in vogue in landscape epidemiology. The former offer two major advantages: (1) simple smoothing (i.e., fitting "models" in **G** only) will miss the fine-resolution complexities that are probably universal in determining the fine points of where species are and are not found to occur, and

(2) the ability to transfer geographically via niche-based insights is completely absent in purely spatial approaches. Making these points clear to the epidemiology and public health communities has begun (e.g., Peterson 2006a), but will still require considerable additional effort and argument.

Finally, the particular interests and needs of the disease world challenge the niche modeling community to develop particular functionalities of specific utility in disease applications. Foremost among these applications are what could be termed "time-specific" ecological niche models that could begin to capture the essence of the temporal dynamics of species' distributions. In this case, occurrence data would be characterized in latitude, longitude, *and* time, and the occurrences would be related to environmental datasets that are similarly specific in time to produce models for a particular point in time. These models could then, in theory, be projected to other time periods to anticipate time-specific dynamics of species' distributions. One initial exploration has already been developed successfully (Peterson et al. 2005a), but considerable additional exploration is needed.

CHAPTER FIFTEEN

Linking Niches with Evolutionary Processes

As methods for modeling and understanding ecological niches and geographic distributions of species have become increasingly robust and well-understood, evolutionary biologists have begun to pay attention. That is because a critical dimension of the evolutionary biology of species is precisely their ecological requirements, as biogeography, distribution, and genetic variation all hinge rather critically on the ecological niche. Evolutionary studies of ecological niches have thus begun to appear in numbers, amplifying the diversity of challenges to which these techniques have been applied.

CHANGES IN THE AVAILABLE ENVIRONMENT

Since the envelope of environmental space available to a species [i.e., environments represented within $\mathbf{M}$ or $\eta(\mathbf{M})$] changes through time, to avoid extinction a species must either track the geographic extent of its scenopoetic existing fundamental niche, or be able to change it via evolutionary responses in physiological or behavioral traits (Holt 1990). One of the important advantages of expressing Grinnellian niches as subsets of an E-space is that the issue of constancy or change of the environmental substrate on which the niches are manifested becomes apparent (Jackson and Overpeck 2000, Ackerly 2003). For example, Jackson and Overpeck (2000) showed changes in what they termed the "Realized Environmental Space," our $\eta(\mathbf{G})$ or $\mathbf{E}$ available in the study region, measured using two extreme temperatures in a $\mathbf{G}$ covering all of North America, from modern times back to 21,000 years before the present (figure 15.1). The actual, existing environmental combinations of a species' niche will shift in spatial location and extent, and a species must track suitable conditions, adapt to suboptimal ones, or go extinct.

The way in which species respond to these challenges is varied. Ackerly (2003) pointed out that the "leading" and "trailing" range edges may pose con-

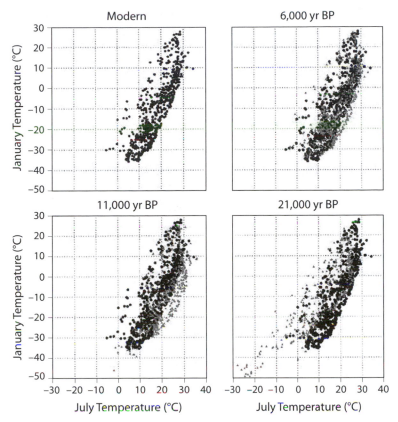

FIGURE 15.1. Illustration of the changing climate conditions in North America at four points in time over the past 21,000 years. The black points in all panels show current-day conditions, while the gray points indicate the conditions at the given earlier period. Redrawn from Jackson and Overpeck (2000).

trasting selective pressures during episodes of change. Imagine the retreat of glaciers in the Northern Hemisphere and the associated northward advance of vegetation. Along the northern edge of the range, populations of these plant species would encounter habitats with few competitors but novel environmental conditions; along the trailing edge, however, populations experience already-known environmental conditions and combinations of species already coadapted but with environmental conditions becoming unsuitable (Ackerly 2003, Brown et al. 2003). The selective pressures are bound to be different in these scenarios. An extremely important question is whether populations can adapt quickly to changes in E-space or whether they must geographically "track" sites with the right conditions. When several correlated environmental variables that are indeed important for the species are considered, tracking them simultaneously may become impossible. Even without extinctions, when the scenopoetic fundamental niches of members of a community of species correlate in different ways with a suite of environmental variables, a change in climate may lead to wholesale rearrangements of species assemblies along gradients (Graham et al. 1996), as illustrated in figure 15.2.

The preceding ideas illustrate the importance of understanding how fast species can adapt to environmental changes in **G**, which can take place over time spans of a few thousand years or even much shorter, over centuries or even decades (Balanyá et al. 2006). This process is that of niche evolution, although niches can evolve for other reasons not related to adaptation (e.g., owing to genetic drift or linkage with other traits under selection). This task would appear to be relatively innocuous: physiological tolerances and habitat associations are clearly features of the evolved phenotype of organisms (Angilletta et al. 2002). However, the concept of niche evolution as a consequence of the evolution of the broader phenotype forms the basis for many key insights from ecological niche modeling, and indeed, since some sort of conservatism in niche features would be required to make possible most of the predictions treated in the last several chapters of this book (Peterson 2006c), the success of those applications appears to provide evidence for conservatism (Peterson et al. 1999). Niche conservatism has many implications (Wiens and Graham 2005), such as the feasibility of forecasting the geography of species' invasions (Peterson 2003a) and effects of climate change on species' occupied and potential distributional areas (Peterson et al. 2005b, Araújo and Rahbek 2006), and for understanding speciation processes (Wiens 2004).

Clearly, though, once we conceive of niches as evolving as part of the overall phenotype of the organism, considerable interest will focus on the circumstances under which niches *have* evolved. That is, if all niches were conserved strictly, then all of life would have the same ecological niche, which is clearly far from the case. Rather, ecological niches of species *do* evolve, and this diversification has been key in structuring life on Earth. Understanding the process of ecological niche evolution and diversification would thus offer key insights into ecology, biogeography, and biodiversity. In this chapter, we offer comments and ideas regarding how the notions of this book can be oriented toward addressing this challenge.

NICHE CONSERVATISM

A body of theoretical ecological work offers a framework in which to consider ecological niche evolution (Brown and Pavlovic 1992, Holt and Gaines 1992,

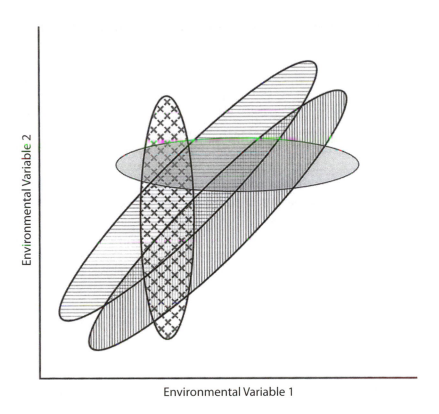

FIGURE 15.2. Illustration of how environmental change can affect species associations. At time 1, existing fundamental niches of species 1 and 2 overlap, so the two can potentially coexist within that intersection. However, at time 2, the existing fundamental niches of the two species do not overlap, so they would not be able to coexist. Redrawn from Jackson and Overpeck (2000).

Kawecki and Stearns 1993, Kawecki 1995, Holt 1996a and 1996b, Holt and Gomulkiewicz 1996, Holt 2003). The essence of these arguments is captured by the following idea: populations outside of the biotically reduced niche (here, $\mathbf{E}_p$) of a species are "sink populations" (Pulliam 1988) that will eventually go extinct without immigration or adaptation (by definition, fitness w is

242

less than unity). Other areas outside of **M** will be uninhabited, but could be colonized by invasive populations (i.e., in $\mathbf{G}_I$).

However, adaptation may take place as well. Using a series of models of rather different scenarios, Holt and Gomulkiewicz (1996) have shown that the adaptive process is, generally speaking, slower than extinction dynamics. Therefore, adaptation seldom rescues sink populations from extinction—in which case, evolution mostly takes place within $\mathbf{N}_F$. These models predict that a combination of large initial populations, small degrees of maladaptation and limited immigration rates from source populations are required for populations to adapt to new niche conditions (Holt 2009, Sexton et al. 2009). The spatial structure of the selective pressures is also predicted as an important factor driving niche evolution (Bell and Gonzalez 2009).

Empirical testing of niche conservatism and broader contemplation of the concept has now begun to fill out the picture considerably: niches act as long-term stable, evolved constraints on species' physiological tolerances and needs, as well as on their geographic distributions (Peterson et al. 1999, Martínez-Meyer et al. 2004a, Nogués-Bravo et al. 2008b). The evidence for conservatism in ecological niche characteristics comes from diverse studies: geographic variation in niche characteristics across species' ranges (Peterson and Holt 2003); comparisons of native and invaded ranges of invasive species (Peterson 2003a); longitudinal (i.e., over time) comparisons of niche characteristics within species (Martínez-Meyer et al. 2004a, Martínez-Meyer and Peterson 2006, Waltari et al. 2007, Nogués-Bravo et al. 2008b); cross-phylogeny comparisons (Ackerly 2003, Martínez-Meyer et al. 2004b, Eaton et al. 2008); and transplant experiments showing that fitness is lower at the margins of distributions (Crozier 2004, Angert and Schemske 2005). These diverse perspectives on niche conservatism are reviewed later.

Obviously, despite their evolution, ecological niches are not *wildly* variable over evolutionary time periods. That is, physiologically challenging environmental realms such as the air and land have been invaded only a relatively few times (Gordon and Olson 1994, Padian and Chiappe 1998, Larson 1982). On the flip side of the coin, however, Eltonian and Grinnellian niches, and major morphological and physiological traits that determine, for example, trophic position or other fundamental ecological adaptations, have obviously changed on *some* scale: marine organisms invaded land at several points in evolutionary history, and terrestrial clades have even invaded back into marine environments (e.g., sea snakes and whales). In this sense, ecological niches are not especially static, and do evolve, just not frequently or wildly (Holt 2009).

Of course, in most ecological niche modeling studies, more subtle ecological changes are where the interest lies—that is, as with the broader panorama

of macroevolution versus microevolution (Wright 1982), what we now perceive as macroevolutionary changes occurred far in the past, deep in evolutionary history, yet through microevolutionary processes (Lande 1986). As such,
microevolutionary changes occurring in the recent past are much more tan-
gible, and certainly far more accessible to study and analysis, especially given
the poor and uneven fossil record available for most groups. Microevolutionary
ecological niche change, which we can define for the purposes of this discussion as relatively minor changes in ecological niche parameters that permit the
occupation of new environmental situations and new distributional areas, are
the focus of the remainder of this chapter.

TESTS OF CONSERVATISM

WITHIN DISTRIBUTIONAL AREAS

The simplest and most frequently applied test of niche conservatism is that of
testing whether niche characteristics are constant across species' geographic
distributions. This idea quite simply builds niche models based on one sector
of the occupied distributional area of a species and tests whether those niche
models are predictive with respect to distributions in other sectors (i.e., equivalent to spatially stratified validation; Peterson and Holt 2003). Although the
first formal presentation of the approach as a test of conservatism (Peterson
and Holt 2003) showed examples of nonconservatism, the vast bulk of the examples examined to date have shown conservatism—that is, different sectors
of species' distributions can reliably be predicted based on the remainder of
the species' distributions (Peterson 2001, Peterson 2005a), although counterexamples exist (Raxworthy et al. 2008). A note of caution is that these tests
must consider carefully the range of environments over which the model is
calibrated, to assure that it is representative of the areas onto which the model
is projected (see the discussion of transferability and extrapolation in chapter
7), because if not, niche comparisons may not be valid (see chapters 7 and 9;
Kambhampati and Peterson 2007, Peterson and Nakazawa 2008).

SPECIES' INVASIONS

Invasive species offer an additional level of complexity to testing ecological
niche conservatism. Here, in addition to the element of spatial diversity provided by testing across different sectors of species' geographic distributions,
introduced ranges generally also present distinct biotic environments. In the
context of the BAM diagram, if native- and introduced-range occupied niches
$\mathbf{E}_O$ are highly similar, then either $\mathbf{A} \subset \mathbf{B}$ (the Eltonian Noise Hypothesis) or at

least features of **B** correlate closely with the variables used to define **A**. In such situations, a model based only on scenopoetic variables can recover a meaningful and consistent biological signal about **A** ∩ **B** across very diverse regions (Peterson 2003a). Many such tests have been developed, and most with positive results—that is, native-range ecological characteristics generally have excellent predictive power regarding invaded-range distributions (Richardson and McMahon 1992, Martin 1996, Peterson and Vieglais 2001, Peterson et al. 2003a, Peterson and Robins 2003, Iguchi et al. 2004, Hinojosa-Díaz et al. 2005, Roura-Pascual et al. 2005, Thuiller et al. 2005b, Nyári et al. 2006, Zambrano et al. 2006, Benedict et al. 2007, Peterson et al. 2007d). Apparent exceptions to this predictive nature of species' invasions are treated in chapter 13.

Chapter 13 presents invasive species applications of ecological niche modeling in greater detail. The coincidence of the spatial extent of predictions of G_I with the areas actually invaded by the species is impressive. Results of these studies suggest that (1) niches are frequently conserved across species' invasions in ecological time, and (2) biotic interactions do not shift dramatically among distributional areas to the extent that predictivity is negated. (We note, however, that the species that do invade may be a nonrandom selection from the overall pool of possible invaders, though this possibility will require creative exploration to resolve.) Some recent studies ostensibly documenting negative results in this realm are discussed in chapter 13 and in the following.

SINGLE LINEAGES THROUGH TIME

Perhaps the most direct tests of niche conservatism available are situations in which models can be developed and tested "before and after" some period of time in which change could be manifested. Here, the same dimensions of the ecological niche are estimated and compared in the same region, so many of the caveats of other tests are avoided. The drawback, however, is that opportunities for such tests are relatively rare, so only a few such studies have been developed to date.

Martínez-Meyer and Peterson (2006) developed so-called longitudinal tests for eight plant taxa discernable at least to genus from pollen samples across North America. Ecological niche models were developed for the present-day distributions of each taxon and tested using occurrence data from within 3000 years of the Last Glacial Maximum (21,000 years ago), and vice versa. Of the total of 16 reciprocal tests conducted, all models predicted the independent evaluation data from the other time period better than would be expected at random. In general (see figure 15.3), the cross-time predictions matched distributional expectations quite closely, suggesting that these plant species were tracking a highly conserved ecological niche closely over the past 21,000 years. Other such "before and after" studies include an analysis of mammal distributions in the present and in the Pleistocene (Martínez-Meyer et al. 2004a), and detailed analysis of the extinction of Woolly Mammoths (*Mammuthus primigenius*) from Europe at the end of the Pleistocene (Nogués-Bravo et al. 2008b).

A particularly intriguing study was that which failed to predict the historical distribution of the Spotted Hyena (*Crocuta crocuta*; Varela et al. 2009). In that analysis, the current distribution of the species was accurately predicted, implying that the species is in equilibrium with its environment, and also that climatic variables used for modeling were adequate; however, projecting the niche model back to the Last Interglacial period (126,000 years ago) failed to predict the spatial distribution of records from across western Eurasia, where extensive fossil records are known for that period, suggesting that its current occupied distributional area represents a subset of its scenopoetic fundamental niche.

Nogués-Bravo (2009) reviewed published studies in which niche models were projected to past scenarios. An important point to make is that such tests of conservatism are unidirectional. In other words, one can find support for the hypothesis of conservatism by finding that present and past distributions are environmentally coincident, but failure to demonstrate such overlap does not provide support for the alternative hypothesis of no conservatism. This asymmetric nature of the test results because several ecological and methodological reasons can be invoked to explain *non*overlap between occupied niches in time (Peterson 2011).

SISTER-SPECIES COMPARISONS

The limitation of the longitudinal approach, of course, is that few situations lend themselves readily to such testing, particularly over longer periods of time, primarily due to the paucity of past occurrence records. As a consequence, Peterson et al. (1999) explored the possibility of building such tests over evolutionary time, comparing the ecological niche requirements of sister-species pairs, in essence asking the question of whether ecological niche characteristics had been conserved over a time period of twice the time since the advent of allopatric conditions for sets of previously contiguous conspecific populations. They assessed 37 sister-species pairs of birds, mammals, and butterflies that are distributed on either side of the Isthmus of Tehuantepec in southern Mexico—in all, 74 reciprocal comparisons, testing whether the occupied niche characteristics of one of the species ($\mathbf{E}_O$) were able to predict the geographic distribution ($\mathbf{G}_O$) of its sister-species better than random expectations. Parallel analyses not of sister-species, but rather of confamilial species for each of

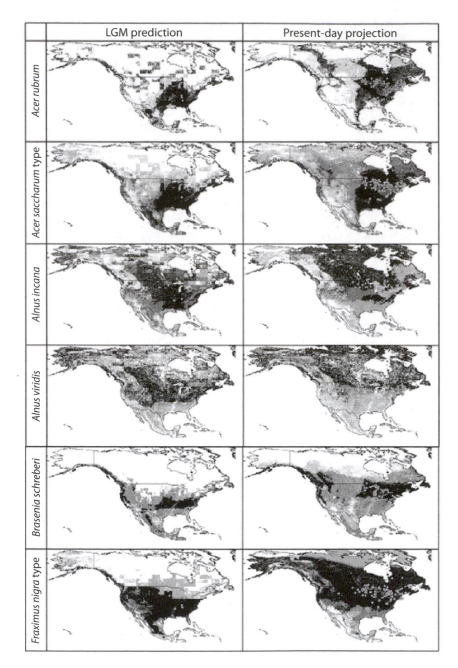

the focal species, showed very low levels of predictive ability, suggesting that ecological niche conservatism is strong between sister-species, but breaks down over longer timescales. Reanalyses of these results (Warren et al. 2008) confirmed that the niches modeled for these species pairs indeed are generally more similar than would be expected at random, although they are basically never identical.

TRACING NICHE CHARACTERISTICS OVER PHYLOGENY

A still more general approach to the question of ecological niche conservatism is to extend analysis back over phylogenies more complex than simple 2-taxon statements (i.e., sister-species pairs). Later, we will treat some more complex questions that can be addressed within this framework. However, for the moment, we focus on the question of niche conservatism, and how it can be studied using phylogenies.

A study that illustrates such approaches is an analysis of ecological niche diversity across the Icteridae (figure 15.4; Eaton et al. 2008)—the American blackbirds—a diverse clade for which detailed phylogenies are increasingly available (Lanyon 1994, Johnson and Lanyon 1999, Lanyon and Omland 1999, Omland and Lanyon 2000, Price and Lanyon 2002 and 2004). In this analysis, ecological niches and potential distributions (G_p) were estimated for each of the >100 species of blackbirds, centroids of species' niches were calculated and compared in environmental space, and species' occupied and potential distributional areas were characterized in geographic space. Results showed that ecological niches were dramatically differentiated only between relatively distantly related species (see figure 15.4B), but that convergent evolution can make distant relatives appear rather similar in environmental space (see figure 15.4D). Most interesting, perhaps, is the point that very close relatives are invariably only slightly differentiated in ecological niche characteristics. Other such studies, in which niche characteristics (expressed in binary format) are mapped onto phylogenetic trees, are beginning to appear in greater numbers (Prinzing et al. 2001, Graham et al. 2004b, Hoffmann 2005, Knouft et al. 2006), although these studies are divided in their conclusions regarding the

---

FIGURE 15.3. Ecological niche models derived from relating Last Glacial Maximum (LGM) detections of pollen of eight tree species to general circulation model reconstructions of LGM climatic parameters. Shown are the LGM reconstructed distribution (potential distributional area G_p; left-hand column) and the projection to present-day climate conditions (G_p; right-hand column). Darker shading indicates greater estimated suitability for the species; independent occurrence data for model evaluation are shown for the present-day projections. Adapted from Martínez-Meyer and Peterson (2006).

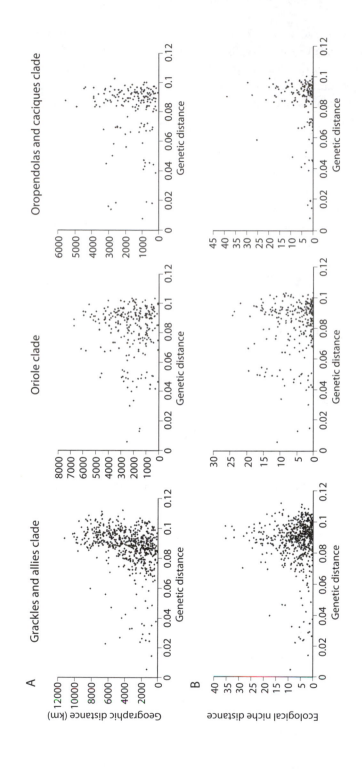

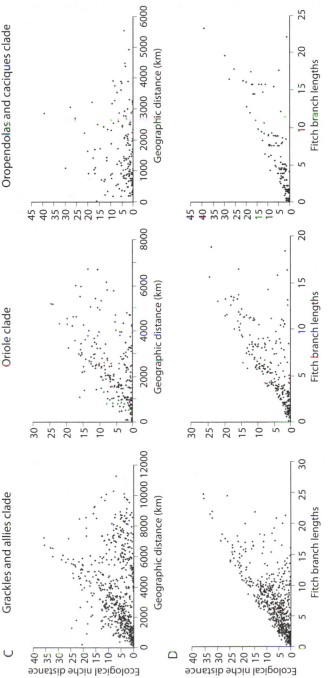

FIGURE 15.4. Illustration of phylogenetic approaches to studying ecological niche divergence and nondivergence in New World blackbird lineages. This figure shows plots of all pairwise species comparisons within each of three blackbird lineages (vertical columns), comparing (A) genetic distance versus geographic distance, (B) genetic distance versus ecological niche distance, (C) geographic distance versus ecological niche distance, and (D) phylogenetic (patristic; Fitch branch lengths) distance versus ecological niche distance. Redrawn from Eaton et al. (2008).

generality of niche conservatism and must confront serious methodological challenges regarding how best to reconstruct ancestral character states of non-binary ecological characteristics. Warren et al. (2008) presented novel randomization tools that will prove useful in such studies.

CONTEXT

The question of evolutionary conservatism of ecological niches of species is in many senses independent of the conceptual framework laid out in the introductory chapters of this book. That is, our conceptual framework is described as two associated spaces—geographic and environmental—at a single point in time. A common assumption that will certainly bear closer examination is that among-population variation in inherited niche characteristics is negligible. This chapter attempts to lay out the panorama of ecological niche change *through time*, which represents yet a third dimension to the question. We now address how temporal change in niche dimensions occurs, how it can be studied, and what can be learned.

The ecological niches of greatest interest initially in this chapter are clearly Grinnellian in nature (see chapter 2)—that is, at a first level, we are most likely to be intrigued by how species' requirements in scenopoetic niche dimensions either change or remain static. As such, we have focused on coarse-resolution, largely climatic scenopoetic environmental dimensions, and have generally neglected the Eltonian, interactive dimensions that may depend critically on other species. Nonetheless, clearly, the evolution of interactions between species is also of potentially great interest—how have the diverse interactions among elements of biodiversity come to be? For example, how did Mallophaga (feather lice) colonize bird feathers, and how did that association become so obligatory over time? Such questions can—in time and with thought—be addressed within these frameworks as well.

LEARNING MORE ABOUT ECOLOGICAL NICHE EVOLUTION

A recent review argued that the question of conservatism of ecological niche characteristics is not particularly interesting (since we *know* that niches evolve), but rather that the important and interesting questions regard how often, how much, and under what circumstances they evolve (Wiens and Graham 2005). We agree. That is not to say that niche conservatism is not *important*: if and only if ecological niches are relatively conserved in a particular lineage can many of the predictive approaches outlined in this book be informative. If, on the other hand, the ecological niche of a species were to vary wildly through time and across space, then even the simple idea of predicting an occupied distributional area would not be likely to succeed, so the conservatism question is extremely relevant. As mentioned earlier, a contribution by Warren et al. (2008) presents a useful clarification of the null hypotheses being tested in a diversity of studies of ecological niche conservatism versus evolution, and may be able to reconcile the different points of view that have been presented in the literature. Nonetheless, *any* evolved character will tend to be more similar among close relatives than among distant relatives. Wiens and colleagues (Wiens 2004, Wiens and Graham 2005, Kozak and Wiens 2006) have correctly pointed out that while "conservatism" sounds like no change, ecological niche conservatism can be the agent of distributional constraint and therefore isolation, divergence, and even speciation.

More interesting, however, is the diversity of questions regarding evolutionary biology of ecological niches that can be addressed using phylogenetic frameworks. Given information regarding the present-day diversity of niche characteristics among species in a clade, understanding the ecology of ancestral forms becomes feasible by means of phylogenetic methods that allow reconstruction of ancestral character states (Cunningham et al. 1998, Martins 2000, Pagel et al. 2004), although these methods are not without uncertainties. In particular, reconstructions of continuous character states (e.g., preferences with respect to temperature and rainfall) have been complicated by assumptions of averaging of ancestral character states, and generally are quite imprecise (Garland et al. 1999).

Graham et al. (2004b) applied these approaches to understanding speciation in dendrobatid frogs in Ecuador. They used both maximum likelihood and least-squares approaches to estimate ancestral niche dimensions as maximum and minimum values for each climatic variable. The result was a detailed view of ecological niche (putatively $\mathbf{E}_A$, but probably between $\mathbf{E}_A$ and $\mathbf{E}_O$) shifts between ancestors and present-day forms (figure 15.5), which permitted reconstruction of ancestral distributional areas (albeit based on present-day climate conditions). No other studies have taken this general approach to understanding ecological niche evolution, to our knowledge.

Another interesting line of inquiry is that of viewing ecological niche characteristics on a phylogenetic framework to detect lineages along which ecological niches have changed dramatically or have remained constant. An example is an analysis of niches across the Neotropical manakins (Pipridae) undertaken by estimating amounts of ecological niche change in comparison to branch lengths on an independent phylogenetic framework (Anciães and Peterson 2009).

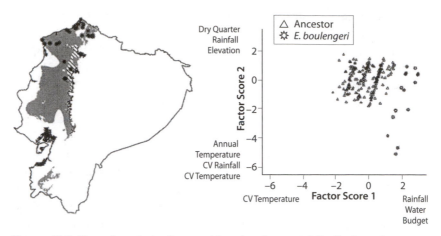

FIGURE 15.5. Example analysis of geographic and environmental distributions of species at present and in the past. From Graham et al. (2004b), this figure shows analyses of the dendrobatid frog lineage that includes *Epipedobates boulengeri*, *E.* sp., *E. tricolor*, *Colostethus machalilla*, and *E. anthonyi*. The left-hand panel shows the modeled potential distributional area $\mathbf{G}_P$ for *E. boulengeri* and for the ancestor of *E. anthonyi*, *C. machalilla*, *E. tricolor*, and *E.* sp., where gray shows the distribution of *E. boulengeri* and black shows the ancestor. (Areas with diagonal shading are areas of overlap between past and present potential distributional areas.) The right-hand panel shows the distribution of *E. boulengeri* and the ancestral taxon in a principal components manipulated E-space.

This study found that niches had been generally conservative, but identified a few lineages (see, e.g., *Chiroxiphia boliviana* in figure 15.6) in which niches have changed dramatically. Such studies have the potential to reconstruct patterns of ecological innovation and to permit insight into when and under what circumstances ecological characteristics do change, and many more are now being developed and published (Martínez-Meyer et al. 2004b, Eaton et al. 2008).

The recent development of paleoclimatic reconstructions and readily available digital data layers describing environmental variables in the past opens doors to new understandings of the geography of speciation and the paleogeography of species. In particular, once ecological niche conservatism has been tested and established in a particular lineage over a particular span of time, it becomes possible to reconstruct past potential distributional patterns, such as

FIGURE 15.6. Analysis of niche characteristics across the Neotropical manakins (Pipridae), depicted as branch lengths summarizing amounts of ecological niche change on an independent phylogenetic framework. Adapted from Anciães and Peterson (2009).

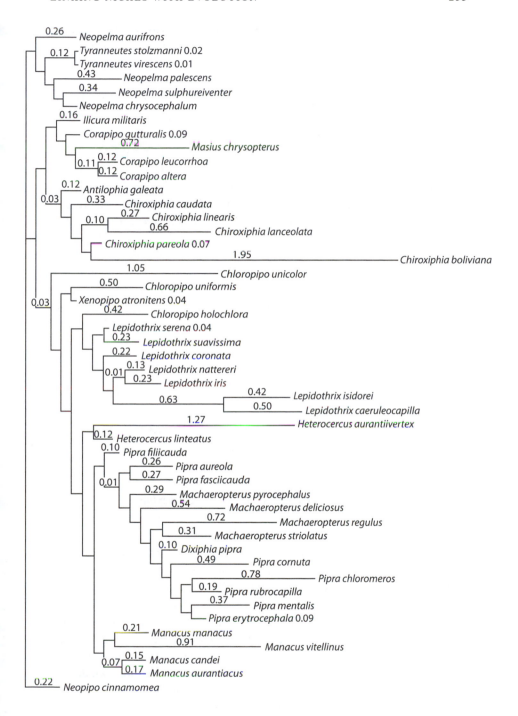

0.26 Neopelma aurifrons
0.12 Tyranneutes stolzmanni 0.02
Tyranneutes virescens 0.01
0.43 Neopelma palescens
0.34 Neopelma sulphureiventer
Neopelma chrysocephalum
0.16 Ilicura militaris
Corapipo gutturalis 0.09
0.72 Masius chrysopterus
0.11 0.12 Corapipo leucorrhoa
0.12 Corapipo altera
0.12 Antilophia galeata
0.03 0.33 Chiroxiphia caudata
0.10 0.27 Chiroxiphia linearis
0.66 Chiroxiphia lanceolata
Chiroxiphia pareola 0.07
1.95 Chiroxiphia boliviana
1.05 Chloropipo unicolor
0.50 Chloropipo uniformis
0.03 Xenopipo atronitens 0.04
0.42 Chloropipo holochlora
Lepidothrix serena 0.04
0.23 Lepidothrix suavissima
0.22 Lepidothrix coronata
0.01 0.13 Lepidothrix nattereri
0.23 Lepidothrix iris
0.42 Lepidothrix isidorei
0.63 0.50 Lepidothrix caeruleocapilla
1.27 Heterocercus aurantiivertex
0.12 Heterocercus linteatus
0.10 Pipra filiicauda
0.26 Pipra aureola
0.27 Pipra fasciicauda
0.01 0.29 Machaeropterus pyrocephalus
0.54 Machaeropterus deliciosus
0.72 Machaeropterus regulus
0.31 Machaeropterus striolatus
0.10 Dixiphia pipra
0.49 Pipra cornuta
0.78 Pipra chloromeros
0.19 Pipra rubrocapilla
0.37 Pipra mentalis
Pipra erytrocephala 0.09
0.21 Manacus manacus
0.91 Manacus vitellinus
0.07 0.15 Manacus candei
0.17 Manacus aurantiacus
0.22 Neopipo cinnamomea

Pleistocene refugia or dispersal corridors (Waltari et al. 2007). For example, Peterson and Nyári (2007) used ecological niche models projected onto climatic information from the Last Glacial Maximum to reconstruct putative Pleistocene refugia for the Thrush-like Mourner (*Schiffornis turdina*) across the Neotropics. They tested the degree to which these patterns of connectivity and isolation correspond to present-day genetic differentiation and found a close correspondence—in other words, Pleistocene connectivity estimated using ecological niche modeling techniques proved very informative regarding present-day genetic breaks in this highly structured species. These hypotheses, which are independent of the usual suites of molecular data that have traditionally been applied to such questions (i.e., the field of phylogeography), can in turn be used as hypotheses to be tested with those other datasets (Carstens and Richards 2007, Knowles et al. 2007, Strasburg et al. 2007, Jakob et al. 2009).

In the last few years, paleodistributions generated with niche modeling techniques have been combined with phylogeographic methods to test hypotheses regarding patterns of population-genetic structure of different taxa (Carstens and Richards 2007, Knowles et al. 2007). In this new approach, niche models based on current distribution data are "hindcasted" to identify potential refugia and likely routes of dispersal, which are then contrasted with population-genetic signals of demographic processes (e.g., bottleneck effects; Jakob et al. 2007, Buckley et al. 2010). Combination of these two independent lines of evidence has proven useful to understanding the roles of geography and ecological constraints on species in responses to climate change (Cordellier and Pfenninger 2009). Although only in initial stages, this fusion of fields promises fruitful future insights (Hickerson et al. 2010).

FUTURE DIRECTIONS AND CHALLENGES

This realm of applications of ecological niche modeling is only beginning to open and be explored rigorously. Several important questions remain for analysis. A potentially fruitful realm concerns the question of how ecological niches interact with processes of evolution, geography, and environmental history to produce biological diversification, including how the available environmental space itself may change through time. An important area of inquiry not yet explored in detail is how dynamic ecological landscapes (i.e., changing available environmental spaces) may constrain the geographic and evolutionary possibilities of species (Jackson and Overpeck 2000). These questions would certainly rank among some of the most fundamental and fascinating in systematics and evolutionary biology, so exciting insights lie ahead.

A second suite of "next steps" looks back to the question of ecological niche conservatism. Here, as we have reviewed earlier, a growing mass of evidence points to the generality of conservatism of ecological traits over short-to-moderate periods of evolutionary time (e.g., Peterson et al. 1999), particularly in the context of species' invasions (Peterson 2003a). We point to exceptions to the conservatism "norm" as being some of the most intriguing and exciting situations encountered, and consider that more research is required to assert the generality of these results, and most interestingly, to explore under what circumstances ecological niche change *does* occur.

Finally, we offer the contrast between Grinnellian and Eltonian niches as an additional fertile field of exploration. Do scenopoetic dimensions of species' ecological niches show greater evolutionary conservatism than linked, interactive dimensions? The existence of many long-term ecological associations that show impressive constancy over evolutionary time periods (DiMichele et al. 2004, McGill et al. 2005), coupled with growing evidence about the fast pace of local adaptations in Eltonian variables (Thompson 2005), makes these questions relevant and interesting.

CHAPTER SIXTEEN

Conclusions

This book is the result of years of discussion, debate, and exploration among seven authors, each of whom has had a distinct trajectory of research efforts that have led to an interest in species' niches and distributions. As a consequence, this book represents a consensus, and sometimes détente, of diverse viewpoints and approaches. What we hope we have achieved, nonetheless, is a step toward a comprehensive framework for thinking about the geography and ecology of where species are and are not distributed.

Our central idea in this book, particularly looking back over the years of discussion and development that it has required of us, is that a firm conceptual framework is critical to further progress in this field. Indeed, in some senses, we believe that an appropriate conceptual framework will prove more important than choosing the best and most accurate modeling algorithms—rather, we suspect that more inaccuracy is introduced into results from incorrect assumptions and nonrepresentative samples. We hope that this book offers a synthesis that provides a firm conceptual foundation for varied work in the field of niches and distributions.

A source of consternation to some to whom we have presented these ideas is our rather radical reworking of traditional concepts in ecology. In reality, we attempted whenever possible not to abandon the classical framework of the ideas of Hutchinson and MacArthur, but several key points were not treated sufficiently by them. In those cases, it has been necessary to revisit concepts, rework nomenclature, and add concepts to clarify. We suspect that the added detail that we have perceived is perceptible chiefly because new data and software tools exist that the previous generations of ecologists and biogeographers did not have available. Regardless, we have attempted to link our new concepts to the preceding suite of concepts, such that the lineage of thinking and discussion is not lost.

Clearly, however, much more work is needed in this field. The preceding pages that constitute this book are replete with unfinished thinking, incomplete development, and challenges for future work. Such is, we believe, the situation when a field of great promise is nonetheless only a decade or two old. Major challenges that we perceive include (1) full integration of the BAM framework with central concepts of population biology and statistical theory; (2) greater methodological clarity regarding model evaluation in relation to the specific quantitites (niches and distributions) being estimated; (3) better clarity in thinking regarding niche conservatism versus evolution as regards scenopoetic versus bionomic environmental dimensions; and (4) much-improved linkage between correlational and mechanistic approaches to estimating and understanding ecological niches. Each of these realms represents a major suite of challenges that requires integration of careful conceptual thinking with detailed empirical exploration—we hope that this book will open doors to such concrete advances.

More generally, we perceive that the syntheses that we present in this book had *not* been achieved previously owing to the tendency toward reductionism in organismal biology. Ecologists, biogeographers, and evolutionary biologists were not "talking to" one another sufficiently in recent decades, and as a consequence did not explore these areas of overlapping interest in sufficient detail. Ecological niche modeling offers an exciting suite of novel tools that have already proven to be of interest across disciplines—the further growth and maturation of this field will require conceptual linkages that can come only from integrated thinking across scales ranging from the ecological to the biogeographic. Such advances will, we hope, be increasingly feasible as a common language and conceptual framework are presented (as we have attempted to do) and adopted as a platform for cross-disciplinary discussions, debates, and syntheses.

APPENDICES

Glossary of Symbols Used

This section summarizes the mathematical notation used throughout this book.

$|A|$	The cardinality of set A.
A^C	The complement of set A.
$\hat{A}$	An estimate of a set A.
$\hat{f}$	An estimate of a function f.
$\hat{Y}$	An estimate or prediction of a variable Y.
$E\{X\}$	Expected value of random variable X.
$P(A)$	Probability of event A.
$P(A|B)$	Conditional probability of event A given B.

The following list summarizes specific symbols used to express concepts in this book.

a, b, c, d	Entries of confusion matrix.
$a_{i,l}$	Per-unit time probability of finding resource l.
$\mathbf{A}$	Set of cells in geographic space where intrinsic growth rates are positive (abiotically suitable).
$b(g)$	Probability of a collector visiting cell g, or sampling bias if heterogeneous.
$b_{i,g}$	Bionomic parameter.
B	Number of bootstrap samples in bootstrapping scheme.
$\mathbf{B}$	Set of cells in $\mathbf{G}$ where biotic conditions are favorable for presence of a species.
$\hat{c}(\mathbf{X})$	A continuous output estimate of function f based on the result of an algorithm $\mu(\mathbf{G}_{data}, \mathbf{E})$.
$c_{i,g}$	Mean field biotic interaction parameter
d	Number of excluded (deleted) datapoints in jackknife scheme.
D	Binary random variable to denote detection of species by collector.
$D(g)$	Conditional probability of detection of species by collector.

$d_i(\vec{e}_g)$ — Environmentally determined death rate of species i at environment $\vec{e}_g$.

$\vec{e}_g = (e_1, e_2, \dots, e_v)_g$ — Vector of v environmental variables at cell g. A generic element of $\mathbf{E}$.

E-space — Multidimensional space of scenopoetic variables. Mathematically, it is $\mathbf{E}$.

$\mathbf{E}$ — Environmental space of scenopoetic environmental variables. Colloquially, it is referred to as "E-space."

$\mathbf{E}'$ — A generic subset of $\mathbf{E}$ obtained from the mapping function η.

$\mathbf{E}_A$ — Scenopoetic existing fundamental niche, defined by $\eta(\mathbf{G}_A)$.

$\mathbf{E}_I$ — Invadable niche space, defined by $\eta(\mathbf{G}_I)$.

$\mathbf{E}_O$ — Occupied niche, defined by $\eta(\mathbf{G}_O)$.

$\mathbf{E}_P$ — Biotically reduced niche, defined by $\eta(\mathbf{G}_P)$.

E_{val} — An estimate of expected loss, or average validation error, prediction error, or testing error.

E_{ver} — An estimate of verification error, or calibration error, or training error.

E_{val}^K — An estimate of expected loss, using K-fold cross validation.

$E_{\text{val}}^{\text{boot}}$ — An estimate of expected loss, using bootstrap samples.

η — Function that maps geography into environment.

$\eta(A)$ — For $A \subseteq \mathbf{G}$, the direct image of set A, or the set $\{\eta(g) | g \in A\}$.

$\eta^{-1}(A)$ — For $A \subseteq \mathbf{E}$, the inverse image of set A, or the set $\{g \in \mathbf{G} | \eta(g) \in A\}$.

$f(\mathbf{X}, \mathbf{Z})$ — Nature's response mechanism relating variables $\mathbf{X}$, $\mathbf{Y}$ to response $\mathbf{Y}$.

$f(\mathbf{X})$ — Idealized approximation of nature's response mechanism relating variables $\mathbf{X}$ to response $\mathbf{Y}$, disregarding effects of $\mathbf{Z}$.

$\hat{f}(\mathbf{X})$ — A binary estimate of function f based on the result of an algorithm $\mu(\mathbf{G}_{\text{data}}, \mathbf{E})$.

$\hat{f}_u$ — Threshold-dependent binary model obtained by thresholding at value u.

$\hat{f}_{-i}$ — Model calibrated by setting aside ith observation.

$\hat{f}^{*i}(\mathbf{X})$ — Bootstrap replicates of fitted models.

g — A generic element of $\mathbf{G}$ (i.e., a single grid cell).

G-space — A set area of geography. Mathematically, it is $\mathbf{G}$.

G	Geographic space composed of cells or pixels, generally two-dimensional. Colloquially, it is "G-space."
G′	A generic spatial subset of space **G**.
G$_A$	The abiotically suitable area, defined by $\eta^{-1}(\mathbf{E}_A)$.
G$_B$	Biotically suitable area; generally referred to simply as **B**.
G$_M$	Accessible area, based on the species' present and historical movements; generally referred to simply as **M**.
G$_I$	The invadable distributional area, defined as $\mathbf{A} \cap \mathbf{B} \cap \mathbf{M}^C$.
G$_O$	The occupied distributional area, defined as $\mathbf{A} \cap \mathbf{B} \cap \mathbf{M}$.
G$_P$	The potential distributional area, defined as $\mathbf{G}_O \cup \mathbf{G}_I$.
G$_{\text{data}}$	Data; set of observations (presences, and, if existing, true absences).
G$_+$	Occurrence data documenting presences of species.
Δ**G**$_O$	Change in occupied distributional area before and after some change event.
h	Number of cells that compose **G**, or $|\mathbf{G}|$.
I	Binary random variable to denote species' access to a site.
J	Binary random variable to denote abiotic suitability.
K	Binary random variable to denote biotic suitability.
K	Number of equal-sized pools for K-fold cross-validation.
k	Number of subsets in data-splitting scheme.
k	Negative binomial parameter in abundance estimation.
$k(i)$	Index notation for K-fold cross-validation.
$L(\mathbf{Y}, \hat{\mathbf{Y}})$	Loss function for quantifying error committed when predicting **Y** with an estimate $\hat{\mathbf{Y}}$.
m	Number of data points used for validating a model.
M	Movement set of geographic cells that have been accessible to a species within a given time span.
$\mu(\mathbf{G}_{\text{data}}, \mathbf{E})$	Result of applying an algorithm to data given an environmental space **E** to estimate subsets of **G**.
n	Number of data points used for calibrating a model.
$\hat{N}$	Estimate of total number of individuals.
N	Available niche space, or $\eta(\mathbf{G}) \subseteq \mathbf{E}$ (used in appendix B only).
N$_F$	Scenopoetic fundamental niche.

$\hat{\mathbf{N}}$	Niche estimated by whatever method.
O	Binary random variable to denote observation of species by collector.
p	Number of spatially structured partitions.
$P_A(g)$	Conditional probability of abiotic suitability at site g.
$P_B(g)$	Conditional probability of biotic suitability at site g.
$P_M(g)$	Conditional probability of species' access at site g.
$P_{AB}(g)$	Product of $P_A(g)$ and $P_B(g)$.
$P_{BAM}(g)$	Product of $P_A(g)$, $P_M(g)$, and $P_B(g)$.
$\varphi_{i,g}(\vec{e}_{i,g}, \vec{R}_{i,g}; \vec{x}_g)$	Regulation term in phenomenological equations.
$\psi(\mathbf{M}_i; \vec{x}_i)$	Movement function in phenomenological equations.
q	Weight in loss function for omission-commission.
r	Generic growth rate of a species.
$\bar{r}$	Long-run growth rate of a species.
r_g	Intrinsic growth rate of a generic species in cell g.
$r_{i,g}(\vec{e}_g, \vec{R}_{i,g})$	Intrinsic growth rate of species i in cell g as a function of the scenopoetic variables ($\vec{e}_g$) and the relevant resource-consumer parameters ($\vec{R}_{i,g}$).
$\vec{R}_{i,g}$	Equation parameters related to interactions with other species.
R_l^*	Level of resource l at equilibrium.
s	The number of interacting species that inhabit $\mathbf{G}$.
S	An arbitrary species, as in Hutchinson (1957).
t	Generic time.
$\mathbf{T}_i$	Transition matrix expressing instantaneous probabilities of intercell movements for species i.
u	Threshold of occurrence used for converting continuous or ordinal output to binary output.
v	Number of environmental variables defined over $\mathbf{G}$, or the number of dimensions of the niche hypervolume.
V_i	Span of i-th environmental variable (used only in appendix B).
w	Fitness.
$w_{i,l}$	Conversion parameters to transform resource-encounter rates to units of population growth rate.
x	Longitude.
x_g	Density of a generic species in cell g.
$x_{i,g}$	Density of species i in cell g.
$\vec{x}_i$	Vector of population densities of all i species in every cell at a given time.

X	Variable to denote random visit by observer to a site.
$\mathbf{X}_i$	Observed values of environmental variables interpreted as values of $\vec{e}_g$.
$\mathbf{X}_i^*$	Observed values of environmental variables different from observations used for calibrating a model.
$\mathbf{X}$	Generic observable explanatory variables.
y	Latitude.
$\mathbf{Y}$	Generic response variable issued by nature.
$\mathbf{Y}_g$	Specific response variable issued by nature at location g.
$\hat{\mathbf{Y}}_i$	Predicted values of $\mathbf{Y}_i$ for observed data, equivalent to $\hat{f}(\mathbf{X}_i)$.
$\hat{\mathbf{Y}}_i^*$	Predicted values of $\mathbf{Y}_i$ for additional data for model evaluation, equivalent to $\hat{f}(\mathbf{X}_i^*)$.
$\mathbf{Z}$	Generic nonobservable explanatory variables.

Set Theory for G- and E-Space

In this appendix, we present some set theoretical operations that are mathematically valid for subsets of G-space and E-space.

1. $\mathbf{E} = e_1 \times e_2 \times \ldots e_m$, the environmental space, is the Cartesian product of all the sets of possible values of the m environmental variables.
2. $\mathbf{G}$ is the set of all cells (defined by their coordinates) existing in the selected region of the world.
3. The set $\eta(\mathbf{G}) = \mathbf{N} \subseteq \mathbf{E}$ is the available niche space, i.e., the subset of $\mathbf{E}$ that actually exists in the region of the world under consideration (Jackson and Overpeck 2000).
4. The difference between the sets $\mathbf{A}$ and $\mathbf{B}$ is defined as $\mathbf{A} - \mathbf{B} = \mathbf{A} \cap \mathbf{B}^C$

The following is a list of valid operations in E-space.

$\eta(\mathbf{G}_1 \cup \mathbf{G}_2) = \eta(\mathbf{G}_1) \cup \eta(\mathbf{G}_2)$	The environment of a union of areas is the union of the environments of each separate area.
$\eta(\mathbf{G}_1 \cap \mathbf{G}_2) \subseteq \eta(\mathbf{G}_1) \cap \eta(\mathbf{G}_2)$	The environment of an intersection of areas is contained in the intersection of the environments of the separate areas.
$\eta(\mathbf{G}_1 - \mathbf{G}_2) \supseteq \eta(\mathbf{G}_1) - \eta(\mathbf{G}_2) = \eta(\mathbf{G}_1) \cap \eta(\mathbf{G}_2)^C$	The environment of the difference of two areas is contained in the intersection of the environment of the first with the complement of the environment of the second.
$\eta(\mathbf{G}_1^C) \supseteq \mathbf{N} - \eta(\mathbf{G}_1) = \mathbf{N} \cap \eta(\mathbf{G}_1)^C$	The environment of the complement of an area is contained in the intersection of the niche space with the complement of the environment of the area.
$\mathbf{G}_1 \subseteq \mathbf{G}_2 \Rightarrow \eta(\mathbf{G}_1) \subseteq \eta(\mathbf{G}_2)$	If an area is contained in another, its environment is contained in the environment of the other.

If no repeated elements are present in $\mathbf{E}$, then the preceding symbols for containing become equalities.

The following is a list of valid G-space operations.

$\eta^{-1}(\mathbf{N}) = \mathbf{G}$	The area corresponding to the entire niche space is the region in consideration.
$\eta^{-1}(\mathbf{E}_1 \cup \mathbf{E}_2) = \eta^{-1}(\mathbf{E}_1) \cup \eta^{-1}(\mathbf{E}_2)$	The area corresponding to the union of two environmental sets is equal to the union of the areas of the two environmental sets, taken separately.
$\eta^{-1}(\mathbf{E}_1 \cap \mathbf{E}_2) = \eta^{-1}(\mathbf{E}_1) \cap \eta^{-1}(\mathbf{E}_2)$	The area corresponding to the intersection of two environmental sets equals the intersection of the areas of the two environmental sets, taken separately.
$\eta^{-1}(\mathbf{N} - \mathbf{E}_1) = \mathbf{G} - \eta^{-1}(\mathbf{E}_1)$	The area corresponding to the complement (with respect to the available niche space) equals the geographic region minus the area corresponding to the environmental subset in question.

These operations are mathematically valid. However, their biological interpretation contains several subtleties, as is discussed in the corresponding chapters. In addition, we stress that in general although both the operations for obtaining the environments of sets of geographic cells $\eta(\mathbf{G}') = \mathbf{E}'$ and the geographic cells corresponding to sets of environmental vectors $\eta^{-1}(\mathbf{E}') = \mathbf{G}'$ can be implemented in a GIS, and they are in a sense inverse operations, one cannot simply assume that $\eta^{-1}[\eta(\mathbf{G}')] = \eta^{-1}(\mathbf{E}') = \mathbf{G}'$. For this equality to be valid, a one-to-one correspondence must exist between environment vectors and geographic cells. If different geographic cells present the same environmental conditions, then the equality may not be valid.

Glossary

A

Abiotic niche—The set of environments in which abiotic conditions are favorable for the species. In practice, largely equivalent to the *scenopoetic niche*. Contrast with *biotically reduced niche*.

Abiotically suitable area—The geographic region where, in the absence of competitors and other negatively interacting species, and given unlimited dispersal abilities, the abiotic environment is favorable for the species. Contrast with *occupied distributional area* and *potential distributional area*.

Absence data (see also *presence/absence data*)—Datasets containing "records" of places where sampling has occurred but the species has not been documented. Contrast with *presence-only data*.

Algorithm—A specific sequence of instructions for solving a problem or developing a task. Examples of algorithms used to model species niches and distributions include BIOCLIM, desktop GARP, Maxent, and so on. Contrast with *model*.

Ancillary data—As used here, additional factors not included in the modeling algorithms and that influence distributions of species beyond the simple effects of *scenopoetic variables* and *bionomic variables*. Examples include geographic barriers and historical events. Ancillary data normally are used for postprocessing results of an algorithm. See *abiotically suitable area* and *occupied distributional area*; contrast with *metadata*.

Apparent commission error—A kind of *commission error* that is not real, but rather derives from misinformative *evaluation data* (see also *absence data*), inappropriate selection of the *study region* for evaluation, or both. See *nonequilibrium distributions*.

Area under the curve (= AUC)—A statistic generated from a *receiver operating characteristic plot* (ROC), the area under the curve (AUC) represents an overall measure of *model performance* across all *thresholds* and strengths of a prediction. AUC is a nonparametric measure that ranges 0 –1 (random expectation of AUC is 0.5), and summarizes the model's ability to rank presence records higher than absence records (or higher than a sample from the background, in the case of *presence–background data*); it does not evaluate the model's *goodness-of-fit*. See *model evaluation*.

Artifactual absence—A situation when a species is not truly absent, but rather the lack of a record is an artifact of inadequate or nonexistent sampling. See *absence data*.

Asymmetric loss—A loss function that combines the *omission error* and *commission error*, but not with equal weight. Contrast with *symmetric loss*.

AUC—See *area under the curve*.

B

Background data — Information on environmental variation across the *study area* (the "background"), whether or not sampling has occurred or whether or not the species of interest has been found there. See also *presence/pseudoabsence data.*

BAM diagram — A Venn diagram that displays the joint fulfillment in geographic space (*G-space*) of three sets of conditions that together determine a species' distribution: *B*, for biotic conditions; *A*, for abiotic conditions; and *M*, for movement of the species.

Bias — See *sampling bias.*

Binomial test — A test employed when each independent result is one of two possible outcomes. In the current context, it is often applied to determine whether *evaluation data* fall into regions of a binary geographic prediction (usually after applying a *threshold* to a continuous or an ordinal output) more often than expected by chance, constituting a one-tailed test of *model significance.* See *model evaluation.*

Biogeographic regions — Portions of *G-space* delimited by patterns of spatial coincidence in the ranges (*occupied distributional area*) of large numbers of species. Biogeographic regions are usually related to current and/or past dispersal limitations, and to some degree environmental characteristics. Note that this concept differs markedly from that of a biome, which depends almost entirely on environmental characteristics and does not directly include contingent effects of history.

Bionomic variables — Variables that are dynamically linked to the occurrence of a species, such as competitors, prey, predators. Contrast with *scenopoetic variables,* noticing that the distinction is not absolute.

Biotic interactions — Interactions between and among species — for example, competition, mutualism, predation. See *BAM diagram.*

Biotically reduced niche — The set of environments in which the abiotic environment is favorable for the species and in which negative interactors are not capable of excluding the populations of the species of interest. See *potential distributional area;* contrast with *scenopoetic niche.*

Biotope — Equivalent to *G-space,* defined here as the geographic space composed of cells or pixels covering a particular region. See *extent, grain,* and *study region.*

C

Calibration — See *model calibration.*

Calibration data — The *primary occurrence data* used to calibrate the model. See *model calibration;* contrast with *evaluation data.*

Commission error — A measure of *model performance* based on the *confusion matrix.* As a rate that ranges from 0 to 1, it indicates the proportion of negative *evaluation data* (localities of known or assumed absence for the species; see *absence data*) that fall in pixels of predicted presence for the species (typically after applying a *threshold* to a continuous or an ordinal prediction); it is equivalent to the *false positive rate* and equals 1 minus *specificity.* See *apparent commission error;* contrast with *omission error.*

Confusion matrix — A matrix relating rows summarizing predicted presence and absence of a species (via a binary prediction) to columns indicating the true (or assumed) status (from occurrence records of the species, as well as absence, pseudoabsence, or background data). The four cells of the matrix thus indicate distinct combinations of prediction versus reality and are commonly used to calculate the *omission error* rate and *commission error* rate (including both true and *apparent commission error*).

D

Data partitioning—See *data splitting*.

Data splitting—A partitioning of occurrence data (typically in some random manner). Can be used to refer to splits internal to *model calibration*; however, some authors use it to refer to random division of *primary occurrence data* into *calibration data* and *evaluation data*.

Detectability—A measure of the degree to which members of the species are apparent to observers using a given set of techniques, quantified as a random reduction in detection probability from unity to something below that in any site across the *occupied distributional area*. (Note, however, possible heterogeneities in detectability across geography.)

Direct image—The result of applying a function to a set of elements in its domain; herein, the function η maps cells in geographic space to their corresponding environments. If G' is a set of cells in geographic space, its direct image $E' = η(G')$ is the corresponding set of environments occurring in G'.

Direct variables—Variables that affect organisms physiologically but that are not consumed by them; equivalent to Hutchinson's *scenopoetic variables* (note, not synonymous with *proximal variables*).

Dispersal limitation—A factor that impedes dispersal (movement) by individuals of a species. See *BAM diagram*.

Distal variables—Variables to which an organism responds only via multiple causal links (not synonymous with *indirect variables*).

Distributional area—A portion of *G-space*, typically that inhabited by the species. See other important distributional areas, such as the *abiotically suitable area, potential distributional area*, and *occupied distributional area*.

Distributional equilibrium—The situation in which a species inhabits the full spatial extent of its *abiotically suitable area*. See *BAM diagram*; contrast with *nonequilibrium distribution*.

E

E-space (= *environmental space*)—A multidimensional space **E** of *scenopoetic variables*. See *environmental data;* contrast with *G-space*.

Ecological niche modeling—Estimation of the different niches (fundamental, existing, potential, occupied), particularly those defined using scenopoetic conditions. In practice, carried out via estimation of abiotically suitable conditions from observations of the presence of a species; such models can be used to estimate different distributional areas (the *abiotically suitable area, potential distributional area*, and *occupied distributional area*) by stating assumptions about factors in **B** and **M**, the latter area being the goal of *species distribution modeling*. See *BAM diagram, E-space*, and *G-space*.

Eltonian niche (= *functional niche*)—A niche concept oriented toward community-ecology questions, defined at small spatial extents at which experimental manipulations are feasible, emphasizing the functional role of species in communities, and including models of resource consumption and impacts. Contrast with *Grinnellian niche*.

Eltonian Noise Hypothesis—The hypothesis that $\mathbf{A} \approx \mathbf{B}$ (see *BAM diagram*)—that is, that the cells of *G-space* that are abiotically suitable for a species generally are biotically suitable as well. This hypothesis is not true when important biotic interactions (e.g., competition, predation, mutalism) are spatially distributed in large, homogeneous regions across areas that are abiotically suitable for the species.

Ensemble prediction—A consensus prediction of a niche or a distributional area made by combining results of different methods, alternative parameterizations of the same method, or multiple iterations of stochastic methods, to generate a composite value of *suitability*.

Environmental data—Values for environmental variables (generally *scenopoetic variables*) used in *ecological niche modeling*. Typically, these variables must be a coincident raster grid for the *study region* employed in *model calibration*. See *E-space*.

Environmental envelope—Simple methods for *ecological niche modeling* by identifying shapes in multidimensional *E-space* that enclose environments associated with known presence data (e.g., BIOCLIM). Note, however, that this term has been applied to other, more complex, techniques in some literature.

Environmental niche (see also *Grinnellian niche*)—A term used in some literature to refer essentially to what we call Grinnellian niches—that is, niches based on a space of mostly abiotic, nonlinked variables.

Environmental space—See *E-space*.

Evaluation—See *model evaluation*.

Evaluation data (see also *model evaluation*)—Occurrence data used to evaluate *model performance* and/or *model significance*. Evaluation data can be *presence-only data*, as well as often *presence/absence data*, *presence/background data*, or *presence/pseudoabsence data*. Contrast with *calibration data*.

Existing environmental space—The *E-space* that is represented across the *study region* at the time in question.

Expected loss—The average *loss function* over typical values of *evaluation data*; for example, under zero-one loss, the expected loss is simply the probability that prediction and observation disagree. See *asymmetric loss* and *symmetric loss*.

Extent (= *spatial extent*)—The size and placement of the study region in geographic space (*G-space*). See *scale*; contrast with *grain*.

Extrapolation—Prediction into environmental values beyond the range (in *E-space*) of the area on which the model was calibrated (common when a model is applied to cross-time or cross-space situations; see also the related but distinct concept of *transferability*). Contrast with *interpolation*.

F

False negative rate (= *omission error rate*)—The rate at which a model incorrectly predicts absence—that is, the proportion of data for the species falling outside the area predicted present for the species; ranges from 0 to 1 and equals 1 minus *sensitivity*. Contrast with *false positive rate*.

False positive rate (= *commission error rate*)—The rate at which a model incorrectly predicts presence—that is, the proportion of absence data for the species falling in areas predicted present for the species; ranges from 0 to 1 and equals 1 minus *specificity*. Contrast with *false negative rate*.

Functional niche—The meaning of niche associated with the Eltonian interpretation, stressing the impact (as opposed to the requirements) of a species on its habitat, resources, and interacting species, mostly at local scales. See *Eltonian niche*.

Fundamental niche—The set of all environmental states that permit a species to exist. Contrast with *realized niche*. Herein, we distinguish Eltonian fundamental niches from Grinnellian fundamental niches. The latter is the set of scenopoetic (noninteracting and nonlinked) conditions that the species can tolerate. The former is defined by Chase and Leibold (2003) in terms of interacting variables.

G

G-space—A set area of geography (**G**). See *study region*; contrast with *E-space*.

Geographic distribution—See *occupied distributional area*.

Geographic niche—See *Grinnellian niche*.

Geographic space—See *G-space*.

Georeferencing—Here, coordinates of latitude and longitude (or another system) indicating the position of a point in space. High-quality georeferenced *primary occurrence data* also should include an estimate of *uncertainty*, as well as *metadata* documenting the record and the source/method of georeferencing.

Goodness-of-fit—A parametric test evaluating *model performance* and sometimes also *model significance* via an assessment of how well model output matches the likelihood of the species' presence, for example via calculation of a Pearson product-moment correlation between the model prediction and evaluation data, typically with *presence/absence data* (but sometimes *presence/background data* or *presence/pseudoabsence data*). Contrast with *area under the curve*.

Grain (= *resolution*, = *spatial resolution*)—The size of the cells, sometimes called pixels, of the raster grid in the *study region* in geographic space (*G-space*). See *scale*; contrast with *extent*.

Grinnellian niche—Niche concepts defined on the basis of environmental space of scenopoetic (noninteracting and nonlinked) environmental variables and focused on geographic scales and requirements. Contrast with *Eltonian niche*.

H

Hutchinsonian Duality—The linked nature of the niche and distribution in *E-space* and *G-space*.

I

Idealized variables—A category of environmental variables focusing on the degree to which they have direct physiological effects on organisms.

Impacts—One of the two basic elements in niche definitions, impacts are the effects of a species on the elements of its environment (e.g., resources, habitat structure, population densities of interactors). See *Eltonian niche*; contrast with *requirements*.

Indirect variables—Variables with no causal physiological effects on individuals but that have a correlation with species' occurrences because of correlations with other factors. Contrast with *direct variables*.

Interpolation—Prediction between known values of an independent variable. Herein, for example, between environmental values within the range (in *E-space*) of *occurrence data* with which the model was calibrated. See also *spatial interpolation*; contrast with *extrapolation*.

Invadable distributional area—The additional area that a species *could* occupy if present distributional constraints were to be overcome. See *BAM diagram*; contrast with *occupied distributional area* and *potential distributional area*.

Invadable niche space—The subset of *E-space* corresponding to the elements of *G-space* that the species could occupy if distributional constraints were to be overcome. See *invadable distributional area*.

J

Jackknife—A statistical approach employing repeated rounds of subsampling from a dataset with systematic deletion of a set number of observations. In *ecological niche modeling*, it has been applied to *primary occurrence data* (when few records are available) and to *environmental data* (to determine the contribution of various variables). A general formula for jackknifes is $n - d$, where n is the overall number of elements available, and d is the number deleted in each iteration. To date, $n - 1$ jackknives have been most common.

K

K-fold cross-validation—An approach to *data splitting* in which *primary occurrence data* are split into k roughly equal-sized subsets ($k > 2$), and each subset is held out successively (for *model evaluation*) while the other $k - 1$ parts are used for *model calibration*. Note that this approach does not fulfill the conditions of true *model validation*.

L

Loss function—An evaluation criterion for quantifying error (discrepancy between prediction and observation); no loss is incurred in cases in which prediction and observation agree. See also *uncertainty*.

M

Metadata — As used here, additional data describing details of the format and source of data — for example, for *primary occurrence data* and *environmental data*. Contrast with *ancillary data*.

Misclassification rate — A measure of model performance derived from the *confusion matrix* that combines *omission error* and *commission error* into a single index.

Model — A simplified representation of some aspects of nature for the purpose of research. Contrast with *algorithm*.

Model calibration (= *calibration*, = *training*) — The step or steps involved in forming a model, here one that estimates a species' niche based on *primary occurrence data* and values of environmental variables.

Model evaluation (= *evaluation*, = *testing*) — Refers collectively to the diversity of testing situations: use of *evaluation data* that are not independent of the *calibration data* (see *model verification*), semi-independent, or fully independent (see *model validation*). Such evaluation quantifies how successful the model is in predicting observations used for evaluation.

Model performance (= *performance*) — Characterization of how well or poorly a model achieves a particular goal related to prediction, including quantifications of *omission error* and/or *commission error*, but not necessarily including statistical assessment of *model significance*. Measures of model performance may be *threshold-dependent* or *threshold-independent*, and either parametric (e.g., for *goodness-of-fit*) or nonparametric (e.g., ranking in the *area under the curve*).

Model prediction — The output of *ecological niche modeling*, generally depicted in *G-space*, and often after processing — for example, after applying a *threshold* to convert a continuous or an ordinal prediction to a binary one.

Model significance (= *significance*, = *statistical significance*) — Determination via statistical tests whether predictions of *evaluation data* differ from a random null hypothesis with a particular level of probabilistic confidence. Often based on some measure of *model performance*, tests of model significance typically assess whether the model predicts evaluation data better than random expectations (one-tailed hypothesis).

Model testing — See *testing* and *model evaluation*.

Model training — See *training* and *model calibration*.

Model validation (= *validation*) — A term referring to one kind of *model evaluation*, specifically with *evaluation data* that are fully independent of the *calibration data*. Contrast with *model verification*.

Model verification (= *verification*) — A term referring to one kind of *model evaluation*, specifically with *evaluation data* that are not independent of the *calibration data*. Contrast with *model validation*.

Modeling algorithm — See *algorithm*.

Modifiable Areal Unit Problem — The dilemma that changing the *grain* (= *spatial resolution*) and/or the location of a raster grid can lead to different values of spatial statistics — for example, changing the grid may lead to different estimates of the niche of a species.

N

Niche concept—One of several related and often confused concepts used to address the environmental requirements (biotic and abiotic) that need to be fulfilled for a population to survive, together with the impacts the population has on these environmental factors. Different schools place the emphasis on different aspects of the preceding idea, leading to a variety of definitions and concepts. See *Eltonian niche* and *Grinnellian niche*.

Niche conservatism—The propensity of species to maintain inherited niche characteristics over evolutionary time, e.g., with closely related species showing similar niches. See also *niche identity* and *niche similarity*; contrast with *niche evolution*.

Niche evolution—The process of evolutionary change (by adaptation or other processes) in the inherited niche characteristics of a species over time. Contrast with *niche conservatism*.

Niche identity—One of two null hypotheses regarding *niche conservatism* versus evolution in which the idea is tested that two niches are identical (Warren et al. 2008). Contrast with *niche similarity*.

Niche similarity—The other of two null hypotheses regarding *niche conservatism* versus evolution in which the idea is tested that two niches are more similar to one another than would be expected at random (Warren et al. 2008). Contrast with *niche identity*.

Nonequilibrium distribution—The pattern present when a species inhabits less than its full *abiotically suitable area*, owing to dispersal limitations and/or biotic interactions.

O

Observational data—See *primary occurrence data*.

Occupied distributional area (= *geographic distribution*, = *range*)—The subset of regions accessible to a species in which both abiotic and biotic conditions are favorable for it to maintain populations, and to which it has been able to disperse.

Occupied niche space—The subset of *E-space* that the species inhabits; it is equivalent to the set of environments in the *occupied distributional area*.

Occurrence data—See *primary occurrence data*.

Omission error—A measure of model of performance. As a rate, the proportion of grid cells of known occurrence of the species that are predicted absent by the model. Typically calculated based on *evaluation data*, but also may be relevant for *calibration data*.

Overfitting—The situation when model complexity becomes excessive and a model shows close fit to *calibration data* but is less able to predict independent or even semi-independent *evaluation data*. Note that overfitting can be to noise and/or to *sampling bias*.

Overprediction—Predicting too broad an area in *G-space* for a species. See *commission error*, but note the difference between real and *apparent commission error*.

P

Performance—See *model performance.*

Postprocessing—The step or steps involved in taking into account factors that cause *nonequilibrium distributions.* Specifically, *ecological niche modeling* results initially representing species' *abiotically suitable areas* can be processed post hoc, leading to closer approximations of the species' *occupied distributional area.* See *ancillary data.*

Potential distributional area—The union of the *occupied distributional area* and *invadable distributional area* for a species—that is, the regions where the abiotic and biotic conditions are suitable. (Note, however, that much literature uses potential distribution in a different way, as a synonym of what we term the *abiotically suitable area.*) See *biotically reduced niche.*

Potential geographic distribution—See *potential distributional area.*

Potential niche—The intersection of a fundamental scenopoetic niche with the set of environments actually existing at a given time. The name was proposed by Jackson and Overpeck (2000); since it is confusing, we use *scenopoetic existing fundamental niche* as a more effective synonym.

Presence data—See *presence-only data.*

Presence/absence data (see also *absence data*)—Datasets containing records of where a species has been observed to be present, as well as sites where it is absent, or assumed to be, despite sampling efforts (but note that the species may actually inhabit these latter sites, if sampling is present but inadequate). Contrast with *presence/background data, presence-only data,* and *presence/pseudoabsence data.*

Presence/background data—Datasets containing records of where a species has been observed to be present, as well as information regarding environmental variation across the study area (the "background"), whether or not sampling has occurred there (and if so, whether or not records of the species exist from those regions). Contrast with *presence/absence data, presence-only data,* and *presence/pseudoabsence data.*

Presence-only data—Datasets containing records of where a species has been observed to be present, but lacking any information regarding sites where it is absent. Contrast with *presence/absence data, presence/background data,* and *presence/pseudoabsence data.*

Presence/pseudoabsence data—Datasets containing records of where a species has been observed to be present, as well as sites where it has not been observed (but note that the species may actually inhabit these latter sites, which can lack records of it due to nonexistent or inadequate sampling). Contrast with *presence/absence data, presence/background data,* and *presence-only data.*

Primary biodiversity data—See *primary occurrence data.*

Primary occurrence data (= *occurrence data*)—Records of species' presence (and sometimes absence), especially *voucher specimens* in natural history museums and herbaria, but also including observational records from visual observations and auditory records (e.g., of birds, amphibians, bats). Contrast with *secondary data.*

Proximal variables—Variables to which organisms respond directly (note, not synonymous with *direct variables*).

Pseudoabsence data—See *presence/pseudoabsence data.*

R

Range — See *occupied distributional area.*

Realized niche — The set of all environmental states that would permit a species to exist in the presence of competitors or other negatively interacting species and restrictive factors. Contrast with *fundamental niche.*

Receiver operating characteristic plot (= ROC) — A plot of two measures derived from the *confusion matrix*: *sensitivity* and *specificity*—sensitivity on the y axis versus 1 – specificity on the x axis. The *area under the curve* (AUC) of the ROC plot is commonly used as a *threshold-independent evaluation* of *model performance*. See *commission error* and *omission error.*

Requirements — The set of environmental factors without which a population cannot have positive growth rates.

Resolution — See *grain, temporal resolution.*

Resource variables — Variables consumed by organisms; equivalent to Hutchinson's *bionomic variables.*

Response curves — The relationship between the species' occurrence (dependent variable) and individual environmental variables (independent variable), describing probabilities of presence for a species across the range of values of an environmental variable.

Road bias — The pattern in which most biological sampling takes place near roadways. See *sampling bias.*

ROC — See *receiver operating characteristic plot.*

S

Sampling bias — Variation in the probability that a site will be sampled by biologists. Often, such bias corresponds to accessibility (in *G-space*) and often also leads to sampling bias in *E-space.*

Sampling effort — Generally, the strength or intensity of sampling by biologists, but may also include consideration of the suite of techniques employed.

Scale — The broad and more or less nebulous concept that includes both *extent* and *grain.*

Scenopoetic fundamental niche — The combination of scenopoetic variables that permits a species to have positive intrinsic growth rates, in the absence of competitors. See *abiotic niche, fundamental niche, Grinnellian niche.*

Scenopoetic niche — A niche in an *E-space* of scenopoetic variables. See *abiotic niche.*

Scenopoetic variables — Variables that are not consumed or affected by individuals of a species. See *E-space* and *scenopoetic niche*; contrast with *bionomic variables.*

Secondary data (= secondary occurrence data) — Summaries or syntheses of *primary occurrence data*, typically via subjective processes and at a coarse resolution (e.g., range maps).

Secondary occurrence data — See *secondary data.*

Sensitivity — The absence of *omission error*. See *confusion matrix.*

Significance — See *model significance.*

Sink populations—Populations in which population growth is insufficient to maintain populations without immigration. Contrast with *source populations*.

Source populations—Populations in which population growth is sufficient to maintain populations without immigration. Contrast with *sink populations*.

Spatial autocorrelation—The pattern of entities being nonindependent in space—for example, in clusters of occurrences of a species. Spatial autocorrelation also occurs for environmental variables. Spatially autocorrelated occurrence data do not represent independent samples for *model calibration* or *model evaluation*.

Spatial extent—See *extent*.

Spatial interpolation—The process of estimating values of a variable merely by consideration of places nearby in *G-space*. See also *interpolation*.

Spatial resolution—See *grain*.

Spatial transferability—See *transferability*.

Spatially structured data partitioning—See *spatially structured data splitting*.

Spatially structured data splitting (= *spatially structured data partitioning*)—A kind of *data splitting* where occurrence data are divided not randomly (as is typical) but rather spatially; hence, *calibration data* and *evaluation data* are more likely to be fully independent of each other. See *sampling bias* and *overfitting*.

Species distribution modeling—Application of niche theory to questions about real spatial distributions of species, typically in the present—specifically, via estimation of the *occupied distributional area* from occurrence information for a species via its relationship to environmental characteristics and their correlations with *dispersal limitation* and *biotic interactions*. Contrast with *ecological niche modeling*, which aims to estimate the *abiotically suitable area*, from which other distributions (e.g., the *potential distributional area* and *occupied distributional area*) can be derived.

Specificity—The absence of *commission error*. See *confusion matrix*; note problems that arise because of *apparent commission error*.

Statistical significance—See *model significance*.

Study region—Defined by its *extent*, the area in geographic space (*G-space*) chosen for the analyses of a particular study. Generally, the study region for *model calibration* is the same as for *model evaluation*, but not necessarily. See *background data* and *scale*.

Suitability—The degree to which the environment is appropriate for the species in question. May be equal to probability of presence if certain assumptions are met (i.e., no *dispersal limitations* or limiting *biotic interactions*; see *BAM diagram*). Note also that evaluations of *goodness-of-fit* assess the degree to which model output matches suitability (in contrast with nonparametric ranking approaches, such as the *receiver operating characteristic plot*).

Symmetric loss—A loss function that combines the *omission error* and *commission error* with equal weight. Contrast with *asymmetric loss*.

T

Temporal resolution—The time span covered by a particular parameter (e.g., an *environmental variable*). See also *grain*.

Testing—See *model evaluation*.

Test dataset—See *evaluation data*.

Threshold—Herein, the threshold of occurrence, the value at or above which a model is deemed to predict presence for the species. Various rules can be employed for selecting a threshold, but consideration must be given to the kind of *primary occurrence data* involved (e.g., *presence-only data* versus *presence/absence data*). See *thresholding*.

Threshold-dependent—A strategy and methodology for evaluating *model performance* or *model significance* based on a binary prediction, typically obtained by applying a *threshold* to a continuous or an ordinal prediction of suitability. Contrast with *threshold-independent*.

Threshold-independent—A strategy and methodology for evaluating *model performance* or *model significance* without applying a *threshold* to convert a continuous or an ordinal prediction of *suitability* to a binary one of presence versus absence. Contrast with *threshold-dependent*.

Thresholding—Herein, the process selecting a *threshold* of occurrence, for converting continuous or ordinal model output to a binary prediction of "present" versus "absent."

Training—See *model calibration*.

Training data—See *calibration data*.

Transferability—The application of a model (calibrated in one region) to another place in geography (*G-space*) and/or to another time period. Contrast with *extrapolation*.

U

Uncertainty—The likely or possible level of error regarding data or a prediction (here, relevant for *primary occurrence data*, *environmental data*, and *model prediction*).

V

Validation—See *model validation*.

Variable selection—Herein, decisions regarding which *environmental data* to use in modeling.

Verification—See *model verification*.

Voucher specimen—A physical specimen, typically deposited in a natural history museum or herbarium, that documents the presence of a species. Vouchered records represent the ideal kind of *primary occurrence data*.

Bibliography

Ackerly, D. D. 2003. Community assembly, niche conservatism, and adaptive evolution in changing environments. International Journal of Plant Sciences **164**:S165–S184.

Almeida, C. E., E. Folly-Ramos, A. T. Peterson, V. Lima-Neiva, M. Gumiel, R. Duarte, M. Locks, M. Beltrão, and J. Costa. 2009. Could the bug *Triatoma sherlocki* be vectoring Chagas disease in small mining communities in Bahia, Brazil? Medical and Veterinary Entomology **23**:410–417.

Amarasekare, P. 2003. Competitive coexistence in spatially structured environments: A synthesis. Ecology Letters **6**:1109–1122.

Anciães, M., and A. T. Peterson. 2006. Climate change effects on Neotropical manakin diversity based on ecological niche modeling. Condor **108**:778–791.

———. 2009. Ecological niches and their evolution among Neotropical manakins (Aves: Pipridae). Journal of Avian Biology **40**:591–604.

Anderson, B. J., H. R. Akçakaya, M. B. Araújo, D. A. Fordham, E. Martínez-Meyer, W. Thuiller, and B. W. Brook. 2009. Dynamics of range margins for metapopulations under climate change. Proceedings of the Royal Society B **276**:1415–1420.

Anderson, M. J. 2001. A new method for non-parametric multivariate analysis of variance. Austral Ecology **26**:32–46.

Anderson, R. P. 2003. Real vs. artefactual absences in species distributions: Tests for *Oryzomys albigularis* (Rodentia: Muridae) in Venezuela. Journal of Biogeography **30**:591–605.

Anderson, R. P., M. Gómez-Laverde, and A. T. Peterson. 2002a. Geographical distributions of spiny pocket mice in South America: Insights from predictive models. Global Ecology and Biogeography **11**:131–141.

Anderson, R. P., D. Lew, and A. T. Peterson. 2003. Evaluating predictive models of species' distributions: Criteria for selecting optimal models. Ecological Modelling **162**:211–232.

Anderson, R. P., and E. Martínez-Meyer. 2004. Modeling species' geographic distributions for preliminary conservation assessments: An implementation with the spiny pocket mice (*Heteromys*) of Ecuador. Biological Conservation **116**:167–179.

Anderson, R. P., A. T. Peterson, and S. L. Egbert. 2006. Vegetation-index models predict areas vulnerable to purple loosestrife (*Lythrum salicaria*) invasion in Kansas. Southwestern Naturalist **51**:471–480.

Anderson, R. P., A. T. Peterson, and M. Gómez-Laverde. 2002b. Using niche-based GIS modeling to test geographic predictions of competitive exclusion and competitive release in South American pocket mice. Oikos **98**:3–16.

Anderson, R. P., and A. Raza. 2010. The effect of the extent of the study region on GIS models of species geographic distributions and estimates of niche evolution:

Preliminary tests with montane rodents (genus *Nephelomys*) in Venezuela. Journal of Biogeography **37**:1378–1393.

Angert, A. L., and D. W. Schemske. 2005. The evolution of species' distributions: reciprocal transplants across the elevation ranges of *Mimulus cardinalis* and *M. lewisii*. Evolution **59**:1671–1684.

Angilletta, M. J., P. H. Niewiarowski, and C. A. Navas. 2002. The evolution of thermal physiology in ectotherms. Journal of Thermal Biology **27**:249–268.

Araújo, M. B., M. Cabeza, W. Thuiller, L. Hannah, and P. H. Williams. 2004. Would climate change drive species out of reserves? An assessment of existing reserve selection methods. Global Change Biology **10**:1618–1626.

Araújo, M. B., and A. Guisan. 2006. Five (or so) challenges for species' distribution modelling. Journal of Biogeography **33**:1677–1688.

Araújo, M. B., C. J. Humphries, P. J. Densham, R. Lampinen, W.J.M. Hagemeijer, A. J. Mitchell-Jones, and J. P. Gasc. 2001. Would environmental diversity be a good surrogate for species diversity? Ecography **24**:103–110.

Araújo, M. B., and M. Luoto. 2007. The importance of biotic interactions for modelling species distributions under climate change. Global Ecology and Biogeography **16**:743–753.

Araújo, M. B., and M. New. 2007. Ensemble forecasting of species distributions. Trends in Ecology and Evolution **22**:42–47.

Araújo, M. B., and R. G. Pearson. 2005. Equilibrium of species' distributions with climate. Ecography **28**:693–695.

Araújo, M. B., R. G. Pearson, W. Thuiller, and M. Erhard. 2005a. Validation of species-climate impact models under climate change. Global Change Biology **11**:1504–1513.

Araújo, M. B., and C. Rahbek. 2006. How does climate change affect biodiversity? Science **313**:1396–1397.

Araújo, M. B., W. Thuiller, and R. G. Pearson. 2006. Climate warming and the decline of amphibians and reptiles in Europe. Journal of Biogeography **33**:1712–1728.

Araújo, M. B., W. Thuiller, P. H. Williams, and I. Reginster. 2005b. Downscaling European species atlas distributions to a finer resolution: implications for conservation planning. Global Ecology and Biogeography **14**:17–30.

Araújo, M. B., W. Thuiller, and N. G. Yoccoz. 2009. Reopening the climate envelope reveals macroscale associations with climate in European birds. Proceedings of the National Academy of Sciences USA **106**:E45–E46.

Araújo, M. B., R. J. Whittaker, R. J. Ladle, and M. Erhard. 2005c. Reducing uncertainty in projections of extinction risk from climate change. Global Ecology and Biogeography **14**:529–538.

Araújo, M. B., and P. H. Williams. 2000. Selecting areas for species persistence using occurrence data. Biological Conservation **96**:331–345.

Araújo, M. B., P. H. Williams, and R. J. Fuller. 2002. Dynamics of extinction and the selection of nature reserves. Proceedings of the Royal Society B **269**:1971–1980.

Argáez, J. A., J. Andrés Christen, M. Nakamura, and J. Soberón. 2005. Prediction of potential areas of species distributions based on presence-only data. Environmental and Ecological Statistics **12**:27–44.

Aspinall, R. J., and B. G. Lees. 1994. Sampling and analysis of spatial environmental data. In Advances in Spatial Data Handling, T. C. Waugh and R. G. Healey, editors, pp. 1066–1097. University of Edinburgh, Edinburgh, UK.

Austin, M. P. 1980. Searching for a model to use in vegetation analysis. Vegetatio **42**: 11–21.

———. 1985. Continuum concept, ordination methods, and niche theory. Annual Review of Ecology and Systematics **16**:39–61.

———. 2002. Spatial prediction of species distribution: An interface between ecological theory and statistical modelling. Ecological Modelling **157**:101–118.

Austin, M. P., L. Belbin, J. A. Meyers, M. D. Doherty, and M. Luoto. 2006. Evaluation of statistical models used for predicting plant species distributions: Role of artificial data and theory. Ecological Modelling **199**:197–216.

Austin, M. P., A. O. Nicholls, and C. R. Margules. 1990. Measurement of the realized qualitative niche: Environmental niches of five *Eucalyptus* species. Ecological Monographs **60**:161–177.

Austin, M. P., and T. M. Smith. 1989. A new model for the continuum concept. Vegetatio **83**:35–47.

Australian Weed Committee. 2008. Weeds Australia, http://www.weeds.org.au/. Australian Weed Committee, Canberra.

Avise, J. C. 2000. *Phylogeography: The History and Formation of Species.* Harvard University Press, Cambridge, MA.

Bahn, V., and B. J. McGill. 2007. Can niche-based distribution models outperform spatial interpolation? Global Ecology and Biogeography **16**:733–742.

Bailey, N.T.J. 1964. *The Elements of Stochastic Processes.* John Wiley & Sons, New York.

Balanyá, J., J. M. Oller, R. B. Huey, G. W. Gilchrist, and L. Serra. 2006. Global genetic change tracks global climate warming in *Drosophila subobscura*. Science **313**: 1773–1775.

Barry, S., and J. Elith. 2006. Error and uncertainty in habitat models. Journal of Applied Ecology **43**:413–423.

Barve, N., V. Barve, A. Jiménez-Valverde, A. Lira-Noriega, S. P. Maher, A. T. Peterson, J. Soberón, and F. Villalobos. 2011. The crucial role of the accessible area in ecological niche modeling and species distribution modeling. Ecological Modelling: doi:10.1016/j.ecolmodel.2011.02.011.

Baselga, A., and M. B. Araújo. 2009. Individualistic vs. community modelling of species distributions under climate change. Ecography **32**:55–65.

Basille, M., C. Calenge, E. Marboutin, R. Andersen, and J.-M. Gaillard. 2008. Assessing habitat selection using multivariate statistics: Some refinements of the ecological-niche factor analysis. Ecological Modelling **211**:233–240.

Bates, J. M., and C. W. Granger. 1969. The combination of forecasts. Operational Research Quarterly **20**:451–468.

BBS. 2008. North American Breeding Bird Survey. U.S. Geological Survey, http://www.mbr-pwrc.usgs.gov/bbs/, Washington, DC.

Beale, C. M., J. J. Lennon, and A. Gimona. 2008. Opening the climate envelope reveals no macroscale associations with climate in European birds. Proceedings of the National Academy of Sciences USA **105**:14908–14912.

Beard, C. B., G. Pye, F. J. Steurer, R. Rodríguez, R. Campman, A. T. Peterson, J. Ramsey, R. A. Wirtz, and L. E. Robinson. 2003. Chagas disease in a domestic transmission cycle in southern Texas, USA. Emerging Infectious Diseases 9: 103–105.

Beaumont, L. J., R. V. Gallagher, W. Thuiller, P. O. Downey, M. R. Leishman, and L. Hughes. 2009. Different climatic envelopes among invasive populations may lead to underestimations of current and future biological invasions. Diversity and Distributions 15:409–420.

Beaumont, L. J., L. Hughes, and M. Poulsen. 2005. Predicting species distributions: Use of climatic parameters in BIOCLIM and its impact on predictions of species' current and future distributions. Ecological Modelling 186:251–270.

Beaumont, L. J., A. J. Pitman, M. Poulsen, and L. Hughes. 2007. Where will species go? Incorporating new advances in climate modelling into projections of species distributions. Global Change Biology 13:1368–1385.

Beerling, D. J., B. Huntley, and J. P. Bailey. 1995. Climate and the distribution of *Fallopia japonica*: Use of an introduced species to test the predictive capacity of response surfaces. Journal of Vegetation Science 6:269–282.

Begon, M., C. R. Townsend, and J. L. Harper. 2006. *Ecology: From Individuals to Ecosystems*, 4th ed. Blackwell Publishing, Oxford, UK.

Bell, G. 1982. *The Masterpiece of Nature: The Evolution and Genetics of Sexuality.* Croom Helm, London.

Bell, G., and A. Gonzalez. 2009. Evolutionary rescue can prevent extinction following environmental change. Ecology Letters 12:942–948.

Benedict, M. Q., R. S. Levine, W. A. Hawley, and L. P. Lounibos. 2007. Spread of the tiger: Global risk of invasion by the mosquito *Aedes albopictus*. Vector-Borne and Zoonotic Diseases 7:76–85.

Berry, P. M., T. P. Dawson, P. A. Harrison, and R. G. Pearson. 2002. Modelling potential impacts of climate change on the bioclimatic envelope of species in Britain and Ireland. Global Ecology and Biogeography 11:453–462.

Bhalla, D. K., and D. B. Warheit. 2004. Biological agents with potential for misuse: A historical perspective and defensive measures. Toxicology and Applied Pharmacology 199:71–84.

Birch, L. C. 1953. Experimental background to the study of the distribution and abundance of insects. I. The influence of temperature, moisture, and food on the innate capacity for increase of three grain beetles. Ecology 34:698–711.

Bojórquez-Tapia, L. A., I. Azuara, E. Ezcurra, and O. Flores-Villela. 1995. Identifying conservation priorities in Mexico through geographic information systems and modeling. Ecological Applications 5:215–231.

Boman, S., A. Grapputo, L. Lindstrom, A. Lyytinen, and J. Mappes. 2008. Quantitative genetic approach for assessing invasiveness: Geographic and genetic variation in life-history traits. Biological Invasions 10:1135–1145.

Boulinier, T., J. D. Nichols, J. R. Sauer, J. E. Hines, and K. H. Pollock. 1998. Estimating species richness: The importance of heterogeneity in species detectability. Ecology 79:1018–1028.

Bourg, N. A., W. J. McShea, and D. E. Gill. 2005. Putting a CART before the search: Successful habitat prediction for a rare forest herb. Ecology 86:2793–2804.

Boyce, M. S., P. R. Vernier, S. E. Nielsen, and F.K.A. Schmiegelow. 2002. Evaluating resource selection functions. Ecological Modelling 157:281–300.

Bradley, B. A., and E. Fleishman. 2008. Can remote sensing of land cover improve species distribution modeling? Journal of Biogeography 35:1158–1159.

Brankston, G., L. Gitterman, Z. Hirji, C. Lemieux, and M. Gardam. 2007. Transmission of influenza A in human beings. Lancet Infectious Diseases 7:257–265.

Breiman, L. 2001. Statistic modeling: The two cultures. Statistical Science 16:199–215.

Breiman, R. F., M. R. Evans, W. Preiser, J. Maguire, A. Schnur, A. Li, H. Bekedam, and J. S. MacKenzie. 2003. Role of China in the quest to define and control Severe Acute Respiratory Syndrome. Emerging Infectious Diseases 9:1037–1041.

Bright, P. W., and T. J. Smithson. 2001. Biological invasions provide a framework for reintroductions: Selecting areas in England for pine marten releases. Biodiversity and Conservation 10:1247–1265.

Broennimann, O., U. A. Treier, H. Müller-Schaerer, W. Thuiller, A. T. Peterson, and A. Guisan. 2007. Evidence of climatic niche shift during biological invasion. Ecology Letters 10:701–709.

Brooks, T. M., S. L. Pimm, and N. J. Collar. 1997. Deforestation predicts the number of threatened birds in insular Southeast Asia. Conservation Biology 11:382–394.

Brotons, L., W. Thuiller, M. B. Araújo, and A. H. Hirzel. 2004. Presence/absence versus presence-only modelling methods for predicting bird habitat suitability. Ecography 27:437–448.

Broussard, L. A. 2001. Biological agents: Weapons of warfare and bioterrorism. Molecular Diagnosis 6:323–333.

Brown, J. H. 1971. Mechanisms of competitive exclusion between two species of chipmunks. Ecology 52:305–311.

———. 1995. Macroecology. University of Chicago Press, Chicago.

Brown, J. H., G. C. Stevens, and D. M. Kaufman. 2003. The geographic range: Size, shape, boundaries, and internal structure. Annual Review of Ecology and Systematics 27:597–623.

Brown, J. S., and N. B. Pavlovic. 1992. Evolution in heterogeneous environments: Effects of migration on habitat specialization. Evolutionary Ecology 6:360–382.

Buckland, S. T., A. E. Magurran, R. E. Green, and R. M. Fewster. 2005. Monitoring change in biodiversity through composite indices. Philosophical Transactions of the Royal Society B 360:243–254.

Buckley, L. B. 2008. Linking traits to energetics and population dynamics to predict lizard ranges in changing environments. American Naturalist 171:E1–E19.

Buckley, T. R., K. Marske, and D. Attanayake. 2010. Phylogeography and ecological niche modelling of the New Zealand stick insect Clitarchus hookeri (White) support survival in multiple coastal refugia. Journal of Biogeography 37:682–695.

Buermann, W., S. Saatchi, T. B. Smith, B. R. Zutta, J. A. Chaves, B. Mila, and C. H. Graham. 2008. Predicting species distributions across the Amazonian and Andean regions using remotely sensed data. Journal of Biogeography 35:1160–1176.

Bullock, J. M., R. J. Edwards, P. D. Carey, and R. J. Rose. 2000. Geographical separation of two Ulex species at three spatial scales: Does competition limit species' ranges? Ecography 23:257–271.

Bunn, A. G., D. L. Urban, and T. H. Keitt. 2000. Landscape connectivity: A conservation application of graph theory. Journal of Environmental Management **59**: 265–278.

Burg, T. M., and J. P. Croxall. 2004. Global population structure and taxonomy of the wandering albatross species complex. Molecular Ecology **13**:2345–2355.

Busby, J. R. 1991. BIOCLIM—A bioclimate analysis and prediction system. Plant Protection Quarterly **6**:8–9.

Cade, T. J., and S. A. Temple. 1995. Management of threatened bird species: Evaluation of the hands-on approach. Ibis **137**:S161–S172.

Cadena, C. D., and A. M. Cuervo. 2009. Molecules, ecology, morphology, and songs in concert: How many species is *Arremon torquatus* (Aves: Emberizidae)? Biological Journal of the Linnaean Society **99**:152–176.

Calenge, C., G. Darmon, M. Basille, A. Loison, and J.-M. Jullien. 2008. The factorial decomposition of the Mahalanobis distances in habitat selection studies. Ecology **89**:555–566.

Carnes, B., and N. A. Slade. 1982. Some comments on niche in canonical space. Ecology **63**:888–893.

Carpenter, G., A. N. Gillson, and J. Winter. 1993. DOMAIN: A flexible modelling procedure for mapping potential distributions of plants and animals. Biodiversity and Conservation **2**:667–680.

Carroll, C., M. K. Phillips, N. H. Schumaker, and D. W. Smith. 2003. Impacts of landscape change on wolf restoration success: Planning a reintroduction program based on static and dynamic spatial models. Conservation Biology **17**:536–548.

Carstens, B. C., and C. L. Richards. 2007. Integrating coalescent and ecological niche modeling in comparative phylogeography. Evolution **61**:1439–1454.

Castro-Esau, K. L., G. A. Sánchez-Azofeifa, and T. Caelli. 2004. Discrimination of lianas and trees with leaf-level hyperspectral data. Remote Sensing of Environment **90**:353–372.

Ceballos, G., and P. R. Ehrlich. 2006. Global mammal distributions, biodiversity hotspots, and conservation. Proceedings of the National Academy of Sciences USA **103**:19374–19379.

Chalmers, N. R. 1996. Monitoring and inventorying biodiversity: Collections, data, and training. In *Biodiversity, Science and Development: Towards a New Partnership*, F. di Castri and T. Younes, editors, pp. 171–179. CAB International, Wallingford, UK.

Chapin, F. S., III, E. S. Zavaleta, V. T. Eviner, R. L. Naylor, P. M. Vitousek, H. L. Reynolds, D. U. Hooper, S. Lavorel, O. E. Sala, S. E. Hobbie, M. C. Mack, and S. Diaz. 2000. Consequences of changing biodiversity. Nature **405**:234–242.

Chapman, A. D. 2005. Principles of Data Quality, Version 1.0. Global Biodiversity Information Facility, Copenhagen.

Chase, J. M., and M. A. Leibold. 2003. *Ecological Niches: Linking Classical and Contemporary Approaches*. University of Chicago Press, Chicago.

Chen, G., and A. T. Peterson. 2002. Prioritization of areas in China for the conservation of endangered birds using modelled geographical distributions. Bird Conservation International **12**:197–209.

Chesmore, D. 2004. Automated bioacoustic identification of species. Anais da Academia Brasileira de Ciências **76**:436–440.

Chesmore, E. D., and E. Ohya. 2007. Automated identification of field-recorded songs of four British grasshoppers using bioacoustic signal recognition. Bulletin of Entomological Research **94**:319–330.

Chesson, P. 2000. General theory of competitive coexistence in spatially varying environments. Theoretical Population Biology **58**:211–237.

Clark, M. L., D. A. Roberts, and D. B. Clark. 2005. Hyperspectral discrimination of tropical rain forest tree species at leaf to crown scales. Remote Sensing of Environment **96**:375–398.

Clavero, M., and E. García-Berthou. 2005. Invasive species are a leading cause of animal extinctions. Trends in Ecology and Evolution **20**:110.

Clemen, R. T. 1989. Combining forecasts: A review and annotated bibliography. International Journal of Forecasting **5**:559–583.

Cody, M. L. 1974. *Competition and the Structure of Bird Communities*. Princeton University Press, Princeton, NJ.

Collingham, Y. C., M. O. Hill, and B. Huntley. 1996. The migration of sessile organisms: A simulation model with measurable parameters. Journal of Vegetation Science **7**:831–846.

Colwell, R. K. 1992. Niche: A bifurcation in the conceptual lineage of the term. In *Keywords in Evolutionary Biology*, E. F. Keller and E. A. Lloyd, editors, pp. 241–248. Harvard University Press, Cambridge, MA.

Colwell, R. K., and J. A. Coddington. 1994. Estimating terrestrial biodiversity through extrapolation. Philosophical Transactions of the Royal Society of London B **345**:101–118.

Colwell, R. K., and E. R. Fuentes. 1975. Experimental studies of the niches. Annual Review of Ecology and Systematics **6**:281–310.

Colwell, R. K., and D. J. Futuyma. 1971. On the measurement of niche breadth and overlap. Ecology **52**:567–576.

Colwell, R. K., and T. F. Rangel. 2009. Hutchinson's duality: The once and future niche. Proceedings of the National Academy of Sciences USA **106**:19651–19658.

CONABIO. 2009. Comisión Nacional para el Conocimiento y Uso de la Biodiversidad, http://www.conabio.gob.mx. Mexico City.

Cordellier, M., and M. Pfenninger. 2009. Inferring the past to predict the future: Climate modelling predictions and phylogeography for the freshwater gastropod *Radix balthica* (Pulmonata, Basommatophora). Molecular Ecology **18**:534–544.

Costa, G. C., C. Wolfe, D. B. Shepard, J. P. Caldwell, and L. J. Vitt. 2008. Detecting the influence of climatic variables on species distributions: A test using GIS niche-based models along a steep longitudinal environmental gradient. Journal of Biogeography **35**:637–646.

Costa, J., M.G.R. Freitas-Sibajev, V. Marchón-Silva, M. Quinhoñes-Pires, and R. S. Pacheco. 1997. Isoenzymes detect variation in populations of *Triatoma brasiliensis* (Hemiptera: Reduviidae: Triatominae). Memórias do Instituto Oswaldo Cruz **92**:459–464.

Costa, J., A. T. Peterson, and C. B. Beard. 2002. Ecological niche modeling and differentiation of populations of *Triatoma brasiliensis* Neiva, 1911, the most important Chagas disease vector in northeastern Brazil (Hemiptera, Reduviidae, Triatominae). American Journal of Tropical Medicine and Hygiene **67**:516–520.

Cramer, J. S. 2003. *Logit Models: From Economics and Other Fields*. Cambridge University Press, Cambridge, UK.

Crisci, J. V., L. Katinas, and P. Posadas. 2003. *Historical Biogeography: An Introduction*. Harvard University Press, Cambridge, MA.

Crooks, K. R., and M. E. Soulé. 1999. Mesopredator release and avifaunal extinctions in a fragmented system. Nature **400**:563–566.

Crozier, L. G. 2004. Field transplants reveal summer constraints on a butterfly range expansion. Oecologia **141**:148–157.

Crozier, L. G., and G. Dwyer. 2006. Combining population-dynamic and ecophysiological models to predict climate-induced insect range shifts. American Naturalist **167**:853–866.

Cunha, B. A. 2002. Anthrax, tularemia, plague, ebola, or smallpox as agents of bioterrorism: Recognition in the emergency room. Clinical Microbiology and Infection **8**:489–503.

Cunningham, C. W., K. E. Omland, and T. H. Oakley. 1998. Reconstructing ancestral character states: A critical reappraisal. Trends in Ecology and Evolution **13**:361–366.

Davies, R. G., C.D.L. Orme, V. Olson, G. H. Thomas, S. G. Ross, T.-S. Ding, P. C. Rasmussen, A. J. Stattersfield, P. M. Bennett, T. M. Blackburn, I.P.F. Owens, and K. J. Gaston. 2006. Human impacts and the global distribution of extinction risk. Proceedings of the Royal Society B **273**:2127–2133.

Davis, A. J., L. S. Jenkinson, J. H. Lawton, B. Shorrocks, and S. Wood. 1998. Making mistakes when predicting shifts in species range in response to global warming. Nature **391**:783–786.

De'ath, G., and K. E. Fabricius. 2000. Classification and regression trees: A powerful yet simple technique for ecological data analysis. Ecology **81**:3178–3192.

Deutsch, C. A., J. J. Tewksbury, R. B. Huey, K. S. Sheldon, C. K. Ghalambor, D. C. Haak, and P. R. Martin. 2008. Impacts of climate warming on terrestrial ectotherms across latitude. Proceedings of the National Academy of Sciences USA **105**: 6668–6672.

DiMichele, W. A., A. K. Behrensmeyer, T. D. Olszewski, C. C. Labandiera, J. M. Pandolfi, S. L. Wing, and R. Bobe. 2004. Long-term stasis in ecological assemblages: Evidence from the fossil record. Annual Review of Ecology, Evolution, and Systematics **35**:285–322.

Diniz-Filho, J. A. F., L. M. Bini, T. F. Rangel, R. D. Loyola, C. Hof, D. Nogués-Bravo, and M. B. Araújo. 2009. Partitioning and mapping uncertainties in ensembles of forecasts of species turnover under climate change. Ecography **32**:897–906.

Dobson, A. P., A. Jolly, and D. Rubenstein. 1989. The greenhouse effect and biological diversity. Trends in Ecology and Evolution **4**:64–68.

Dodd, L. E., and M. S. Pepe. 2003. Partial AUC estimation and regression. Biometrics **59**:614–623.

Doledec, S., D. Chessel, and C. Gimaret-Carpentier. 2000. Niche separation in community analysis: A new method. Ecology **81**:2914–2927.

Dormann, C. F., J. M. McPherson, M. B. Araújo, R. Bivand, J. Bolliger, G. Carl, R. G. Davies, A. Hirzel, W. Jetz, W. D. Kissling, I. Kühn, R. Ohlemüller, P. R. Peres-Neto, B. Reineking, B. Schröder, F. M. Schurr, and R. Wilson. 2007. Methods to account for spatial autocorrelation in the analysis of species distributional data: A review. Ecography **30**:609–628.

Dray, S., D. Chessel, and J. Thioulouse. 2003. Co-inertia analysis and the linking of ecological data tables. Ecology **84**:3078–3089.

Drezner, T. D., and C. M. Garrity. 2003. Saguaro distribution under nurse plants in Arizona's Sonoran Desert: Directional and microclimate influences. Professional Geographer **55**:505–512.

Dueser, R. D., and H. H. Shuggart, Jr. 1979. Niche pattern in a forest-floor small-mammal fauna. Ecology **60**:108–118.

Dullinger, S., I. Kleinbauer, J. Peterseil, M. Smolik, and F. Essl. 2009. Niche-based distribution modelling of an invasive alien plant: Effects of population status, propagule pressure, and invasion history. Biological Invasions **11**:2401–2414.

Eaton, M. D., J. Soberón, and A. T. Peterson. 2008. Phylogenetic perspective on ecological niche evolution in American blackbirds (Family Icteridae). Biological Journal of the Linnaean Society **94**:869–878.

Edwards, J. L. 2004. Research and societal benefits of the global biodiversity information facility. BioScience **54**:485–486.

Efron, B. 1987. *The Jackknife, the Bootstrap, and Other Resampling Plans*. Society for Industrial and Applied Mathematics, Philadelphia.

Efron, B., and R. J. Tibshirani. 1993. *An Introduction to the Bootstrap*. Chapman and Hall, New York.

Egbert, S. L., E. Martínez-Meyer, M. A. Ortega-Huerta, and A. T. Peterson. 2002. Use of datasets derived from time-series AVHRR imagery as surrogates for land cover maps in predicting species' distributions. Proceedings IEEE 2002 International Geoscience and Remote Sensing Symposium (IGARSS) **4**:2337–2339.

Eisen, L., and R. J. Eisen. 2007. Need for improved methods to collect and present spatial epidemiologic data for vectorborne diseases. Emerging Infectious Diseases **13**:1816–1820.

Elith, J., C. H. Graham, R. P. Anderson, M. Dudík, S. Ferrier, A. Guisan, R. J. Hijmans, F. Huettmann, J. R. Leathwick, A. Lehmann, J. Li, L. G. Lohmann, B. A. Loiselle, G. Manion, C. Moritz, M. Nakamura, Y. Nakazawa, J. M. Overton, A. T. Peterson, S. J. Phillips, K. Richardson, R. Scachetti-Pereira, R. E. Schapire, J. Soberón, S. Williams, M. S. Wisz, and N. E. Zimmermann. 2006. Novel methods improve prediction of species' distributions from occurrence data. Ecography **29**:129–151.

Elith, J., and J. R. Leathwick. 2007. Predicting species distributions from museum and herbarium records using multiresponse models fitted with multivariate adaptive regression splines. Diversity and Distributions **13**:265–275.

Elith, J., J. R. Leathwick, and T. Hastie. 2008. A working guide to boosted regression trees. Journal of Animal Ecology **77**:802–813.

Elliott, P., J. C. Wakefield, N. G. Best, and D. J. Briggs. 2000. *Spatial Epidemiology: Methods and Applications*. Oxford University Press, Oxford, UK.

Elliott, P., and D. Wartenberg. 2004. Spatial epidemiology: Current approaches and future challenges. Environmental Health Perspectives **112**:998–1006.

Elkinton, J. S., and A. M. Liebhold. 1990. Population dynamics of gypsy moth in North America. Annual Review of Ecology and Systematics **35**:571–596.

Elton, C. S. 1927. *Animal Ecology*. Sidgwick and Jackson, London.

Engler, R., and A. Guisan. 2009. MigClim: Predicting plant distribution and dispersal in a changing climate. Diversity and Distributions **15**:590–601.

Erasmus, B.F.N., A. S. Van Jaarsveld, S. L. Chown, M. Kshatriya, and K. J. Wessels. 2002. Vulnerability of South African animal taxa to climate change. Global Change Biology **8**:679–693.

Erickson, R. O. 1945. The *Clematis fremontii* var. *riehlii* population of the Ozarks. Annals of the Missouri Botanical Garden **32**:413–460.

Erwin, T. L. 1991. How many species are there? Revisited. Conservation Biology **5**: 330–333.

Faith, D. P. 1993. Biodiversity and systematics: The use and misuse of divergence information in assessing taxonomic diversity. Pacific Conservation Biology **1**: 53–57.

Farber, O., and R. Kadmon. 2003. Assessment of alternative approaches for bioclimatic modeling with special emphasis on the Mahalanobis distance. Ecological Modelling **160**:115–130.

Feria, T. P., and A. T. Peterson. 2002. Prediction of bird community composition based on point-occurrence data and inferential algorithms: A valuable tool in biodiversity assessments. Diversity and Distributions **8**:49–56.

Ferrier, S., G.V.N. Powell, K. S. Richardson, G. Manion, J. M. Overton, T. F. Allnutt, S. E. Cameron, K. Mantle, N. D. Burgess, D. P. Faith, J. F. Lamoreux, G. Kier, R. J. Hijmans, V. A. Funk, G. A. Cassis, B. L. Fisher, P. Flemons, D. Lees, J. C. Lovett, and R.S.A.R. Van Rompaey. 2004. Mapping more of terrestrial biodiversity for global conservation assessment. BioScience **54**:1101–1109.

Ferrier, S., G. Watson, J. Pearce, and M. Drielsma. 2002. Extended statistical approaches to modelling spatial pattern in biodiversity in northeast New South Wales. I. Species-level modelling. Biodiversity and Conservation **11**:2275–2307.

Fielding, A. H., and J. F. Bell. 1997. A review of methods for the assessment of prediction errors in conservation presence/absence models. Environmental Conservation **24**:38–49.

Fielding, A. H., and P. F. Haworth. 1995. Testing the generality of bird-habitat models. Conservation Biology **9**:1466–1481.

Fitzpatrick, M. C., and J. F. Weltzin. 2005. Ecological niche models and the geography of biological invasions: A review and novel application. In *Invasive Plants: Ecological and Agricultural Aspects*, S. Inderjit, editor, pp. 45–60. Birkhauser Basel, Switzerland.

Fitzpatrick, M. C., J. F. Weltzin, N. J. Sanders, and R. R. Dunn. 2007. The biogeography of prediction error: Why does the introduced range of the fire ant over-predict its native range? Global Ecology and Biogeography **16**:24–33.

Foden, W., G. F. Midgley, G. Hughes, W. J. Bond, W. Thuiller, M. T. Hoffman, P. Kaleme, L. G. Underhill, A. Rebelo, and L. Hannah. 2007. A changing climate is eroding the geographical range of the Namib Desert tree *Aloe* through population declines and dispersal lags. Diversity and Distributions **13**:645–653.

Fotheringham, A. S., C. Brunsdon, and M. Charlton. 2000. *Quantitative Geography: Perspectives on Spatial Data Analysis*. SAGE Publications, London.

Franklin, J. 2010. *Mapping Species Distributions: Spatial Inference and Prediction*. Cambridge University Press, Cambridge, UK.

Freeman, E. A., and G. G. Moisen. 2008. A comparison of the performance of threshold criteria for binary classification in terms of predicted prevalence and Kappa. Ecological Modelling **217**:48–58.

Funk, V. A., and K. S. Richardson. 2002. Systematic data in biodiversity studies: Use it or lose it. Systematic Biology **51**:303–316.

Ganeshaiah, K. N., N. Barve, N. Nath, K. Chandrashekara, M. Swamy, and R. U. Shaanker. 2003. Predicting the potential geographical distribution of the sugarcane woolly aphid using GARP and *DIVA*-GIS. Current Science **85**:1526–1528.

Garland, T., Jr., P. E. Midford, and A. R. Ives. 1999. An introduction to phylogenetically based statistical methods, with a new method for confidence intervals on ancestral values. American Zoologist **39**:374–388.

Gaston, K. J. 2003. *The Structure and Dynamics of Geographic Ranges*. Oxford University Press, Oxford, UK.

Gaubert, P., M. Papeş, and A. T. Peterson. 2006. Natural history collections and the conservation of poorly known taxa: Ecological niche modeling in central African rainforest genets (*Genetta* spp.). Biological Conservation **130**:106–117.

Gause, G. F. 1936. *The Struggle for Existence*. Williams and Wilkins, Baltimore.

Getz, W. M., and C. C. Wilmers. 2004. A local nearest-neighbor convex-hull construction of home ranges and utilization distributions. Ecography **27**:489–505.

Godown, M. E., and A. T. Peterson. 2000. Preliminary distributional analysis of U.S. endangered bird species. Biodiversity and Conservation **9**:1313–1322.

Godsoe, W. 2010. I can't define the niche but I know it when I see it: A formal link between statistical theory and the ecological niche. Oikos **119**:53–60.

Golubov, J., M. C. Mandujano, and J. Soberón. 2001. La posible invasión de *Cactoblastis cactorum* Berg en México. Cactáceas y Suculentas Mexicanas **46**:75–78.

González, C., O. Wang, S. E. Strutz, C. González-Salazar, V. Sánchez-Cordero, and S. Sarkar. 2010. Climate change and risk of leishmaniasis in North America: Predictions from ecological niche models of vector and reservoir species. PLoS Neglected Tropical Diseases **4**:e585.

Good, R. D. 1931. A theory of plant geography. New Phytologist **30**:149–171.

Gordon, M. S., and E. C. Olson. 1994. *Invasions of the Land: The Transitions of Organisms from Aquatic to Terrestrial Life*. Columbia University Press, New York.

Gower, J. C. 1971. A general coefficient of similarity and some of its properties. Biometrics **27**:857–871.

Graham, C. H., J. Elith, R. J. Hijmans, A. Guisan, A. T. Peterson, B. A. Loiselle, and the NCEAS Predicting Species Distributions Working Group. 2007. The influence of spatial errors in species occurrence data used in distribution models. Journal of Applied Ecology **45**:239–247.

Graham, C. H., S. Ferrier, F. Huettman, C. Moritz, and A. T. Peterson. 2004a. New developments in museum-based informatics and applications in biodiversity analysis. Trends in Ecology and Evolution **19**:497–503.

Graham, C. H., and R. J. Hijmans. 2006. A comparison of methods for mapping species ranges and species richness. Global Ecology and Biogeography **15**:578–587.

Graham, C. H., S. R. Ron, J. C. Santos, C. J. Schneider, and C. Moritz. 2004b. Integrating phylogenetics and environmental niche models to explore speciation mechanisms in dendrobatid frogs. Evolution **58**:1781–1793.

Graham, R. W., E. L. Lundelius, M. A. Graham, E. K. Schroeder, R. S. Toomey, E. Anderson, A. D. Barnosky, J. A. Burns, C. S. Churcher, D. K. Grayson, R. D. Guthrie, C. R. Harington, G. T. Jefferson, L. D. Martin, H. G. McDonald, R. E. Morlan, H. A. Semken, S. D. Webb, L. Werdelin, and M. C. Wilson. 1996. Spatial response of mammals to late quaternary environmental fluctuations. Science 272:1601–1606.

Green, R. E., Y. C. Collingham, S. G. Willis, R. D. Gregory, K. W. Smith, and B. Huntley. 2008. Performance of climate envelope models in retrodicting recent changes in bird population size from observed climatic change. Biology Letters 4:599–602.

Green, R. H. 1971. A multivariate statistical approach to the Hutchinsonian niche: Bivalve molluscs of central Canada. Ecology 52:544–556.

Grinnell, J. 1917. The niche-relationships of the California Thrasher. Auk 34:427–433.

Guisan, A., O. Broennimann, R. Engler, M. Vust, N. G. Yoccoz, A. Lehman, and N. E. Zimmermann. 2006. Using niche-based models to improve the sampling of rare species. Conservation Biology 20:501–511.

Guisan, A., T. C. Edwards, Jr., and T. Hastie. 2002. Generalized linear and generalized additive models in studies of species distributions: Setting the scene. Ecological Modelling 157:89–100.

Guisan, A., C. H. Graham, J. Elith, F. Huettman, and the NCEAS Predicting Species Distributions Working Group. 2007. Sensitivity of predictive species distribution models to change in grain size. Diversity and Distributions 13:332–340.

Guisan, A., and W. Thuiller. 2005. Predicting species distribution: Offering more than simple habitat models. Ecology Letters 8:993–1009.

Guisan, A., and N. E. Zimmermann. 2000. Predictive habitat distribution models in ecology. Ecological Modelling 135:147–186.

Guo, Q., M. Kelly, and C. H. Graham. 2005. Support vector machines for predicting distribution of sudden oak death in California. Ecological Modelling 182:75–90.

Guralnick, R. P., J. Wieczorek, R. Beaman, R. J. Hijmans, and the BioGeomancer Working Group. 2006. BioGeomancer: Automated georeferencing to map the world's biodiversity data. PLoS Biology 4:e381.

Gyapong, J. O., D. Kyelem, I. Kleinschmidt, K. Agbo, F. Ahouandogbo, J. Gaba, G. Owusu-Banahene, S. Sanou, Y. K. Sodahlon, G. Biswas, O. O. Kale, D. H. Molyneux, J. B. Roungou, M. C. Thomson, and J. Remme. 2002. The use of spatial analysis in mapping the distribution of bancroftian filariasis in four West African countries. Annals of Tropical Medicine and Parasitology 96:695–705.

Hall, B. P., and R. E. Moreau. 1970. An Atlas of Speciation in African Passerine Birds. Trustees of the British Museum (Natural History), London.

Hall, E. R. 1981. The Mammals of North America, 2nd ed. John Wiley & Sons, New York.

Hampe, A. 2004. Bioclimate envelope models: What they detect and what they hide. Global Ecology and Biogeography 13:469–471.

Handley, C. O., Jr. 1976. Mammals of the Smithsonian Venezuelan Project. Brigham Young University Science Bulletin, Biological Series 20:1–91.

Hannah, L., G. Midgley, S. Andelman, M. Araújo, G. Hughes, E. Martínez-Meyer, R. Pearson, and P. Williams. 2007. Protected area needs in a changing climate. Frontiers in Ecology and Environment 5:131–138.

Harper, J. 1977. Population Biology of Plants. Academic Press, London.

Hastie T. J., and R. Tibshirani. 1990. Generalized Additive Models. Chapman and Hall, London.

Hastie, T., R. Tibshirani, and J. Friedman. 2001. The Elements of Statistical Learning: Data Mining, Inference, and Prediction. Springer, New York.

Hay, S. I., J. Cox, D. J. Rogers, S. E. Randolph, D. I. Stern, G. D. Shanks, M. F. Myers, and R. W. Snow. 2002. Climate change and the resurgence of malaria in the East Africa highlands. Nature **415**:905–909.

He, F., and K. J. Gaston. 2000. Occupancy-abundance relationships and sampling scales. Ecography **23**:503–511.

Heikkinen, R. K., M. Luoto, M. B. Araújo, R. Virkkala, W. Thuiller, and M. T. Sykes. 2006. Methods and uncertainties in bioclimatic envelope modelling under climate change. Progress in Physical Geography **30**:751–777.

Heikkinen, R. K., M. Luoto, R. Virkkala, R. G. Pearson, and J.-H. Körber. 2007. Biotic interactions improve prediction of boreal bird distributions at macro-scales. Global Ecology and Biogeography **16**:754–763.

Hernandez, P. A., C. H. Graham, L. L. Master, and D. L. Albert. 2006. The effect of sample size and species characteristics on performance of different species distribution modeling methods. Ecography **29**:773–785.

Hewitt, G. 2000. The genetic legacy of the Quaternary ice ages. Nature **405**:907–913.

Heyer, W. R., J. Coddington, W. J. Kress, P. Acevedo, D. Cole, T. L. Erwin, B. J. Meggers, M. G. Pogue, R. W. Thorington, R. P. Vari, M. J. Weitzman, and S. H. Weitzman. 1999. Amazonian biotic diversity and conservation decisions. Ciência e Cultura **51**:372–385.

Hickerson, M. J., B. C. Carstens, J. Cavender-Bares, K. A. Crandall, C. H. Graham, J. B. Johnson, L. Rissler, P. F. Victoriano, and A. D. Yoder. 2010. Phylogeography's past, present, and future: 10 years after Avise, 2000. Molecular Phylogenetics and Evolution **54**:291–301.

Hickling, R., D. B. Roy, J. K. Hill, R. Fox, and C. D. Thomas. 2006. The distributions of a wide range of taxonomic groups are expanding polewards. Global Change Biology **12**:450–455.

Hidalgo-Mihart, M. G., L. Cantú-Salazar, A. González-Romero, and C. A. López-González. 2004. Historical and present distribution of coyote (*Canis latrans*) in Mexico and Central America. Journal of Biogeography **31**:2025–2038.

Higgins, S. I., D. M. Richardson, R. M. Cowling, and T. H. Trinder-Smith. 1999. Predicting the landscape-scale distribution of alien plants and their threat to plant diversity. Conservation Biology **13**:303–313.

Hijmans, R. J., S. E. Cameron, J. L. Parra, P. G. Jones, and A. Jarvis. 2005. Very high resolution interpolated climate surfaces for global land areas. International Journal of Climatology **25**:1965–1978.

Hijmans, R. J., L. Guarino, M. Cruz, and E. Rojas. 2001. Computer tools for spatial analysis of plant genetic resources data: 1. DIVA-GIS. Plant Genetic Resources Newsletter **127**:15–19.

Hilbert, D. W., and B. Ostendorf. 2001. The utility of artificial neural networks for modelling the distribution of vegetation in past, present and future climates. Ecological Modelling **146**:311–327.

Hinojosa-Díaz, I. A., O. Yáñez-Ordóñez, G. Chen, and A. T. Peterson. 2005. The North American invasion of the Giant Resin Bee (Hymenoptera: Megachilidae). Journal of Hymenoptera Research **14**:69–77.

Hinsinger, P., C. Plassard, C. Tang, and B. Jaillard. 2003. Origins of root-mediated pH changes in the rhizosphere and their responses to environmental constraints: A review. Plant and Soil **248**:43–59.

Hirzel, A. H., J. Hausser, D. Chessel, and N. Perrin. 2002. Ecological-niche factor analysis: How to compute habitat-suitability maps without absence data. Ecology 83:2027–2036.

Hirzel, A. H., and G. Le Lay. 2008. Habitat suitability modelling and niche theory. Journal of Applied Ecology 45:1372–1381.

Hoffmann, M. H. 2001. The distribution of *Senecio vulgaris*: Capacity of climatic range models for predicting adventitious ranges. Flora 196:395–403.

———. 2005. Evolution of the realized climatic niche in the genus *Arabidopsis* (Brassicaceae). Evolution 59:1425–1436.

Holt, R. D. 1990. The microevolutionary consequences of climate change. Trends in Ecology and Evolution 5:311–315.

———. 1996a. Adaptive evolution in source-sink environments: Direct and indirect effects of density-dependence on niche evolution. Oikos 75:182–192.

———. 1996b. Demographic constraints in evolution: Towards unifying the evolutionary theories of senescence and niche conservatism. Evolutionary Ecology 10:1–11.

———. 2003. On the evolutionary ecology of species' ranges. Evolutionary Ecology Research 5:159–178.

———. 2009. Bringing the Hutchinsonian niche into the 21st century: Ecological and evolutionary perspectives. Proceedings of the National Academy of Sciences USA 106:19659–19665.

Holt, R. D., and M. S. Gaines. 1992. Analysis of adaptation in heterogeneous landscapes: Implications for the evolution of fundamental niches. Evolutionary Ecology 6:433–447.

Holt, R. D., and R. Gomulkiewicz. 1996. The evolution of species' niches: A population dynamic perspective. In *Case Studies in Mathematical Modeling—Ecology, Physiology, and Cell Biology*, H. G. Othmer, F. R. Adler, M. A. Lewis, and J. C. Dallon, editors, pp. 25–50. Prentice-Hall, Saddle River, NJ.

Holt, R. D., and T. H. Keitt. 2000. Alternative causes for range limits: A metapopulation perspective. Ecology Letters 3:41–47.

Holt, R. D., T. H. Keitt, M. A. Lewis, B. A. Maurer, and M. L. Taper. 2005. Theoretical models of species' borders: Single species approaches. Oikos 108:18–27.

Holt, R. D., J. H. Lawton, J. K. Gaston, and T. M. Blackburn. 1997. On the relationship between range size and local abundances: Back to the basics. Oikos 78:183–190.

Honig, M. A., R. M. Cowling, and D. M. Richardson. 1992. The invasive potential of Australian banksias in South-African fynbos—A comparison of the reproductive potential of *Banksia ericifolia* and *Leucadendron laureolum*. Australian Journal of Ecology 17:305–314.

Hooper, H. L., R. Connon, A. Callaghan, G. Fryer, S. Yarwood-Buchanan, J. Biggs, S. J. Maund, T. H. Hutchinson, and R. M. Sibly. 2008. The ecological niche of *Daphnia magna* characterized using population growth rate. Ecology 89:1015–1022.

Howell, S.N.G., and S. Webb. 1995. *A Guide to the Birds of Mexico and Northern Central America*. Oxford University Press, Oxford, UK.

Howells, O., and G. Edwards-Jones. 1997. A feasibility study of reintroducing wild boar *Sus scrofa* to Scotland: Are existing woodlands large enough to support minimum viable populations. Biological Conservation 81:77–89.

Huisman, J., and F. J. Weissing. 2001. Fundamental unpredictability in multispecies competition. American Naturalist **157**:488–494.

Hunter, P. R. 2003. Climate change and waterborne and vector-borne disease. Journal of Applied Microbiology **94**:37S–46S.

Huntley, B., P. M. Berry, W. Cramer, and A. P. McDonald. 1995. Modelling present and potential future ranges of some European higher plants using climate response surfaces. Journal of Biogeography **22**:967–1001.

Huntley, B., Y. C. Collingham, S. G. Willis, and R. E. Green. 2008. Potential impacts of climatic change on European breeding birds. PLoS ONE **3**:e1439.

Huntley, B., R. E. Green, Y. C. Collingham, J. K. Hill, S. G. Willis, P. J. Bartlein, W. Cramer, W.J.M. Hagermeijer, and C. J. Thomas. 2004. The performance of models relating species geographic distributions to climate is independent of trophic level. Ecology Letters **7**:417–426.

Hurlbert, A. H., and W. Jetz. 2007. Species richness, hotspots, and the scale dependence of range maps in ecology and conservation. Proceedings of the National Academy of Sciences **104**:13384–13389.

Hutchinson, G. E. 1957. Concluding remarks. Cold Spring Harbor Symposia on Quantitative Biology **22**:415–427.

———. 1978. *An Introduction to Population Ecology*. Yale University Press, New Haven, CT.

ICZN. 1999. International Code of Zoological Nomenclature, 4th ed.; http://www.iczn.org/iczn/index.jsp. International Commission on Zoological Nomenclature, London.

Iguchi, K., K. Matsuura, K. McNyset, A. T. Peterson, R. Scachetti-Pereira, K. A. Powers, D. A. Vieglais, E. O. Wiley, and T. Yodo. 2004. Predicting invasions of North American basses in Japan using native range data and a genetic algorithm. Transactions of the American Fisheries Society **133**:845–854.

INPE. 2009. Instituto Nacional de Pesquisas Espaciais; http://www.inpe.br. Ministério da Ciência y Tecnología, São José dos Campos, São Paulo.

IPCC. 2007. *Climate Change 2007: The Physical Science Basis*. S. Solomon, D. Qin, M. Manning, Z. Chen, M. Marquis, K. B. Averyt, M. Tignor, and H. L. Miller editors. Cambridge University Press, Cambridge, UK.

———. 2009. The IPCC Data Distribution Centre; http://www.ipcc-data.org. Intergovernmental Panel on Climate Change. Geneva, Switzerland.

Iverson, L. R., and A. M. Prasad. 1998. Predicting abundance of 80 tree species following climate change in the eastern United States. Ecological Monographs **68**: 465–485.

Jackson, S. T., and J. T. Overpeck. 2000. Responses of plant populations and communities to environmental changes of the late Quaternary. Paleobiology **26**:194–220.

Jakob, S. S., A. Ihlow, and F. R. Blattner. 2007. Combined ecological niche modelling and molecular phylogeography revealed the evolutionary history of *Hordeum marinum* (Poaceae)—Niche differentiation, loss of genetic diversity, and speciation in Mediterranean Quaternary refugia. Molecular Ecology **16**:1713–1727.

Jakob, S. S., E. Martínez-Meyer, and F. R. Blattner. 2009. Phylogeographic analyses and paleodistribution modeling indicates Pleistocene in situ survival of *Hordeum* species (Poaceae) in southern Patagonia without genetic or spatial restriction. Molecular Biology and Evolution **26**:907–923.

James, F. C. 1971. Ordinations of habitat relationships among breeding birds. The Wilson Bulletin **83**:215–236.

James, F. C., R. F. Johnston, N. O. Wamer, G. J. Niemi, and W. J. Boecklen. 1984. The Grinnellian niche of the Wood Thrush. American Naturalist **124**:17–47.

Jarvis, A., K. Williams, D. Williams, L. Guarino, P. J. Caballero, and G. Mottram. 2005. Use of GIS for optimizing a collecting mission for a rare wild pepper (*Capsicum flexuosum* Sendtn.) in Paraguay. Genetic Resources and Crop Evolution **52**:671–682.

Jiménez-Valverde, A., N. Barve, A. Lira-Noriega, S. P. Maher, Y. Nakazawa, M. Papeş, J. Soberón, J. Sukumaran, and A. T. Peterson. 2010. Dominant climate influences on North American bird distributions. Global Ecology and Biogeography **20**: 114–118.

Jiménez-Valverde, A., J. M. Lobo, and J. Hortal. 2008. Not as good as they seem: The importance of concepts in species distribution modelling. Diversity and Distributions **14**:885–890.

Jiménez-Valverde, A., A. T. Peterson, J. Soberón, J. Overton, P. Aragón, P., and J. M. Lobo. 2011. Use of niche models in invasive species risk assessments. Biological Invasions: doi:10.1007/s10530-011-9963-4.

Johnson, K. P., and S. M. Lanyon. 1999. Molecular systematics of the grackles and allies, and the effect of additional sequence (cyt *b* and ND2). Auk **116**:759–768.

Jones, P. G., and A. Gladkov. 1999. FloraMap: A computer tool for predicting the distribution of plants and other organisms in the wild. Centro Internacional de Agricultura Tropical, Cali, Colombia.

Kadmon, R., O. Farber, and A. Danin. 2004. Effect of roadside bias on the accuracy of predictive maps produced by bioclimatic models. Ecological Applications **14**: 401–413.

Kambhampati, S., and A. T. Peterson. 2007. Ecological niche conservation and differentiation in the wood-feeding cockroaches, *Cryptocercus*, in the United States. Biological Journal of the Linnaean Society **90**:457–466.

Kawecki, T. J. 1995. Demography of source-sink populations and the evolution of ecological niches. Evolutionary Ecology **9**:38–44.

Kawecki, T. J., and S. C. Stearns. 1993. The evolution of life histories in spatially heterogeneous environments: Optimal reaction norms revisited. Evolutionary Ecology **7**:155–174.

Kearney, M. 2006. Habitat, environment, and niche: What are we modelling? Oikos **115**:186–191.

Kearney, M., and W. P. Porter. 2004. Mapping the fundamental niche: Physiology, climate, and the distribution of a nocturnal lizard. Ecology **85**:3119–3131.

Keating, K. A., and S. Cherry. 2004. Use and interpretation of logistic regression in habitat-selection studies. Journal of Wildlife Management **68**:774–789.

Keith, D. A., H. R. Akçakaya, W. Thuiller, G. F. Midgley, R. G. Pearson, S. J. Phillips, H. M. Regan, M. B. Araújo, and T. G. Rebelo. 2008. Predicting extinction risks under climate change: Coupling stochastic population models with dynamic bioclimatic habitat models. Biology Letters **4**:560–563.

Keitt, T. H., O. N. Bjørnstad, P. M. Dixon, and S. Citron-Pousty. 2002. Accounting for spatial pattern when modeling organism-environment interactions. Ecography **25**:616–625.

Kilpatrick, A. M., A. A. Chmura, D. W. Gibbons, R. C. Fleischer, P. P. Marra, and P. Daszak. 2006. Predicting the global spread of H5N1 avian influenza. Proceedings of the National Academy of Sciences USA **103**:19368–19373.

Kirkpatrick, M., and N. H. Barton. 1997. Evolution of a species' range. American Naturalist **150**:1–23.

Kitchener, A. 1991. *The Natural History of the Wild Cats*. Comstock Press, Ithaca, NY.

Kitron, U. 1998. Landscape ecology and epidemiology of vector-borne diseases: Tools for spatial analysis. Journal of Medical Entomology **35**:435–445.

Kluza, D. A., D. A. Vieglais, J. K. Andreasen, and A. T. Peterson. 2007. Sudden oak death: Geographic risk estimates and predictions of origins. Plant Pathology **56**: 580–587.

Knouft, J. H., J. B. Losos, R. E. Glor, and J. J. Kolbe. 2006. Phylogenetic analysis of the evolution of the niche in lizards of the *Anolis sagrei* group. Ecology **87**:S29–S38.

Knowles, L. L., B. C. Carstens, and M. L. Keat. 2007. Coupling genetic and ecological-niche models to examine how past population distributions contribute to divergence. Current Biology **17**:940–946.

Koleff, P., M. Tambutti, I. J. March, R. Esquivel, C. Cantú, and A. Lira-Noriega. 2009. Identificación de prioridades y análisis de vacíos y omisiones en la conservación de la biodiversidad de México. Pages 651–718 in CONABIO, editor. In *Capital Natural de México, Volumen II. Estado de Conservación y Tendencias de Cambio*, CONABIO, J. Soberón, G. Halffter, and J. Llorente, editors, pp. 651–718. Comisión Nacional para el Uso y Conocimiento de la Biodiversidad, Mexico City.

Komar, N. 2003. West Nile virus: Epidemiology and ecology in North America. Advances in Virus Research **61**:185–234.

Koplin, J. R., and R. S. Hoffmann. 1968. Habitat overlap and competitive exclusion in voles (*Microtus*). American Midland Naturalist **80**:494–507.

Körner, C. 2007. The use of "altitude" in ecological research. Trends in Ecology and Evolution **22**:569–574.

Kovats, R. S., D. Campbell-Lendrum, A. J. McMichael, A. Woodward, and J. S. Cox. 2001. Early effects of climate change: Do they include changes in vector-borne disease? Philosophical Transactions of the Royal Society of London B **356**: 1057–1068.

Kozak, K. H., and J. J. Wiens. 2006. Does niche conservatism promote speciation? A case study in North American salamanders. Evolution **60**:2604–2621.

Kremen, C., A. Cameron, A. Moilanen, S. J. Phillips, C. D. Thomas, H. Beentje, J. Dransfield, B. L. Fisher, F. Glaw, T. C. Good, G. J. Harper, R. J. Hijmans, D. C. Lees, E. Louis, Jr., R. A. Nussbaum, C. J. Raxworthy, A. Razafimpahanana, G. E. Schatz, M. Vences, D. R. Vieites, P. C. Wright, and M. L. Zjhra. 2008. Aligning conservation priorities across taxa in Madagascar with high-resolution planning tools. Science **320**:222–226.

Krishtalka, L., and P. S. Humphrey. 2000. Can natural history museums capture the future? BioScience **50**:611–617.

Kutz, S. J., E. P. Hoberg, L. Polley, and E. J. Jenkins. 2005. Global warming is changing the dynamics of Arctic host-parasite systems. Proceedings of the Royal Society B **272**:2571–2576.

Lande, R. 1986. The dynamics of peak shifts and the pattern of morphological evolution. Paleobiology **12**:343–354.

————. 1988. Genetics and demography in biological conservation. Science **241**: 1455–1460.

Lande, R., S. Engen, and B.-E. Saether. 2003. *Stochastic Population Dynamics in Ecology and Conservation*. Oxford University Press, New York.

Lanyon, S. M. 1994. Polyphyly of the blackbird genus *Agelaius* and the importance of assumptions of monophyly in comparative studies. Evolution **48**:679–693.

Lanyon, S. M., and K. E. Omland. 1999. A molecular phylogeny of the blackbirds (Icteridae): Five lineages revealed by cytochrome-*b* sequence data. Auk **116**: 629–639.

Larson, G. 1982. Great Moments in Evolution [Visual image]. *The Far Side*. Far Works, Inc., Los Angeles.

Latimer, A. M., S. Wu, A. E. Gelfand, and J. A. Silander, Jr. 2006. Building statistical models to analyze species distributions. Ecological Applications **16**:33–50.

Lawson, A. B., A. Biggeri, D. Böhning, E. Lesaffre, J.-F. Viel, R. Bertollini, editors. 1999. *Disease Mapping and Risk Assessment for Public Health*. John Wiley & Sons, New York.

Lawton, J. H. 2000. Concluding remarks: A review of some open questions. In *Ecological Consequences of Heterogeneity*, M. J. Hutchings, E. A. John, and A. J. A. Stewart, editors, pp. 401–424. Cambridge University Press, Cambridge, UK.

Leathwick, J. R., and M. P. Austin. 2001. Competitive interactions between tree species in New Zealand's old-growth indigenous forest. Ecology **82**:2560–2573.

Lee, C. E. 2002. Evolutionary genetics of invasive species. Trends in Ecology and Evolution **17**:386–391.

Lee, C. E., J. L. Remfert, and Y.-M. Chang. 2007. Response to selection and evolvability of invasive populations. Genetica **129**:179–192.

Lee, J., and D. Stucky. 1998. On applying viewshed analysis for determining least-cost paths on digital elevation models. International Journal of Geographical Information Science **12**:891–905.

Legendre, P., and L. Legendre. 1998. *Numerical Ecology*, 2nd English ed. Elsevier, Amsterdam.

Leibold, M. A. 1995. The niche concept revisited: Mechanistic models and community context. Ecology **76**:1371–1382.

Levine, R. S., A. T. Peterson, and M. Q. Benedict. 2004. Geographic and ecologic distributions of the *Anopheles gambiae* complex predicted using a genetic algorithm. American Journal of Tropical Medicine and Hygiene **70**:105–109.

Levine, R. S., A. T. Peterson, K. L. Yorita, D. Carroll, I. K. Damon, and M. G. Reynolds. 2007. Ecological niche and geographic distribution of human monkeypox in Africa. PLoS ONE **2**:e176.

Levins, R. 1968. *Evolution in Changing Environments*. Princeton University Press, Princeton, NJ.

Lim, B. K., A. T. Peterson, and M. D. Engstrom. 2002. Robustness of ecological niche modeling algorithms for mammals in Guyana. Biodiversity and Conservation **11**:1237–1246.

Lindenmayer, D. B., H. A. Nix, J. P. McMahon, M. F. Hutchinson, and M. T. Tanton. 1991. The conservation of Leadbeater's possum, *Gymnobelideus leadbeateri*

(McCoy): A case study of the use of bioclimatic modelling. Journal of Biogeography **18**:371–383.

Liu, C., P. M. Berry, T. P. Dawson, and R. G. Pearson. 2005. Selecting thresholds of occurrence in the prediction of species distributions. Ecography **28**:385–393.

Llorente-Bousquets, J. E., L. Oñate-Ocaña, A. Luis-Martínez, and I. Vargas-Fernández. 1997. *Papilionidae y Pieridae de México: Distribución Geográfica e Ilustración.* Comisión Nacional para el Conocimiento y Uso de la Biodiversidad, Mexico City.

Lobo, J. M., A. Jiménez-Valverde, and R. Real. 2007. AUC: A misleading measure of the performance of predictive distribution models. Global Ecology and Biogeography **17**:145–151.

Loiselle, B. A., C. A. Howell, C. H. Graham, J. M. Goerck, T. Brooks, K. G. Smith, and P. H. Williams. 2003. Avoiding pitfalls of using species distribution models in conservation planning. Conservation Biology **17**:1591–1600.

López-Cárdenas, J., F. E. González-Bravo, P. M. Salazar-Schettino, J. C. Gallaga-Solórzano, E. Ramírez-Barba, J. Martínez-Méndez, V. Sánchez-Cordero, A. T. Peterson, and J. M. Ramsey. 2005. Fine-scale predictions of distributions of Chagas disease vectors in the state of Guanajuato, Mexico. Journal of Medical Entomology **42**:1068–1081.

López-Darias, M., and J. M. Lobo. 2008. Factors affecting invasive species abundance: The Barbary Ground Squirrel on Fuerteventura Island, Spain. Zoological Studies **47**:268–281.

MacArthur, R. H. 1972. *Geographical Ecology: Patterns in the Distribution of Species.* Harper & Row, New York.

MacArthur, R. H., and E. O. Wilson. 1967. *The Theory of Island Biogeography.* Princeton University Press, Princeton, NJ.

Mace, G. M., N. J. Collar, K. J. Gaston, C. Hilton-Taylor, H. R. Akçakaya, N. Leader-Williams, E. J. Milner-Gulland, and S. N. Stuart. 2008. Quantification of extinction risk: IUCN's system for classifying threatened species. Conservation Biology **22**:1424–1442.

MacKenzie, D. I., J. D. Nichols, G. B. Lachman, S. Droege, J. A. Royle, and C. A. Langtimm. 2002. Estimating site occupancy rates when detection probabilities are less than one. Ecology **83**:2248–2255.

Mackey, B. G., and D. B. Lindenmayer. 2001. Towards a hierarchical framework for modelling the spatial distribution of animals. Journal of Biogeography **28**:1147–1166.

Maguire, B., Jr. 1973. Niche response structure and the analytical potentials of its relationship to the habitat. American Naturalist **107**:213–246.

Manel, S., H. C. Williams, and S. J. Ormerod. 2001. Evaluating presences-absence models in ecology: The need to account for prevalence. Journal of Applied Ecology **38**:921–931.

Manning, A. D., J. Fischer, A. Felton, B. Newell, W. Steffen, and D. B. Lindenmayer. 2009. Landscape fluidity: A unifying perspective for understanding and adapting to global change. Journal of Biogeography **36**:193–199.

Margules, C. R., and R. L. Pressey. 2000. Systematic conservation planning. Nature **405**:243–253.

Marmion, M., M. Parviainen, M. Luoto, R. K. Heikkinen, and W. Thuiller. 2009. Evaluation of consensus methods in predictive species distribution modelling. Diversity and Distributions **15**:59–69.

Martin, W. K. 1996. The current and potential distribution of the common myna *Acridotheres tristis* in Australia. Emu **96**:166–173.

Martínez-Gordillo, D., O. Rojas-Soto, and A. Espinosa de los Monteros. 2010. Ecological niche modelling as an exploratory tool for identifying species limits: An example based on Mexican muroid rodents. Journal of Evolutionary Biology **23**:259–270.

Martínez-Meyer, E., and A. T. Peterson. 2006. Conservatism of ecological niche characteristics in North American plant species over the Pleistocene-to-Recent transition. Journal of Biogeography **33**:1779–1789.

Martínez-Meyer, E., A. T. Peterson, and W. W. Hargrove. 2004a. Ecological niches as stable distributional constraints on mammal species, with implications for Pleistocene extinctions and climate change projections for biodiversity. Global Ecology and Biogeography **13**:305–314.

Martínez-Meyer, E., A. T. Peterson, and A. G. Navarro-Sigüenza. 2004b. Evolution of seasonal ecological niches in the *Passerina* buntings (Aves: Cardinalidae). Proceedings of the Royal Society B **271**:1151–1157.

Martínez-Meyer, E., A. T. Peterson, J. I. Servín, and L. F. Kiff. 2006. Ecological niche modelling and prioritizing areas for species reintroductions. Oryx **40**:411–418.

Martins, E. P. 2000. Adaptation and the comparative method. Trends in Ecology and Evolution **15**:296–299.

Maurer, B. A., and M. L. Taper. 2002. Connecting geographical distributions with population processes. Ecology Letters **5**:223–231.

McClean, C. J., J. C. Lovett, W. Küper, L. Hannah, J. H. Sommer, W. Barthlott, M. Termansen, G. F. Smith, S. Tokumine, and J.R.D. Taplin. 2005. African plant diversity and climate change. Annals of the Missouri Botanical Garden **92**:139–152.

McCormack, J. E., A. J. Zellmer, and L. L. Knowles. 2010. Does niche divergence accompany allopatric divergence in Aphelocoma jays as predicted under ecological speciation? Insights from tests with niche models. Evolution **64**:1231–1244.

McGill, B. J., B. J. Enquist, E. Weiher, and M. Westoby. 2006. Rebuilding community ecology from functional traits. Trends in Ecology and Evolution **21**:178–185.

McGill, B. J., E. A. Hadley, and B. A. Maurer. 2005. Community inertia of Quaternary small mammal assemblages in North America. Proceedings of the National Academy of Sciences USA **102**:16701–16706.

McNyset, K. M. 2005. Use of ecological niche modelling to predict distributions of freshwater fish species in Kansas. Ecology of Freshwater Fish **14**:243–255.

Medley, K. A. 2010. Niche shifts during the global invasion of the Asian tiger mosquito, *Aedes albopictus* Skuse (Culicidae), revealed by reciprocal distribution models. Global Ecology and Biogeography **19**:122–133.

Meszéna, G., M. Gyllenberg, L. Pásztor, and J.A.J. Metz. 2006. Competitive exclusion and limiting similarity: A unified theory. Theoretical Population Biology **69**: 68–87.

Midgley, G. F., L. Hannah, D. Millar, M. C. Rutherford, and L. W. Powrie. 2002. Assessing the vulnerability of species richness to anthropogenic climate change in a biodiversity hotspot. Global Ecology and Biogeography **11**:445–451.

Midgley, G. F., L. Hannah, D. Millar, W. Thuiller, and A. Booth. 2003. Developing regional and species-level assessments of climate change impacts on biodiversity in the Cape Floristic Region. Biological Conservation **112**:87–97.

Mittermeier, R. A., N. Myers, J. B. Thomsen, G.A.B. da Fonseca, and S. Olivieri. 1998. Biodiversity hotspots and major tropical wilderness areas: Approaches to setting conservation priorities. Conservation Biology **12**:516–520.

Moffett, A., N. Shackelford, and S. Sarkar. 2007. Malaria in Africa: Vector species' niche models and relative risk maps. PLoS ONE **2**:e824.

Mohamed, K. I., M. Papeş, R. Williams, B. W. Benz, and A. T. Peterson. 2006. Global invasive potential of 10 parasitic witchweeds and related Orobanchaceae. AMBIO **35**:281–288.

Moilanen, A. 2005. Reserve selection using nonlinear species distribution models. American Naturalist **165**:695–706.

Molofsky, J., R. Durrett, J. Dushoff, D. Griffeath, and S. Levin. 1999. Local frequency dependence and global coexistence. Theoretical Population Biology **55**:270–282.

Moreno, C. E., and G. Halffter. 2000. Assessing the completeness of bat biodiversity inventories using species accumulation curves. Journal of Applied Ecology **37**: 149–158.

Muñoz, M. de S., R. De Giovanni, M. de Siqueira, T. Sutton, P. Brewer, R. Pereira, D. Canhos, and V. Canhos. 2009. OpenModeller: A generic approach to species' potential distribution modelling. GeoInformatica (online): doi:10.1007/s10707-009-0090-7.

Myers, N. 1979. *The Sinking Ark: A New Look at the Problem of Disappearing Species*. Pergamon Press, Oxford, UK.

Myers, N., R. A. Mittermeier, C. G. Mittermeier, G.A.B. da Fonseca, and J. Kent. 2000. Biodiversity hotspots for conservation priorities. Nature **403**:853–858.

Myneni, R. B., F. G. Hall, P. J. Sellers, and A. L. Marshak. 1995. The interpretation of spectral vegetation indexes. IEEE Transactions on Geoscience and Remote Sensing **33**:481–486.

Nakazawa, Y., R. Williams, A. T. Peterson, P. Mead, E. Staples, and K. L. Gage. 2007. Climate change effects on plague and tularemia in the United States. Vector Borne and Zoonotic Diseases **7**:529–540.

NAS. 2002. Predicting Invasions of Nonindigenous Plants and Plant Pests. National Academy Press, Washington, DC.

Natori, Y., and W. P. Porter. 2007. Model of Japanese serow (*Capricornis crispus*) energetics predicts distribution on Honshu, Japan. Ecological Applications **17**:1441–1459.

Navarro-Sigüenza, A. G., A. T. Peterson, and A. Gordillo-Martínez. 2006. Atlas de las Distribuciones de las Aves de México, unpublished database. Mexico City.

New, M., M. Hulme, and P. D. Jones. 1997. *A 1961–1990 Mean Monthly Climatology of Global Land Areas*. Climatic Research Unit, University of East Anglia, Norwich, UK.

Nix, H. A. 1986. A biogeographic analysis of Australian Elapid Snakes. In *Atlas of Elapid Snakes*, R. Longmore, editor, pp. 4–15. Australian Government Publishing Service, Canberra.

Nogués-Bravo, D. 2009. Predicting the past distribution of species climatic niches. Global Ecology and Biogeography **18**:521–531.

Nogués-Bravo, D., M. B. Araújo, T. Romdal, and C. Rahbek. 2008a. Scale effects and human impact on the elevational species richness gradients. Nature **453**: 216–219.

Nogués-Bravo, D., J. Rodriguez, J. Hortal, P. Batra, and M. B. Araújo. 2008b. Climate change, humans, and the extinction of the woolly mammoth. PLoS Biology 6:e79.

NRSC. 2009. National Remote Sensing Centre; http://www.nrsc.gov.in/. Department of Space, Balanagar, Hyderabad, India.

Nyári, Á., C. Ryall, and A. T. Peterson. 2006. Global invasive potential of the House Crow (*Corvus splendens*) based on ecological niche modelling. Journal of Avian Biology 37:306–311.

Odling-Smee, F. J., K. N. Laland, and M. W. Feldman. 2003. *Niche Construction: The Neglected Process in Evolution*. Princeton University Press, Princeton, NJ.

Omland, K. E., and S. M. Lanyon. 2000. Reconstructing plumage evolution in orioles (*Icterus*): Repeated convergence and reversal in patterns. Evolution 54:2119–2133.

Openshaw, S., and P. J. Taylor. 1981. The modifiable areal unit problem. In *Quantitative Geography: A British View*, N. Wrigley and R. J. Bennett, editors, pp. 60–69. Routledge & Kegan Paul, Ltd., London.

Orme, C.D.L., R. G. Davies, V. A. Olson, G. H. Thomas, T.-S. Ding, P. C. Rasmussen, R. S. Ridgely, A. J. Stattersfield, P. M. Bennett, I.P.F. Owens, T. M. Blackburn, and K. J. Gaston. 2006. Global patterns of geographic range size in birds. PLoS Biology 4:e208.

Ortega-Huerta, M. A. and., and A. T. Peterson. 2004. Modelling spatial patterns of biodiversity for conservation prioritization in north-eastern Mexico. Diversity and Distributions 10:39–54.

Ortiz-Martínez, T., V. Rico-Gray, and E. Martínez-Meyer. 2008. Predicted and verified distributions of *Ateles geoffroyi* and *Alouatta palliata* in Oaxaca, Mexico. Primates 49:186–194.

Ortiz-Pulido, R., A. T. Peterson, M. B. Robbins, R. Díaz, A. G. Navarro-Sigüenza, and G. Escalona-Segura. 2002. The Mexican Sheartail (*Doricha eliza*): Morphology, behavior, distribution, and status. Wilson Bulletin 114:153–160.

Padian, K., and L. M. Chiappe. 1998. The origin and early evolution of birds. Biological Reviews 73:1–42.

Pagel, M., A. Meade, and D. Barker. 2004. Bayesian estimation of ancestral character states on phylogenies. Systematic Biology 53:673–684.

Panetta, F. D., and J. Dodd. 1987. Bioclimatic prediction of the potential distribution of skeleton weed *Chondrilla juncea* L. in Western Australia. Journal of the Australian Institute of Agricultural Science 53:11–16.

Papeş, M., and A. T. Peterson. 2003. Predicting the potential invasive distribution for *Eupatorium adenophorum* Spreng. in China. Journal of Wuhan Botanical Research 21:137–142.

Papeş, M., R. Tupayachi, P. Martínez, A. T. Peterson, and G.V.N. Powell. 2010. Using hyperspectral satellite imagery for regional inventories: A test with tropical emergent trees in the Amazon Basin. Journal of Vegetation Science 21:342–354.

Parmesan, C. 2006. Ecological and evolutionary responses to recent climate change. Annual Review of Ecology, Evolution, and Systematics 37:637–669.

Patten, M. A. 2004. Correlates of species richness in North American bat families. Journal of Biogeography 31:975–985.

Patterson, B. D. 1999. Contingency and determinism in mammalian biogeography: The role of history. Journal of Mammalogy **80**:345–360.

Pearce, J. L., and M. S. Boyce. 2006. Modelling distribution and abundance with presence-only data. Journal of Applied Ecology **43**:405–412.

Pearce, J. L., and D. Lindenmayer. 1998. Bioclimatic analysis to enhance reintroduction biology of the endangered helmeted honeyeater (*Lichenostomus melanops cassidix*) in southeastern Australia. Restoration Ecology **6**:238–243.

Pearson, R. G. 2007. *Species' Distribution Modeling for Conservation Educators and Practitioners: Synthesis.* American Museum of Natural History, New York. Available at http://ncep.amnh.org.

Pearson, R. G., and T. P. Dawson. 2003. Predicting the impacts of climate change on the distribution of species: Are bioclimate envelope models useful? Global Ecology and Biogeography **12**:361–371.

———. 2004. Bioclimate envelope models: What they detect and what they hide—Response to Hampe (2004). Global Ecology and Biogeography **13**:471–473.

———. 2005. Long-distance plant dispersal and habitat fragmentation: Identifying conservation targets for spatial landscape planning under climate change. Biological Conservation **123**:389–401.

Pearson, R. G., T. P. Dawson, P. M. Berry, and P. A. Harrison. 2002. SPECIES: A spatial evaluation of climate impact on the envelope of species. Ecological Modelling **154**:289–300.

Pearson, R. G., C. J. Raxworthy, M. Nakamura, and A. T. Peterson. 2007. Predicting species' distributions from small numbers of occurrence records: A test case using cryptic geckos in Madagascar. Journal of Biogeography **34**:102–117.

Pearson, R. G., W. Thuiller, M. B. Araújo, E. Martínez-Meyer, L. Brotons, C. McClean, L. Miles, P. Segurado, T. P. Dawson, and D. C. Lees. 2006. Model-based uncertainty in species' range prediction. Journal of Biogeography **33**:1704–1711.

Peters, R. L., and J. P. Myers. 1991–1992. Preserving biodiversity in a changing climate. Issues in Science and Technology **8**:66–72.

Peterson, A. T. 2001. Predicting species' geographic distributions based on ecological niche modeling. Condor **103**:599–605.

———. 2003a. Predicting the geography of species' invasions via ecological niche modeling. Quarterly Review of Biology **78**:419–433.

———. 2003b. Projected climate change effects on Rocky Mountain and Great Plains birds: Generalities of biodiversity consequences. Global Change Biology **9**:647–655.

———. 2005a. Kansas Gap Analysis: The importance of validating distributional models before using them. Southwestern Naturalist **50**:230–236.

———. 2005b. Predicting potential geographic distributions of invading species. Current Science **89**:9.

———. 2006a. Ecological niche modeling and spatial patterns of disease transmission. Emerging Infectious Diseases **12**:1822–1826.

———. 2006b. Taxonomy *is* important in conservation: A preliminary reassessment of Philippine species-level bird taxonomy. Bird Conservation International **16**:155–173.

———. 2006c. Uses and requirements of ecological niche models and related distributional models. Biodiversity Informatics **3**:59–72.

———. 2007a. Ecological niche modelling and understanding the geography of disease transmission. Veterinaria Italiana **43**:393–400.

———. 2007b. Why not WhyWhere: The need for more complex models of simpler environmental spaces. Ecological Modelling **203**:527–530.

———. 2008a. Biogeography of diseases: A framework for analysis. Naturwissenschaften **95**:483–491.

———. 2008b. Improving methods for reporting spatial epidemiologic data. Emerging Infectious Diseases **14**:1335–1336.

———. 2009. Shifting suitability for malaria vectors across Africa with warming climates. BMC Infectious Diseases **9**:59; doi:10.1186/1471-2334-9-59.

———. 2011. Ecological niche conservatism: A time-structured review of evidence. Journal of Biogeography. In press; doi:10.1111/j.1365-2699.2010.02456.x.

Peterson, A. T., L. G. Ball, and K. P. Cohoon. 2002a. Predicting distributions of Mexican birds using ecological niche modelling methods. Ibis **144**:e27–e32.

Peterson, A. T., N. Barve, L. M. Bini, J. A. Diniz-Filho, A. Jiménez-Valverde, A. Lira-Noriega, J. Lobo, S. Maher, P. de Marco, Jr., E. Martínez-Meyer, Y. Nakazawa, and J. Soberón. 2009a. The climate envelope may not be empty. Proceedings of the National Academy of Sciences USA **106**:e47.

Peterson, A. T., J. T. Bauer, and J. N. Mills. 2004a. Ecological and geographic distribution of filovirus disease. Emerging Infectious Diseases **10**:40–47.

Peterson, A. T., B. W. Benz, and M. Papeş. 2007a. Highly pathogenic H5N1 avian influenza: Entry pathways into North America via bird migration. PLoS ONE **2**:e261.

Peterson, A. T., D. S. Carroll, J. N. Mills, and K. M. Johnson. 2004b. Potential mammalian filovirus reservoirs. Emerging Infectious Diseases **10**:2073–2081.

Peterson, A. T., and K. P. Cohoon. 1999. Sensitivity of distributional predictive algorithms to geographic data completeness. Ecological Modelling **117**:159–164.

Peterson, A. T., S. L. Egbert, V. Sánchez-Cordero, and K. P. Price. 2000. Geographic analysis of conservation priority: Endemic birds and mammals in Veracruz, Mexico. Biological Conservation **93**:85–94.

Peterson, A. T., and R. D. Holt. 2003. Niche differentiation in Mexican birds: Using point occurrences to detect ecological innovation. Ecology Letters **6**:774–782.

Peterson, A. T., and D. A. Kluza. 2003. New distributional modelling approaches for gap analysis. Animal Conservation **6**:47–54.

Peterson, A. T., R. R. Lash, D. S. Carroll, and K. M. Johnson. 2006a. Geographic potential for outbreaks of Marburg hemorrhagic fever. American Journal of Tropical Medicine and Hygiene **75**:9–15.

Peterson, A. T., C. Martínez-Campos, Y. Nakazawa, and E. Martínez-Meyer. 2005a. Time-specific ecological niche modeling predicts spatial dynamics of vector insects and human dengue cases. Transactions of the Royal Society of Tropical Medicine and Hygiene **99**:647–655.

Peterson, A. T., and E. Martínez-Meyer. 2007. Geographic evaluation of conservation status of African forest squirrels (Sciuridae) considering land use change and climate change: The importance of point data. Biodiversity and Conservation **16**:3939–3950.

Peterson, A. T., E. Martínez-Meyer, C. González-Salazar, and P. W. Hall. 2004c. Modeled climate change effects on distributions of Canadian butterfly species. Canadian Journal of Zoology **82**:851–858.

Peterson, A. T., and Y. Nakazawa. 2008. Environmental data sets matter in ecological niche modelling: An example with *Solenopsis invicta* and *Solenopsis richteri*. Global Ecology and Biogeography **17**:135–144.

Peterson, A. T., and A. G. Navarro-Sigüenza. 2009. Making biodiversity discovery more efficient: An exploratory test using Mexican birds. Zootaxa **2246**:58–66.

Peterson, A. T., and Á. S. Nyári. 2007. Ecological niche conservatism and Pleistocene refugia in the Thrush-like Mourner, *Schiffornis* sp., in the Neotropics. Evolution **62**:173–183.

Peterson, A. T., M. A. Ortega-Huerta, J. Bartley, V. Sánchez-Cordero, J. Soberón, R. H. Buddemeier, and D.R.B. Stockwell. 2002b. Future projections for Mexican faunas under global climate change scenarios. Nature **416**:626–629.

Peterson, A. T., and M. Papeş. 2007. Potential geographic distribution of the Bugun Liocichla *Liocichla bugunorum*, a poorly known species from north-eastern India. Indian Birds **2**:146–149.

Peterson, A. T., M. Papeş, D. S. Carroll, H. Leirs, and K. M. Johnson. 2007b. Mammal taxa constituting potential coevolved reservoirs of filoviruses. Journal of Mammalogy **88**:1544–1554.

Peterson, A. T., M. Papeş, and M. Eaton. 2007c. Transferability and model evaluation in ecological niche modeling: A comparison of GARP and Maxent. Ecography **30**:550–560.

Peterson, A. T., M. Papeş, and D. A. Kluza. 2003a. Predicting the potential invasive distributions of four alien plant species in North America. Weed Science **51**: 863–868.

Peterson, A. T., M. Papeş, and J. Soberón. 2008a. Rethinking receiver operating characteristic analysis applications in ecological niche modeling. Ecological Modelling **213**:63–72.

Peterson, A. T., R. S. Pereira, and V. F. Neves. 2004d. Using epidemiological survey data to infer geographic distributions of leishmania vector species. Revista da Sociedade Brasileira de Medicina Tropical **37**:10–14.

Peterson, A. T., A. Robbins, R. Restifo, J. Howell, and R. Nasci. 2009b. Predictable ecology and geography of West Nile virus transmission in the central United States. Journal of Vector Ecology **33**:342–352.

Peterson, A. T., and C. R. Robins. 2003. Using ecological-niche modeling to predict barred owl invasions with implications for spotted owl conservation. Conservation Biology **17**:1161–1165.

Peterson, A. T., V. Sánchez-Cordero, C. B. Beard, and J. M. Ramsey. 2002c. Ecologic niche modeling and potential reservoirs for Chagas disease, Mexico. Emerging Infectious Diseases **8**:662–667.

Peterson, A. T., V. Sánchez-Cordero, E. Martínez-Meyer, and A. G. Navarro-Sigüenza. 2006b. Tracking population extirpations via melding ecological niche modeling with land-cover information. Ecological Modelling **195**:229–236.

Peterson, A. T., V. Sánchez-Cordero, J. Soberón, J. Bartley, R. H. Buddemeier, and A. G. Navarro-Sigüenza. 2001. Effects of global climate change on geographic distributions of Mexican Cracidae. Ecological Modelling **144**:21–30.

Peterson, A. T., R. Scachetti-Pereira, and D. A. Kluza. 2003b. Assessment of invasive potential of *Homalodisca coagulata* in western North America and South America. Biota Neotropica **3**:1–7.

Peterson, A. T., and J. Shaw. 2003. *Lutzomyia* vectors for cutaneous leishmaniasis in southern Brazil: Ecological niche models, predicted geographic distributions, and climate change effects. International Journal of Parasitology **33**:919–931.

Peterson, A. T., J. Soberón, and V. Sánchez-Cordero. 1999. Conservatism of ecological niches in evolutionary time. Science **285**:1265–1267.

Peterson, A. T., A. Stewart, K. I. Mohamed, and M. B. Araújo. 2008b. Shifting global invasive potential of European plants with climate change. PLoS ONE **3**:e2441.

Peterson, A. T., D.R.B. Stockwell, and D. A. Kluza. 2002d. Distributional prediction based on ecological niche modeling of primary occurrence data. In *Predicting Species Occurrences: Issues of Accuracy and Scale*, J. M. Scott, P. J. Heglund, and M. L. Morrison, editors, pp. 617–623. Island Press, Washington, DC.

Peterson, A. T., H. Tian, E. Martínez-Meyer, J. Soberón, V. Sánchez-Cordero, and B. Huntley. 2005b. Modeling distributional shifts of individual species and biomes. In *Climate Change and Biodiversity*, T. E. Lovejoy and L. Hannah, editors, pp. 211–228. Yale University Press, New Haven, CT.

Peterson, A. T., and D. A. Vieglais. 2001. Predicting species invasions using ecological niche modeling: New approaches from bioinformatics attack a pressing problem. BioScience **51**:363–371.

Peterson, A. T., R. Williams, and G. Chen. 2007d. Modeled global invasive potential of Asian gypsy moths, *Lymantria dispar*. Entomologia Experimentalis et Applicata **125**:39–44.

Phillips, S. J. 2008. Transferability, sample selection bias, and background data in presence-only modelling: A response to Peterson et al. (2007). Ecography **31**: 272–278.

Phillips, S. J., R. P. Anderson, and R. E. Schapire. 2006. Maximum entropy modeling of species geographic distributions. Ecological Modelling **190**:231–259.

Phillips, S. J., and M. Dudík. 2008. Modeling of species distributions with Maxent: New extensions and a comprehensive evaluation. Ecography **31**:161–175.

Phillips, S. J., M. Dudík, J. Elith, C. H. Graham, A. Lehmann, J. Leathwick, and S. Ferrier. 2009. Sample selection bias and presence-only distribution models: Implications for background and pseudo-absence data. Ecological Applications **19**:181–197.

Phillips, S. J., P. Williams, G. Midgley, and A. Archer. 2008. Optimizing dispersal corridors for the Cape Proteaceae using network flow. Ecological Applications **18**: 1200–1211.

Pickett, S.T.A., and F. A. Bazzaz. 1978. Organization of an assemblage of early successional species on a soil moisture gradient. Ecology **59**:1248–1255.

Pielou, E. C. 1984. *The Interpretation of Ecological Data: A Primer on Classification and Ordination*. John Wiley and Sons, New York.

Pimentel, D., R. Zuñiga, and D. Morrison. 2004. Update on the environmental and economic costs associated with alien-invasive species in the United States. Ecological Economics **52**:273–288.

Porter, W. P., J. L. Sabo, C. R. Tracy, O. J. Reichman, and N. Ramankutty. 2002. Physiology on a landscape scale: Plant-animal interactions. Integrative and Comparative Biology **42**:431–453.

Prendergast, J. R., S. N. Wood, J. H. Lawton, and B. C. Eversham. 1993. Correcting for variation in recording effort in analyses of diversity hotspots. Biodiversity Letters **1**:39–53.

Price, J. J., and S. M. Lanyon. 2002. A robust phylogeny of the oropendolas: Polyphyly revealed by mitochondrial sequence data. Auk **119**:335–348.

———. 2004. Song and molecular data identify congruent but novel affinities of the Green Oropendola (*Psarocolius viridis*). Auk **121**:224–229.

Primack, R. B. 2006. *Essentials of Conservation Biology*, 4th ed. Sinauer Associates, Inc., Sunderland, MA.

Prinzing, A., W. Durka, S. Klotz, and R. Brandl. 2001. The niche of higher plants: Evidence for phylogenetic conservatism. Proceedings of the Royal Society **268**:2383–2389.

Pulliam, H. R. 1988. Sources, sinks, and population regulation. American Naturalist **132**:652–661.

———. 2000. On the relationship between niche and distribution. Ecology Letters **3**:349–361.

Quattrochi, D. A., and M. F. Goodchild. 1997. *Scale in Remote Sensing and GIS*. Lewis Publishers, Boca Raton, FL.

Rabinowitz, D., S. Cairns, and T. Dillon. 1986. Seven forms of rarity and their frequency in the flora of the British Isles. In *Conservation Biology: The Science of Scarcity and Diversity*, M. E. Soulé, editor, pp. 182–204. Sinauer Associates, Sunderland, MA.

Raes, N., and H. ter Steege. 2007. A null-model for significance testing of presence-only species distribution models. Ecography **30**:727–736.

Ramírez-Bastida, P., A. G. Navarro-Sigüenza, and A. T. Peterson. 2008. Aquatic bird distributions in Mexico: Designing conservation approaches quantitatively. Biodiversity and Conservation **17**:2525–2558.

Randin, C. F., T. Dirnböck, S. Dullinger, N. E. Zimmermann, M. Zappa, and A. Guisan. 2006. Are niche-based species' distribution models transferable in space? Journal of Biogeography **33**:1689–1703.

Raxworthy, C. J., C. M. Ingram, N. Rabibisoa, and R. G. Pearson. 2007. Applications of ecological niche modeling for species delimitation: A review and empirical evaluation using day geckos (*Phelsuma*) from Madagascar. Systematic Biology **56**:907–923.

Raxworthy, C. J., E. Martínez-Meyer, N. Horning, R. A. Nussbaum, G. E. Schneider, M. A. Ortega-Huerta, and A. T. Peterson. 2003. Predicting distributions of known and unknown reptile species in Madagascar. Nature **426**:837–841.

Raxworthy, C. J., R. G. Pearson, B. M. Zimkus, S. Reddy, A. J. Deo, R. A. Nussbaum, and C. M. Ingram. 2008. Continental speciation in the tropics: Contrasting biogeographic patterns of divergence in the *Uroplatus* leaf-tailed gecko radiation of Madagascar. Journal of Zoology **275**:423–440.

Ray, N., A. Lehmann, and P. Joly. 2002. Modeling spatial distribution of amphibian populations: A GIS approach based on habitat matrix permeability. Biodiversity and Conservation **11**:2143–2165.

Reddy, S., and L. M. Dávalos. 2003. Geographical sampling bias and its implications for conservation priorities in Africa. Journal of Biogeography **30**:1719–1727.

Reed, K. D., J. K. Meece, J. R. Archer, and A. T. Peterson. 2008. Ecologic niche modeling of *Blastomyces dermatitidis* in Wisconsin. PLoS ONE **3**:e2034.

Reed, K. D., J. W. Melski, M. B. Graham, R. L. Regnery, M. J. Sotir, M. V. Wegner, J. J. Kazmierczak, E. J. Stratman, Y. Li, J. A. Fairley, G. R. Swain, V. A. Olson, E. K. Sargent, S. C. Kehl, M. A. Frace, R. Kline, S. L. Foldy, J. P. Davis, and

I. K. Damon. 2004. The detection of monkeypox in humans in the Western Hemisphere. New England Journal of Medicine **350**:342–350.

Regal, P. J. 1994. Scientific principles for ecologically based risk assessment of transgenic organisms. Molecular Ecology **3**:5–13.

Rescigno, A., and I. W. Richardson. 1973. The deterministic theory of population dynamics. In *Foundations of Mathematical Biology, Volume 3: Supercellular Systems*, R. Rosen, editor, pp. 238–360. Academic Press, New York.

Richardson, D. M., and J. P. McMahon. 1992. A bioclimatic analysis of *Eucalyptus nitens* to identify potential planting regions in southern Africa. South African Journal of Science **88**:380–387.

Ricklefs, R. E. 2004. A comprehensive framework for global patterns in biodiversity. Ecology Letters **7**:1–15.

Ridgely, R. S., T. F. Allnutt, T. Brooks, D. K. McNicol, D. W. Mehlman, B. E. Young, and J. R. Zook. 2005. Digital Distribution Maps of the Birds of the Western Hemisphere, version 2.1. NatureServe, Arlington, VA.

Rissler, L. J., and J. J. Apodaca. 2007. Adding more ecology into species delimitation: Ecological niche models and phylogeography help define cryptic species in the black salamander (*Aneides flavipunctatus*). Systematic Biology **56**:924–942.

Robertson, M. P., M. H. Villet, and A. R. Palmer. 2004. A fuzzy classification technique for predicting species' distributions: Applications using invasive alien plants and indigenous insects. Diversity and Distributions **10**:461–474.

Rödder, D., S. Schmidtlein, M. Veith, and S. Lötters. 2009. Alien invasive slider turtle in unpredicted habitat: A matter of niche shift or of predictors studied? PLoS ONE **4**:e7843.

Rodrigues, A.S.L., and K. J. Gaston. 2001. How large do reserve networks need to be? Ecology Letters **4**:602–609.

Rogers, D. J., and S. E. Randolph. 1988. Tsetse flies in Africa: Bane or boon? Conservation Biology **2**:57–65.

Rogers, D. J., S. E. Randolph, R. W. Snow, and S. I. Hay. 2002. Satellite imagery in the study and forecast of malaria. Nature **415**:710–715.

Rojas-Soto, O. R., O. Alcántara-Ayala, and A. G. Navarro-Sigüenza. 2003. Regionalization of the avifauna of the Baja California Peninsula, Mexico: A parsimony analysis of endemicity and distributional modelling approach. Journal of Biogeography **30**:449–461.

Rojas-Soto, O. R., E. Martínez-Meyer, A. G. Navarro-Sigüenza, A. Oliveras de Ita, H. Gómez de Silva, and A. T. Peterson. 2008. Modeling distributions of disjunct populations of the Sierra Madre Sparrow. Journal of Field Ornithology **79**:245–253.

Ron, S. R. 2005. Predicting the distribution of the amphibian pathogen *Batrachochytrium dendrobatidis* in the New World. Biotropica **37**:209–221.

Root, T. 1988. Environmental factors associated with avian distributional boundaries. Journal of Biogeography **15**:489–505.

Rotenberry, J. T., K. L. Preston, and S. T. Knick. 2006. GIS-based niche modeling for mapping species' habitat. Ecology **87**:1458–1464.

Rotenberry, J. T., and J. A. Wiens. 1980. Habitat structure, patchiness, and avian communities in North American steppe vegetation: A multivariate analysis. Ecology **61**:1228–1250.

Roura-Pascual, N., A. V. Suarez, C. Gómez, P. Pons, Y. Touyama, A. L. Wild, and A. T. Peterson. 2005. Geographic potential of Argentine ants (*Linepithema humile* Mayr) in the face of global climate change. Proceedings of the Royal Society B **271**:2527–2535.

Roura-Pascual, N., A. V. Suarez, K. M. McNyset, C. Gómez, P. Pons, Y. Touyama, A. L. Wild, F. Gascon, and A. T. Peterson. 2006. Niche differentiation and fine-scale projections for Argentine ants based on remotely sensed data. Ecological Applications **16**:1832–1841.

Sánchez-Cordero, V., P. Illoldi-Rangel, M. Linaje, S. Sarkar, and A. T. Peterson. 2005. Deforestation and extant distributions of Mexican endemic mammals. Biological Conservation **126**:465–473.

Sánchez-Cordero, V., and E. Martínez-Meyer. 2000. Museum specimen data predict crop damage by tropical rodents. Proceedings of the National Academy of Sciences USA **97**:7074–7077.

Scott, J. K., and F. D. Panetta. 1993. Predicting the Australian weed status of southern African plants. Journal of Biogeography **20**:87–93.

Scott, J. M., F. Davis, B. Csuti, R. Noss, B. Butterfield, C. Groves, H. Anderson, S. Caicco, F. D'Erichia, T. C. Edwards, Jr., J. Ulliman, and R. G. Wright. 1993. Gap analysis: A geographic approach to protection of biological diversity. Wildlife Monographs **123**:3–41.

Scott, J. M., T. H. Tear, and F. W. Davis, editors. 1996. *Gap Analysis: A Landscape Approach to Biodiversity Planning*. American Society for Photogrammetry and Remote Sensing, Bethesda, MD.

Segurado, P., and M. B. Araújo. 2004. An evaluation of methods for modelling species distributions. Journal of Biogeography **31**:1555–1568.

Segurado, P., M. B. Araújo, and W. E. Kunin. 2006. Consequences of spatial autocorrelation for niche-based models. Journal of Applied Ecology **43**:433–444.

Sexton, J. P., P. J. McIntyre, A. L. Angert, and K. J. Rice. 2009. Evolution and ecology of species range limits. Annual Review of Ecology, Evolution, and Systematics **40**:415–436.

Shao, J., and C.F.J. Wu. 1989. A general theory for jackknife variance estimation. Annals of Statistics **17**:1176–1197.

Shin, S. I., Z. Liu, B. Otto-Bliesner, E. C. Brady, J. E. Kutzbach, and S. P. Harrison. 2003. A simulation of the Last Glacial Maximum climate using the NCAR-CCSM. Climate Dynamics **20**:127–151.

Silvertown, J. 2004. Plant coexistence and the niche. Trends in Ecology and Evolution **19**:605–611.

Silvertown, J., M. Dodd, D. Gowning, C. Lawson, and K. McConway. 2006. Phylogeny and the hierarchical organization of plant diversity. Ecology **87**:S39–S49.

Siqueira, M. F., G. Durigan, P. de Marco, Jr., and A. T. Peterson. 2009. Something from nothing: Using landscape similarity and ecological niche modeling to find rare plant species. Journal for Nature Conservation **17**:25–32.

Siqueira, M. F., and A. T. Peterson. 2003. Global climate change consequences for cerrado tree species. Biota Neotropica **3**:1–14.

Smith, P. A. 1994. Autocorrelation in logistic regression modelling of species' distribution. Global Ecology and Biogeography Letters **4**:47–61.

Soberón, J. 1999. Linking biodiversity information sources. Trends in Ecology and Evolution **14**:291.

———. 2007. Grinnellian and Eltonian niches and geographic distributions of species. Ecology Letters **10**:1115–1123.

———. 2010. Niche and area of distribution modeling: A population ecology perspective. Ecography **33**:1–9.

Soberón, J., L. Arriaga, and L. Lara. 2002. Issues of quality control in large, mixed-origin entomological databases. In *Towards a Global Biological Information Infrastructure*, H. Saarenmaa and E. Nielsen, editors, pp. 15–22. European Environmental Agency, Copenhagen.

Soberón, J., J. Golubov, and J. Sarukhán. 2001. The importance of *Opuntia* in Mexico and routes of invasion and impact of *Cactoblastis cactorum* (Lepidoptera: Pyralidae). Florida Entomologist **84**:486–492.

Soberón, J., R. Jiménez, J. Golubov, and P. Koleff. 2007. Assessing completeness of biodiversity databases at different spatial scales. Ecography **30**:152–160.

Soberón, J., J. Llorente, and H. Benítez. 1996. An international view of national biological surveys. Annals of the Missouri Botanical Garden **83**:562–573.

Soberón, J., J. B. Llorente, and L. Oñate. 2000. The use of specimen-label databases for conservation purposes: An example using Mexican papilionid and pierid butterflies. Biodiversity and Conservation **9**:1441–1466.

Soberón, J., and M. Nakamura. 2009. Niches and distributional areas: Concepts, methods, and assumptions. Proceedings of the National Academy of Sciences USA **106**:19644–19650.

Soberón, J., and A. T. Peterson. 2004. Biodiversity informatics: Managing and applying primary biodiversity data. Philosophical Transactions of the Royal Society of London B **359**:689–698.

———. 2005. Interpretation of models of fundamental ecological niches and species' distributional areas. Biodiversity Informatics **2**:1–10.

———. 2008. Monitoring biodiversity loss with primary species-occurrence data: Toward national-level indicators for the 2010 Target of the Convention on Biological Diversity. AMBIO **38**:29–34.

Sokal, R. R., and F. J. Rohlf. 1995. *Biometry: The Principles and Practice of Statistics in Biological Research*, 3rd ed. W. H. Freeman, New York.

Solé, R. V., and J. Bascompte. 2006. *Self-organization in Complex Ecosystems*. Princeton University Press, Princeton, NJ.

SSIC. 2009. Rapportsystemet för fåglar; http://www.artportalen.se/birds/. Swedish Species Information Centre.

Stein, B. R., and J. Wieczorek. 2004. Mammals of the world: MaNIS as an example of data integration in a distributed network environment. Biodiversity Informatics **1**:14–22.

Stockwell, D.R.B. 2006. Improving ecological niche models by data mining large environmental datasets for surrogate models. Ecological Modelling **192**:188–196.

———. 2007. *Niche Modeling: Predictions from Statistical Distributions*. Chapman & Hall/CRC, London.

Stockwell, D.R.B., and D. P. Peters. 1999. The GARP modelling system: Problems and solutions to automated spatial prediction. International Journal of Geographical Information Systems **13**:143–158.

Strasburg, J. L., M. Kearney, C. Moritz, and A. R. Templeton. 2007. Combining phylogeography with distribution modeling: Multiple Pleistocene range expansions in
a parthenogenetic gecko from the Australian arid zone. PLoS ONE 2:e760.

Sutherst, R. W. 2001. The vulnerability of animal and human health to parasites under
global change. International Journal for Parasitology 31:933–948.

———. 2003. Prediction of species geographical ranges. Journal of Biogeography 30:
805–816.

Svenning, J.-C., S. Normand, and F. Skov. 2008. Postglacial dispersal limitation of
widespread forest plant species in nemoral Europe. Ecography 31:316–326.

Svenning, J.-C., and F. Skov. 2004. Limited filling of the potential range in European
tree species. Ecology Letters 7:565–573.

Sweeney, A. W., N. W. Beebe, R. D. Cooper, J. T. Bauer, and A. T. Peterson. 2006. Environmental factors associated with distribution and range limits of malaria vector
Anopheles farauti in Australia. Journal of Medical Entomology 43:1068–1075.

Swets, J. A. 1988. Measuring the accuracy of diagnostic systems. Science 240:1285–1293.

Sykes, M. T., I. C. Prentice, and W. Cramer. 1996. A bioclimatic model for the potential
distributions of north European tree species under present and future climates.
Journal of Biogeography 23:203–233.

Tarassenko, L. 1998. *Guide to Neural Computing Applications*, 1st ed. John Wiley &
Sons, New York.

Tax, D.M.J., and R. P. W. Duin. 1999. Support vector domain description. Pattern Recognition Letters 20:1191–1199.

Téllez-Valdés, O., and P. Dávila-Aranda. 2003. Protected areas and climate change: A
case study of the cacti in the Tehuacan-Cuicatlan Biosphere Reserve, Mexico.
Conservation Biology 17:846–853.

Thomas, C. D., A. Cameron, R. E. Green, M. Bakkenes, L. J. Beaumont, Y. C. Collingham, B. F. N. Erasmus, M. F. de Siqueira, A. Grainger, L. Hannah, L. Hughes,
B. Huntley, A. S. van Jaarsveld, G. F. Midgley, L. Miles, M. A. Ortega-Huerta,
A. T. Peterson, O. L. Phillips, and S. E. Williams. 2004a. Extinction risk from
climate change. Nature 427:145–148.

Thomas, C. D., and J. J. Lennon. 1999. Birds extend their ranges northwards. Nature
399:213.

Thomas, J. A., M. G. Telfer, D. B. Roy, C. D. Preston, J.J.D. Greenwood, J. Asher, R. Fox,
R. T. Clarke, and J. H. Lawton. 2004b. Comparative losses of British butterflies,
birds, and plants and the global extinction crisis. Science 303:1879–1881.

Thompson, J. N. 2005. *The Geographic Mosaic of Coevolution*. University of Chicago
Press, Chicago.

Thomson, M. C., D. A. Elnaiem, R. W. Ashford, and S. J. Connor. 1999. Towards a kala
azar risk map for Sudan: Mapping the potential distribution of *Phlebotomus ori-
entalis* using digital data of environmental variables. Tropical Medicine and International Health 4:105–113.

Thomson, M. C., V. Obsomer, M. Dunne, S. J. Connor, and D. H. Molyneux. 2000.
Satellite mapping of loa loa prevalence in relation to ivermectin use in west and
central Africa. Lancet 356:1077–1078.

Thuiller, W. 2003. BIOMOD—Optimizing predictions of species distributions and projecting potential future shifts under global change. Global Change Biology 9:
1353–1362.

———. 2004. Patterns and uncertainties of species' range shifts under climate change. Global Change Biology **10**:2020–2027.

Thuiller, W., M. B. Araújo, and S. Lavorel. 2003. Generalized models versus classification tree analysis: Predicting spatial distributions of plant species at different scales. Journal of Vegetation Science **14**:669–680.

———. 2004a. Do we need land-cover data to model species distributions in Europe? Journal of Biogeography **31**:353–361.

Thuiller, W., M. B. Araújo, R. G. Pearson, R. J. Whittaker, L. Brotons, and S. Lavorel. 2004b. Biodiversity conservation: Uncertainty in predictions of extinction risk. Nature **430**:33.

Thuiller, W., L. Brotons, M. B. Araújo, and S. Lavorel. 2004c. Effects of restricting environmental range of data to project current and future species distributions. Ecography **27**:165–172.

Thuiller, W., B. Lafourcade, R. Engler, and M. Araújo. 2009. BIOMOD—A platform for ensemble forecasting of species distributions. Ecography **32**:369–373.

Thuiller, W., S. Lavorel, M. B. Araújo, M. T. Sykes, and I. C. Prentice. 2005a. Climate change threats to plant diversity in Europe. Proceedings of the National Academy of Sciences USA **102**:8245–8250.

Thuiller, W., D. M. Richardson, P. Pysék, G. F. Midgley, G. O. Hughes, and M. Rouget. 2005b. Niche-based modelling as a tool for predicting the global risk of alien plant invasions. Global Change Biology **11**:2234–2250.

Tiedje, J. M., R. K. Colwell, Y. L. Grossman, R. E. Hodson, R. E. Lenski, R. N. Mack, and P. J. Regal. 1989. The planned introduction of genetically engineered organisms: Ecological considerations and recommendations. Ecology **70**:298–315.

Tilman, D. 1982. *Resource Competition and Community Structure*. Princeton University Press, Princeton, NJ.

Timm, R. M., R. M. Salazar, and A. T. Peterson. 1997. Historical distribution of the extinct tropical seal, *Monachus tropicalis* (Carnivora: Phocidae). Conservation Biology **11**:549–551.

Toribio, M., and A. T. Peterson. 2008. Prioritisation of Mexican lowland rain forests for conservation using modelled geographic distributions of birds. Journal for Nature Conservation **16**:109–116.

Udvardy, M.D.F. 1969. *Dynamic Zoogeography. With Special Reference to Land Animals*. Van Nostrand Reinhold, New York.

Usinger, R. L., P. Wygodzinsky, and R. E. Ryckman. 1966. The biosystematics of Triatominae. Annual Review of Entomology **11**:309–330.

Vaclavik, T., and R. K. Meentemeyer. 2009. Invasive species distribution modeling (iSDM): Are absence data and dispersal constraints needed to predict actual distributions? Ecological Modelling **220**:3248–3258.

Vanak, A. T., M. Irfan-Ullah, and A. T. Peterson. 2008. Gap analysis of Indian Fox conservation using ecological niche modelling. Journal of the Bombay Natural History Society **105**:49–54.

Vandermeer, J. H. 1972. Niche theory. Annual Review of Ecology and Systematics **3**:107–132.

VanDerWal, J., L. P. Shoo, C. N. Johnson, and S. E. Williams. 2009. Abundance and the environmental niche: Environmental suitability estimated from niche models predicts the upper limit of local abundance. American Naturalist **174**:282–291.

Vane-Wright, R. I. 1996. Identification of priorities for the conservation of biodiversity: Systematic biological criteria within a socio-political framework. In *Biodiversity: A Biology of Numbers and Difference*, K. J. Gaston, editor, pp. 309–344. Blackwell Science, Oxford, UK.

Vane-Wright, R. I., C. J. Humphries, and P. H. Williams. 1991. What to protect?—Systematics and the agony of choice. Biological Conservation **55**:235–254.

Van Lieshout, M., R. S. Kovats, M.T.J. Livermore, and P. Martens. 2004. Climate change and malaria: Analysis of the SRES climate and socio-economic scenarios. Global Environmental Change **14**:87–99.

Varela, S., J. Rodríguez, and J. M. Lobo. 2009. Is current climatic equilibrium a guarantee for the transferability of distribution model predictions? A case study of the spotted hyena. Journal of Biogeography **36**:1645–1655.

Veloz, S. D. 2009. Spatially autocorrelated sampling falsely inflates measures of accuracy for presence-only niche models. Journal of Biogeography **36**:2290–2299.

Voss, R. S., and L. H. Emmons. 1996. Mammalian diversity in Neotropical lowland rainforests: A preliminary assessment. Bulletin of the American Museum of Natural History **230**:1–115.

Walker, P. A., and K. D. Cocks. 1991. HABITAT: A procedure for modelling a disjoint environmental envelope for a plant or animal species. Global Ecology and Biogeography Letters **1**:108–118.

Waltari, E., R. J. Hijmans, A. T. Peterson, Á. S. Nyári, S. L. Perkins, and R. P. Guralnick. 2007. Locating Pleistocene refugia: Comparing phylogeographic and ecological niche model predictions. PLoS ONE **2**:e563.

Walther, G.-R., S. Berger, and M. T. Sykes. 2005. An ecological "footprint" of climate change. Proceedings of the Royal Society B **272**:1427–1432.

Ward, G., T. Hastie, S. Barry, J. Elith, and J. R. Leathwick. 2009. Presence-only data and the EM algorithm. Biometrics **65**:554–563.

Warren, D. L., R. E. Glor, and M. Turelli. 2008. Environmental niche equivalency versus conservatism: Quantitative approaches to niche evolution. Evolution **62**: 2868–2883.

Welk, E., K. Schubert, and M. H. Hoffmann. 2002. Present and potential distribution of invasive garlic mustard (*Alliaria petiolata*) in North America. Diversity and Distributions **8**:219–233.

Wells, P. V., and C. D. Jorgensen. 1964. Pleistocene wood rat middens and climatic change in mohave desert: A record of juniper woodlands. Science **143**:1171–1173.

Whittaker, R. H., S. A. Levin, and R. B. Root. 1973. Niche, habitat, and ecotope. American Naturalist **955**:321–338.

Whittaker, R. J., M. B. Araújo, P. Jepson, R. J. Ladle, J.E.M. Watson, and K. J. Willis. 2005. Conservation biogeography: Assessment and prospect. Diversity and Distributions **11**:3–23.

Wieczorek, J., Q. Guo, and R. J. Hijmans. 2004. The point-radius method for georeferencing locality descriptions and calculating associated uncertainty. International Journal of Geographical Information Science **18**:745 –767.

Wiens, J. J. 2004. Speciation and ecology revisited: Phylogenetic niche conservatism and the origin of species. Evolution **58**:193–197.

———. 2007. Species delimitation: New approaches for discovering diversity. Systematic Biology **56**:875–878.

Wiens, J. J., and C. H. Graham. 2005. Niche conservatism: Integrating evolution, ecology, and conservation biology. Annual Review of Ecology, Evolution, and Systematics **36**:519–539.

Wilby, R. L., and T.M.L. Wigley. 1997. Downscaling general circulation model output: A review of methods and limitations. Progress in Physical Geography **21**:530–548.

Wiley, E. O., K. M. McNyset, A. T. Peterson, C. R. Robins, and A. M. Stewart. 2003. Niche modeling and geographic range predictions in the marine environment using a machine-learning algorithm. Oceanography **16**:120–127.

Williams, J. W., and S. T. Jackson. 2007. Novel climates, no-analog communities, and ecological surprises. Frontiers in Ecology and the Environment **5**:475–482.

Williams, P. H. 2001. Complementarity. In *Encyclopedia of Biodiversity*, 2nd ed., S. A. Levin, editor, pp. 813–829. Academic Press, San Diego, CA.

Williams, P. H., L. Hannah, S. Andelman, G. Midgley, M. B. Araújo, G. Hughes, L. Manne, E. Martínez-Meyer, and R. G. Pearson. 2005. Planning for climate change: Identifying minimum-dispersal corridors for the Cape Proteaceae. Conservation Biology **19**:1063–1074.

Williams, R.A.J., F. O. Fasina, and A. T. Peterson. 2008. Predictable ecology and geography of avian influenza (H5N1) transmission in Nigeria and West Africa. Transactions of the Royal Society of Tropical Medicine and Hygiene **102**:471–479.

Williamson, M. 1996. *Biological Invasions*. Chapman & Hall, London.

Wilson, E. O., editor. 1988. *Biodiversity*. National Academy Press, Washington, DC.

Wisz, M. S., R. Hijmans, J. Li, A. T. Peterson, C. H. Graham, A. Guisan, and the NCEAS Predicting Species Distributions Working Group. 2008. Effects of sample size on the performance of species distribution models. Diversity and Distributions **14**:763–773.

Woodward, F. I., and D. J. Beerling. 1997. The dynamics of vegetation change: Health warnings for equilibrium "dodo" models. Global Ecology and Biogeography Letters **6**:413–418.

Wright, S. 1982. The Shifting Balance Theory and macroevolution. Annual Review of Genetics **16**:1–20.

Yeshiwondim, A. K., S. Gopal, A. T. Hailemariam, D. O. Dengela, and H. P. Patel. 2009. Spatial analysis of malaria incidence at the village level in areas with unstable transmission in Ethiopia. International Journal of Health Geographics **8**:5; doi:10.1186/1476-072X-8-5.

Yoshiyama, R. M., and J. Roughgarden. 1977. Species packing in two dimensions. American Naturalist **111**:107–121.

Zambrano, L., E. Martínez-Meyer, N. Menezes, and A. T. Peterson. 2006. Invasive potential of common carp (*Cyprinus carpio*) and Nile tilapia (*Oreochromis niloticus*) in American freshwater systems. Canadian Journal of Fisheries and Aquatic Sciences **63**:1903–1910.

Zaniewski, A. E., A. Lehmann, and J. M. Overton. 2002. Predicting species spatial distributions using presence-only data: A case study of native New Zealand ferns. Ecological Modelling **157**:261–280.

Zavaleta, E. S., R. J. Hobbs, and H. A. Mooney. 2001. Viewing invasive species removal in a whole-ecosystem context. Trends in Ecology and Evolution **16**:454–459.

Zink, R. M., and M. C. McKitrick. 1995. The debate about species concepts and its implications for ornithology. Auk **112**:701–719.

MONOGRAPHS IN POPULATION BIOLOGY
EDITED BY SIMON A. LEVIN AND HENRY S. HORN

1. *The Theory of Island Biogeography*, by Robert H. MacArthur and Edward O. Wilson
2. *Evolution in Changing Environments: Some Theoretical Explorations*, by Richard Levins
3. *Adaptive Geometry of Trees*, by Henry S. Horn
4. *Theoretical Aspects of Population Genetics*, by Motoo Kimura and Tomoko Ohta
5. *Populations in a Seasonal Environment*, by Steven D. Fretwell
6. *Stability and Complexity in Model Ecosystems*, by Robert M. May
7. *Competition and the Structure of Bird Communities*, by Martin L. Cody
8. *Sex and Evolution*, by George C. Williams
9. *Group Selection in Predator-Prey Communities*, by Michael E. Gilpin
10. *Geographic Variation, Speciation, and Clines*, by John A. Endler
11. *Food Webs and Niche Space*, by Joel E. Cohen
12. *Caste and Ecology in the Social Insects*, by George F. Oster and Edward O. Wilson
13. *The Dynamics of Arthropod Predator-Prey Systems*, by Michael P. Hassel
14. *Some Adaptations of Marsh-Nesting Blackbirds*, by Gordon H. Orians
15. *Evolutionary Biology of Parasites*, by Peter W. Price
16. *Cultural Transmission and Evolution: A Quantitative Approach*, by L. L. Cavalli-Sforza and M. W. Feldman
17. *Resource Competition and Community Structure*, by David Tilman
18. *The Theory of Sex Allocation*, by Eric L. Charnov
19. *Mate Choice in Plants: Tactics, Mechanisms, and Consequences*, by Nancy Burley and Mary F. Wilson
20. *The Florida Scrub Jay: Demography of a Cooperative-Breeding Bird*, by Glen E. Woolfenden and John W. Fitzpatrick
21. *Natural Selection in the Wild*, by John A. Endler
22. *Theoretical Studies on Sex Ratio Evolution*, by Samuel Karlin and Sabin Lessard
23. *A Hierarchical Concept of Ecosystems*, by R. V. O'Neill, D. L. DeAngelis, J. B. Waide, and T.F.H. Allen
24. *Population Ecology of the Cooperatively Breeding Acorn Woodpecker*, by Walter D. Koenig and Ronald L. Mumme
25. *Population Ecology of Individuals*, by Adam Lomnicki
26. *Plant Strategies and the Dynamics and Structure of Plant Communities*, by David Tilman

27. *Population Harvesting: Demographic Models of Fish, Forest, and Animal Resources*, by Wayne M. Getz and Robert G. Haight

28. *The Ecological Detective: Confronting Models with Data*, by Ray Hilborn and Marc Mangel

29. *Evolutionary Ecology across Three Trophic Levels: Goldenrods, Gallmakers, and Natural Enemies*, by Warren G. Abrahamson and Arthur E. Weis

30. *Spatial Ecology: The Role of Space in Population Dynamics and Interspecific Interactions*, edited by David Tilman and Peter Kareiva

31. *Stability in Model Populations*, by Laurence D. Mueller and Amitabh Joshi

32. *The Unified Neutral Theory of Biodiversity and Biogeography*, by Stephen P. Hubbell

33. *The Functional Consequences of Biodiversity: Empirical Progress and Theoretical Extensions*, edited by Ann P. Kinzig, Stephen J. Pacala, and David Tilman

34. *Communities and Ecosystems: Linking the Aboveground and Belowground Components*, by David Wardle

35. *Complex Population Dynamics: A Theoretical/Empirical Synthesis*, by Peter Turchin

36. *Consumer-Resource Dynamics*, by William W. Murdoch, Cheryl J. Briggs, and Roger M. Nisbet

37. *Niche Construction: The Neglected Process in Evolution*, by F. John Odling-Smee, Kevin N. Laland, and Marcus W. Feldman

38. *Geographical Genetics*, by Bryan K. Epperson

39. *Consanguinity, Inbreeding, and Genetic Drift in Italy*, by Luigi Luca Cavalli-Sforza, Antonio Moroni, and Gianna Zei

40. *Genetic Structure and Selection in Subdivided Populations*, by François Rousset

41. *Fitness Landscapes and the Origin of Species*, by Sergey Gavrilets

42. *Self-Organization in Complex Ecosystems*, by Ricard V. Solé and Jordi Bascompte

43. *Mechanistic Home Range Analysis*, by Paul R. Moorcroft and Mark A. Lewis

44. *Sex Allocation*, by Stuart West

45. *Scale, Heterogeneity, and the Structure of Diversity of Ecological Communities*, by Mark E. Ritchie

46. *From Populations to Ecosystems: Theoretical Foundations for a New Ecological Synthesis*, by Michel Loreau

47. *Resolving Ecosystem Complexity*, by Oswald J. Schmitz

48. *Adaptive Diversification*, by Michael Doebeli

49. *Ecological Niches and Geographic Distributions*, by A. Townsend Peterson, Jorge Soberón, Richard G. Pearson, Robert P. Anderson, Enrique Martínez-Meyer, Miguel Nakamura, and Miguel Bastos Araújo